AF616460

Effect of IL-10 and anti-TGF-beta antibodies on the morphology of bone marrow stroma cultures
from
Interleukin-10
by
Jan E. DeVries and René de Waal Malefyt
© RG Landes Co. 1995

Molecular Biology Intelligence Unit

Polyamines: Regulation and Molecular Interaction

Robert A. Casero, Jr., Ph.D.

The Johns Hopkins Oncology Center Research Labs
Baltimore, Maryland, U.S.A.

R.G. Landes Company
Austin

Molecular Biology Intelligence Unit

POLYAMINES: REGULATION AND MOLECULAR INTERACTION

R.G. LANDES COMPANY
Austin, Texas, U.S.A.

Submitted: April 1995
Published: July 1995

U.S. and Canada Copyright © 1995 R.G. Landes Company
All rights reserved. Printed in the U.S.A.

Please address all inquiries to the Publisher:
R.G. Landes Company, 909 Pine Street, Georgetown, Texas, U.S.A. 78626
or
P.O. Box 4858, Austin, Texas, U.S.A. 78765
Phone: 512/ 863 7762; FAX: 512/ 863 0081

U.S. and Canada ISBN 1-57059-281-0

International Copyright © 1995 Springer-Verlag, Heidelberg, Germany
All rights reserved.

International ISBN 3-540-60037-X

While the authors, editors and publisher believe that drug selection and dosage and the specifications and usage of equipment and devices, as set forth in this book, are in accord with current recommendations and practice at the time of publication, they make no warranty, expressed or implied, with respect to material described in this book. In view of the ongoing research, equipment development, changes in governmental regulations and the rapid accumulation of information relating to the biomedical sciences, the reader is urged to carefully review and evaluate the information provided herein.

Library of Congress Cataloging-in-Publication Data

Polyamines : regulation and regulation and molecular interaction / edited by Robert A. Casero, Jr.
p. cm. — (Medical intelligence unit)
Includes bibliographical references and index.
ISBN 1-57059-281-0 (alk. paper)
1. Polyamines—Metabolism—Regulation. 2. Polyamines—Pathophysiology.
I. Casero, Robert Anthony. II. Series.
[DNLM: 1. Polyamines—metabolism. QU 61 P781 1995]
QP801.P638P65 1995
612'.0157—dc20
DNLM/DLC 95–14123
for Library of Congress CIP

PUBLISHER'S NOTE

R.G. Landes Company publishes five book series: *Medical Intelligence Unit, Molecular Biology Intelligence Unit, Neuroscience Intelligence Unit, Tissue Engineering Intelligence Unit* and *Biotechnology Intelligence Unit.* The authors of our books are acknowledged leaders in their fields and the topics are unique. Almost without exception, no other similar books exist on these topics.

Our goal is to publish books in important and rapidly changing areas of medicine for sophisticated researchers and clinicians. To achieve this goal, we have accelerated our publishing program to conform to the fast pace in which information grows in biomedical science. Most of our books are published within 90 to 120 days of receipt of the manuscript. We would like to thank our readers for their continuing interest and welcome any comments or suggestions they may have for future books.

Deborah Muir Molsberry
Publications Director
R.G. Landes Company

CONTENTS

Preface ix

1. **Introduction** 1
   Robert A. Casero, Jr.

2. **Transcriptional Control of the ODC Gene** 5
   G. Lynn Law, Run-Sheng Li, David R. Morris
   I. Introduction 5
   II. Factors Determining Basal Promoter Activity 6
   III. Is ODC a Primary Response Gene? 10
   IV. Regulation by cAMP 12
   V. Response to Phorbol Esters and Modulation by Retinoic Acid 14
   VI. Crosstalk Between PKA and PKC 16
   VII. Regulation by Oncogene Products 17

3. **Mammalian S-Adenosylmethionine Decarboxylase Regulation and Processing** 27
   Bruce A. Stanley
   I. Introduction 27
   II. Transcriptional Control of SAMDC 31
   III. Translational Control of SAMDC Expression 33
   IV. Regulation of Rate of Proenzyme Processing 50
   V. Activation of Mature Enzyme 57
   VI. Post-Translational Modifications and Inactivation of Enzyme 59
   VII. Regulation Through Changes in Degradation Rate 61
   VIII. Summary 65

4. **Regulation of Spermidine/Spermine N^1-Acetyltransferase** 77
   Lei Xiao, Robert A. Casero, Jr.
   I. Introduction 77
   II. The Polyamine Metabolic Pathway: a Very Brief Overview 78
   III. The Spermidine/Spermine N^1-Acetyltransferase Protein 79
   IV. Cloning and Characterization of the Spermidine/Spermine N^1-Acetyltransferase Gene 80
   V. Regulation/Deregulation of SSAT Expression and Physiologic Consequences 85
   VI. Summary 92

5. **Biological and Therapeutic Implications of the Effects of Polyamines on Chromatin Condensation 101**
Hirak S. Basu, Laurence J. Marton
I. Summary 101
II. Introduction 102
III. Polyamine-DNA Interactions 102
IV. Polyamine-DNA Interactions and Cell Division 103
V. Possible Therapeutic Significance of Polyamine Analog Induced Changes in Chromatin Structure 109
VI. Unanswered Questions and Unsolved Mysteries 114

6. **Modulation of NMDA Receptors by Polyamines 129**
Keith Williams
I. Overview 129
II. Introduction 129
III. Effects of Polyamines on Native NMDA Receptors 136
IV. Molecular Biology of NMDA Receptors 140
V. Effects of Polyamines on Recombinant NMDA Receptors 145
VI. Implications for the In Vivo Biology of Polyamines 155
VII. Conclusions 158

7. **S-Adenosylmethionine Decarboxylase and Spermidine/Spermine-N^1-Acetyltransferase — Emerging Targets for Rational Inhibitor Design 171**
Patrick M. Woster
I. Introduction 171
II. S-Adenosylmethionine Decarboxylase 173
III. Spermidine/Spermine-N^1-Acetyltransferase 186

8. **The Polyamine Metabolic Pathway as a Target for Chemoprevention 205**
Craig M. Berchtold, Thomas W. Kensler, Robert A. Casero, Jr.
I. Introduction 205
II. Polyamine Metabolic Pathway 208
III. Polyamines in Chemical Carcinogenesis 211
IV. Polyamine Pathway in Clinical Chemoprevention 217
V. Toxicity and Pharmacokinetic Studies of DFMO 220
VI. Polyamine Analogs in Chemoprevention 221
VII. Inflammatory Processes as a Target of Intervention 223
VIII. Conclusion 224

Index 233

EDITOR

Robert A. Casero, Jr., Ph.D.
The Oncology Center
The Johns Hopkins University School of Medicine
Baltimore, Maryland, U.S.A.
chapters 1, 4, 8

CONTRIBUTORS

Hirak S. Basu, Ph.D.
Department of Human Oncology
University of Wisconsin
Madison, Wisconsin, U.S.A.
chapter 5

Craig M. Berchtold, B.S.
Division of Toxicological Sciences
Department of Environmental Health Sciences
Johns Hopkins School of Hygiene and Public Health
Baltimore, Maryland, U.S.A.
chapter 8

Thomas W. Kensler, Ph.D.
Division of Toxicological Sciences
Department of Environmental Health Sciences
Johns Hopkins School of Hygiene and Public Health
Baltimore, Maryland, U.S.A.
chapter 8

G. Lynn Law, Ph.D.
Department of Biochemistry, SJ-70
University of Washington
Seattle, Washington, U.S.A.
chapter 2

Run-Sheng Li, Ph.D.
Department of Biochemistry, SJ-70
University of Washington
Seattle, Washington, U.S.A.
chapter 2

Laurence J. Marton, M.D.
Departments of Pathology/Laboratory Medicine
Oncology and Human Oncology
University of Wisconsin
Madison, Wisconsin, U.S.A.
chapter 5

David R. Morris, Ph.D.
Department of Biochemistry, SJ-70
University of Washington
Seattle, Washington, U.S.A.
chapter 2

CONTRIBUTORS

Bruce A. Stanley, Ph.D.
Department of Cellular and Molecular Physiology
The Pennsylvania State University College of Medicine
Milton S. Hershey Medical Center
Hershey, Pennsylvania, U.S.A.
chapter 3

Keith Williams, Ph.D.
Department of Pharmacology
University of Pennsylvania School of Medicine
Philadelphia, Pennsylvania, U.S.A.
chapter 6

Patrick M. Woster, Ph.D.
Department of Pharmaceutical Sciences
Wayne State University
Detroit, Michigan, U.S.A.
chapter 7

Lei Xiao, Ph.D.
The Oncology Center
The Johns Hopkins University School of Medicine
Baltimore, Maryland, U.S.A.
chapter 4

PREFACE

Throughout the last several years there has been a virtual explosion of work completed in the general field of polyamine research. The recent introduction of several new chemicals and drugs have allowed a more precise study of the metabolic pathway. The past few years have provided molecular probes for each of the regulatory enzymatic steps in the polyamine metabolic pathway and with them the ability to examine the various levels of molecular control of their expression. Each week provides new information as to how these small, but ubiquitous polycations are specifically involved in molecular/cellular interactions, from the regulation of gene expression, to regulation of ion channels, to the self-regulation of the biosynthetic and catabolic enzymes in the pathway. Renewing interest in the field of polyamine metabolism comes, in part, from recent findings which suggest ODC can act as a facilitator of oncogenesis, the unique mechanism of AdoMetDC self-processing, the potential role of SSAT as a mediator of the antineoplastic activity of new polyamine analogues and the production of new inhibitors of each enzymatic step. Although this book is not intended to be a complete review of the field, it will hopefully provide the interested reader a representative view that will entice them into further investigation.

Robert A. Casero, Jr., Ph.D.
The Oncology Center
The Johns Hopkins University
School of Medicine

CHAPTER 1

INTRODUCTION

Robert A. Casero, Jr.

The interest in the field of polyamine metabolism and enzymatic regulation has been increasing steadily over the last several years. The absolute requirement for the polyamines in growth and development in eukaryotic cells has driven the interest in understanding the actual molecular functions of the polyamines to a new level. Several groups have undertaken the task of design and synthesis of several new agents which interfere with polyamine metabolism and thus are used to dissect the important roles which they play in biological processes. The interest in the use of polyamine analogs as therapeutic agents continues to grow with the initiation of clinical trials against several important solid tumor types.[5,8] The recent cloning and characterization of essentially all of the controlling enzymes in the polyamine pathway has allowed a more rigorous examination of both the details of regulation and the precise manner in which the polyamines themselves are involved in the regulation of their own synthesis and degradation. The polyamine metabolic pathway has also been found to be a target for antiparasitic agents and exciting advances continue to be made along those lines.[1] Additionally, the recent flurry of work in the ornithine decarboxylase antizyme field has elegantly demonstrated an apparently unique mechanism for eukaryotic enzymatic regulation (unfortunately not reviewed here, but excellently reviewed by Hayashi and Murakami).[2]

Polyamine metabolism has been reviewed several times over the last 5 years and no attempt will be made to be all inclusive in this work. The interested reader is directed to the several excellent works on polyamine metabolism, the polyamine pathway as a therapeutic target and other works reviewing the major point of interest in the field.[3-8] Here, the major goals are to first examine the regulation of the three rate-limiting enzymes in polyamine metabolism: ornithine decarboxylase (ODC), S-adenosylmethionine decarboxylase (AdoMetDC) and spermidine/spermine N^1-acetyltransferase (SSAT).

Polyamines: Regulation and Molecular Interaction, edited by Robert Casero. © 1995 R.G. Landes Company.

Second, we will examine two examples of how the polyamines are thought to interact with other biomolecules/receptors ultimately effecting structure and function. Polyamine interaction with DNA has long been thought to be an important function of the natural polyamines and as more is learned about the specific interactions and the resultant conformational changes which can be influenced by the polyamine binding to DNA the potential for regional and gene-specific changes are becoming more evident. Further, these proposed interactions may be sufficient, in some cases, to explain the antiproliferative effects of the many new polyamine analogs. The investigations into the specific roles that polyamines play in the functioning of the NMDA receptor provide a specific example of the polyamines potential roles in modifying receptor-mediated events in a specific manner. Although these examples are relatively focused, they may provide a general paradigm which may useful in envisioning general polyamine interactions.

As more is discovered about the polyamines and their functions the need for new compounds to probe further is necessitated. Therefore, the design and synthesis of new agents which interfere with polyamine metabolism is discussed. It is hoped these compounds can be used to finely dissect the molecular interaction in which the polyamines are involved and may further the availability of useful antiproliferative agents.

Finally, although the polyamine pathway continues to be well studied as a target for antineoplastic therapy and will be covered extensively in a future edition of this series (Kenji Nishioka, Editor) much less attention has been focused on the targeting of polyamine metabolism as a strategy for chemoprevention of neoplastic disease. In the last section of this edition, the potential of existing agents and the potential for new compounds that may be useful in chemoprevention will be examined. Chemoprevention in high risk populations is a relatively new field. Inhibitors of polyamine metabolism may possess the necessary qualities of antiprogression activities with acceptably low toxicities for compounds to be administered over a long period to essentially asymptomatic patients. This potential is an intriguing possibility which is at its early stages of investigation.

Again it must be emphasized that the purpose of these chapters is not to be all inclusive and I apologize for not covering other topics of equal importance to those selected here. However, we hope this volume will provide the reader with enough detail in a reasonably broad spectrum of current topics in polyamine research to stimulate further questions and approaches into studying these simple, yet critical biomolecules.

References

1. Bacchi CJ and Yarlett N. Effects of antagonists of polyamine metabolism on African trypanosomes. Acta Tropica 1993;54:225-236.
2. Hayashi SI and Murakami Y. Rapid and regulated degradation of ornithine decarboxylase. Biochem J 1995;306:1-10.
3. Heby O and Persson L. Molecular genetics of polyamine synthesis in eukaryotic cells. TIBS 1990;15:153-158.
4. Janne J, Alhonen L and Leinonen P. Polyamines: From molecular biology to clinical applications. Ann Med 1991;23:241-259.
5. Marton LJ and Pegg AE. Polyamines as targets for therapeutic intervention. Ann Rev Pharmacol 1995;35:55-91.
6. McCann PP and Pegg AE. Ornithine decarboxylase as an enzyme target for therapy. Pharmacol Therapeutics 1992;54:195-215.
7. Pegg AE and McCann PP. S-adenosylmethionine decarboxylase as an enzyme target for therapy. Pharmacol Therapeutics 1992;56:359-377.
8. Porter CW, Regenass U and Bergeron RJ. Polyamine inhibitors and analogs as potential anticancer agents. In: Falk Symposium on Polyamines, Lancaster, UK: Kluwer Press, 1992;301-322.

CHAPTER 2

TRANSCRIPTIONAL CONTROL OF THE ODC GENE

G. Lynn Law, Run-Sheng Li, David R. Morris

I. INTRODUCTION

Ornithine decarboxylase (ODC) catalyzes a key step in the biosynthesis of the polyamines and both the synthesis and degradation of this enzyme are highly regulated. The level of ODC is feedback controlled by the cellular content of free polyamines,[1] possibly to prevent the known toxicity of elevated polyamine levels.[2] Feedback regulation of ODC level is mediated posttranscriptionally, but not by variations in mRNA content.[1] ODC activity also responds rapidly to a variety of exogenous stimuli, including a variety of hormones and growth factors and these changes in enzyme activity are usually the result of elevated mRNA.[3] With a few exceptions,[4-6] the modulation of ODC content by exogenous signals is usually regulated at the transcriptional level. Recent observations that overexpression of ODC leads to neoplastic transformation of fibroblasts[7-9] dramatically demonstrate the importance of precise regulation of this key enzyme.

This chapter focuses specifically on the regulation of transcription of the ODC gene. Our knowledge of the DNA elements and protein factors involved in the basal activity of the promoter is steadily increasing and we have also gained significant knowledge of the details of regulation through cAMP activation of protein kinase (A-kinase, PKA). The exciting new information on the control of ODC transcription by oncogene products, particularly c-Myc, is certainly of interest with respect to the regulation of ODC by growth factors and may also be important to our understanding of the deregulation of ODC that is often seen in tumor cells.

Polyamines: Regulation and Molecular Interaction, edited by Robert Casero. © 1995 R.G. Landes Company.

II. FACTORS DETERMINING BASAL PROMOTER ACTIVITY

The elements regulating basal expression of the ODC gene are embedded within the extremely GC-rich promoter region. The ODC promoter contains a TATAA box and several GC boxes are found within the first 500 bases proximal to the transcriptional start site. Very close to the TATAA box, the promoter has a binding site for members of the CREB/ATF family of transcription factors. This binding site is critical for both basal activity and cAMP-regulated activity of the promoter.

GC boxes are found in many promoter regions of both cellular and viral genes and are thought to be involved in basal transcriptional activity.[10] Although the best characterized transcription factor known to bind to these boxes is Sp1, several other factors, including both activators and inhibitors, have been recently shown to bind to GC-rich regions. One of the GC-rich regions associated with the ODC gene has been shown to be important in determining basal promoter activity.[11] This region is located approximately 70 nucleotides upstream of the TATAA box, at nucleotides -136 to -91, and contains at least two overlapping protein binding sites (Fig. 2.1). Sp1 and/or Sp1-like proteins bind specifically to one site and a new factor, NF-ODC$_1$, binds specifically to the other. This GC box contributes significantly to the basal expression level in a manner that is dependent on cell type. Deletions and point mutations in the ODC promoter that abolish protein binding to this 43bp GC-rich region reduce basal pro-

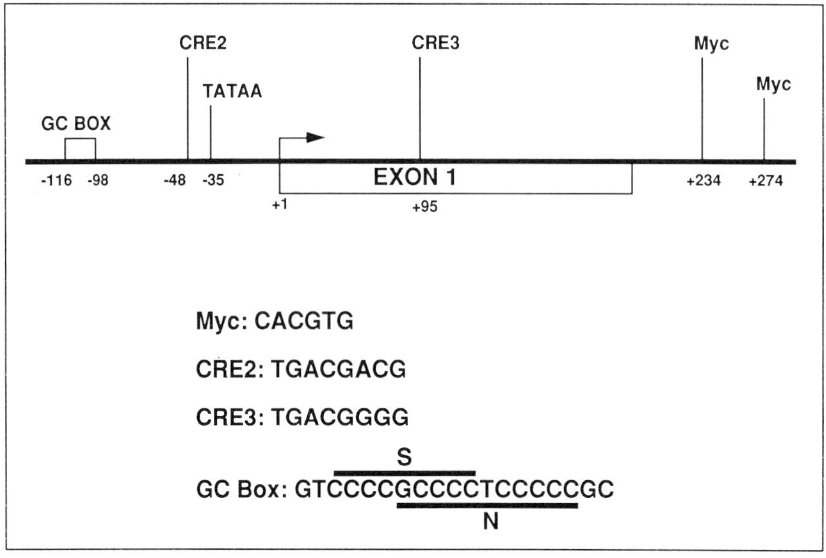

Fig. 2.1. Established regulatory sites in the region of the mouse ODC promoter. The sites are referred to in the text and the sequences are given in the figure. In the sequence of the GC box, the Sp1 binding site is designated "S" and the NF-ODC$_1$ site "N".

moter activity of transfected constructs to 20% of wildtype in the adrenal carcinoma cell line, Y1. In contrast, these same mutations produce no effect on promoter activity in the T-cell line, Jurkat.

A. Transcription Activator Sp1 and Related Proteins

GC-rich regions and the Sp1 binding motif are common promoter elements found in many eukaryotic genes.[10] Sp1 is expressed in most cell types and is involved in basal and induced expression of a large number of genes. This factor has been shown to be involved in both developmental and tissue-specific regulation. Sp1 is believed to exert its activity by interacting, either directly or indirectly, with the general transcription initiation machinery, thereby stabilizing initiation complexes, and these interactions seem to be regulated by coactivator proteins.[12] In some genes, Sp1 activity can also be modulated through competition for binding to the GC box.[13] Overlapping binding sites for Sp1 and other transcription factors have been discovered in several genes,[14-17] including that encoding ODC.[11] The relative levels of Sp1 and proteins that may compete with it for DNA sites and coactivators, as well as the respective binding affinities of these factors, may define Sp1 activity at specific times during development and cell differentiation.

In recent years, several proteins related to Sp1 have been isolated and characterized, including Sp2, Sp3, Sp4, BTEB 1 and BTEB 2. In most cases, the binding affinities of these proteins for the Sp1 consensus site are indistinguishable from that of Sp1 and the proteins are expressed in many different cell types.[18-21] The transactivation domains of these zinc-finger proteins differ significantly in amino acid sequence, suggesting that, although they bind to the same cis element, they may interact with a different set of coactivators and/or transcription factors. This generates a complex regulatory matrix that is capable of mediating a vast array of specific regulatory activities that are needed during cell growth and differentiation. As seems to be true with ODC regulation, the Sp1 binding sites in the promoters of several genes have been shown to regulate expression in ways that differ with cell type.[14,18,22]

The ODC gene contains several likely binding sites for transcription factor Sp1. A GC-rich region approximately 100 bases upstream of the transcriptional initiation site contains a perfect Sp1 site (CCCCGCCCC at nucleotides -114 to -106, Fig. 2.1). This site is conserved in the human, rat and mouse genes. We have shown that this region of the promoter binds strongly to purified Sp1 and forms several specific complexes with proteins in nuclear extracts from different cell types.[11] Formation of all but one of these complexes was specifically competed by the Sp1 consensus binding sequence. (The remaining complex is the result of NF-ODC$_1$ binding and is discussed below.) Interestingly, only one of the complexes was recognized by

Sp1-specific antibody and this complex had the same electrophoretic mobility as that generated with purified Sp1. These data suggest that, in addition to Sp1, other proteins, perhaps related to Sp1, bind to this region of the ODC promoter. Mutations in the Sp1 binding site differentially influenced promoter activity depending on the cell type analyzed. For example, in Y1 cells, HeLa cells, smooth muscle cells and NIH 3T3 cells, these mutations produced a decrease in promoter activity. In contrast, there was no effect of these mutations on expression in two T-cell lines.[11] The relative levels of Sp1 in nuclear extracts did not correlate with this cell-type specificity.

There are also several possible sites of Sp1 binding upstream of the -114/-106 site, but at this time little is known about the roles of these sites in determining ODC promoter activity. Deletion of the region between nucleotides -427 and -208, which contains two potential Sp1 sites (at -280 and -330), reduced ODC promoter strength significantly in NIH 3T3 cells.[23] When an oligonucleotide containing the ODC sequence encompassing these two sites was used in mobility shift assays with nuclear proteins from mouse tissues, protein binding was seen and specific binding was inhibited by addition of excess Sp1 consensus binding sequence[23] (also, our unpublished results). SR Muller and coworkers also found that the region encompassing these two Sp1 sites was involved in basal activity in PC12 cells.[24] In neither instance were the active sequences demonstrated to be the Sp1 sites.

B. Transcription Inhibitor NF-ODC$_1$

In addition to the -114/-106 Sp1 binding site discussed above, we have shown that there is an overlapping site for the newly described nuclear protein, NF-ODC$_1$, in the GC-box region of the human, mouse and rat genes. This protein binds to the sequence (GCCCCTCCCCC), which is located at nucleotides -110 to -100 of the mouse ODC gene (Fig. 2.1). Binding studies, including mobility shift analysis, DNase I footprinting and methylation interference, indicated that this site overlaps the Sp1 binding site and that these two proteins bind the ODC promoter in a mutually exclusive manner.[11] The influence of a mutation that specifically disrupts NF-ODC$_1$ binding on ODC promoter activity differs with cell type. In two T-cell lines, Jurkat and EL-4, this mutation had essentially no effect on basal expression. Conversely, in Y1, HeLa, smooth muscle and NIH 3T3 cells, the mutation blocking NF-ODC$_1$ binding *enhanced* expression to varying degrees.[11] Thus, binding of NF-ODC$_1$ to its site in the ODC promoter seems to inhibit expression and in a cell-type specific manner. NF-ODC$_1$ binding activity was found in nuclear extracts of all cell types we examined, including Jurkat, HeLa, NIH 3T3, Y1, BHK and smooth muscle cells. The relative amounts of binding activity in the various cell types did not correlate with the apparent requirement for the NF-ODC$_1$ binding site for promoter activity, suggesting that other factors such as posttranslational modification or interaction with other proteins may be involved in the repressive activity of this factor.

There are many questions left to be answered about NF-ODC$_1$. This factor may be acting passively to neutralize the activity of Sp1 by simply competing with it at the overlapping binding sites in the promoter. Alternatively, NF-ODC$_1$ may be an active repressor in that it interacts directly with the transcription machinery to decrease the transcription rate. In addition, the activity of NF-ODC$_1$ may be regulated by interactions with cell-specific factors in the same manner postulated for Sp1. Lastly, we know nothing about possible roles of NF-ODC$_1$ in the *regulated* activity of the ODC gene and in the deregulation associated with some neoplastic cells.

C. Other Repressors Binding to GC Boxes

Transcriptional repression is an important mechanism for regulation of normal cell growth and differentiation. Deregulation of transcriptional repressor proteins can lead to neoplastic growth and tumor production. There are several known repressors that bind GC-rich regions. One example is the Wilms tumor suppressor, WT1. Wilms tumor is a pediatric nephroblastoma that is thought to arise when kidney blastemal cells continue to proliferate after birth instead of differentiating.[25] WT1 binds the same GC-rich DNA sequence element as the early growth response (EGR) family of transcription factors.[26] However, unlike members of the EGR family, which stimulate transcription rate, WT1 behaves as a transcriptional repressor in transient transfection experiments. These findings are of great biological significance since small deletions and point mutations in the zinc fingers of WT1, which interfere with DNA binding, have been detected in a number of Wilms tumors. This suggests that WT1 may function to repress the transcription of specific genes that are involved in blastemal cell proliferation. If WT1 repression activity is decreased or eliminated through mutations, normal differentiation of the blastemal cells fails, and there is uncontrolled proliferation of these cells. From the standpoint of this essay, it is interesting that WT1 binds to a GC-rich sequence similar to the binding site for NF-ODC$_1$. However, unpublished electrophoretic mobility shift studies from our laboratory indicate that WT1 and NF-ODC$_1$ are not the same protein.

Another protein factor that appears to repress transcription is called GC factor (GCF). The overexpression of GCF in gastric carcinoma cells resulted in the decreased mRNA levels of several growth factors and receptors including transforming growth factor (TGF-α), insulin-like growth factor II and proto-oncogene, c-*met,* suggesting that GCF is important for the regulation of cell growth and a decrease in its activity may lead to unwanted cell proliferation.[31] The binding sites for GCF and NF-ODC$_1$ differ significantly and the molecular weight of GCF is approximately 100 kDa,[32] whereas we estimate that of purified NF-ODC$_1$

to be around 50 kDa (unpublished results). Therefore, it seems likely that these two binding activities reside in different molecules.

The viral transcriptional protein E2 is another example, unrelated to the ODC gene, in which transcriptional repression through a GC box may be important in tumorigenesis. HPV-16 is a high risk genital papillomavirus that is found in a high percentage of cervical cancers. P97, the principal early promoter of HPV-16, contains three protein binding sites in close proximity to the TATAA box. Sp1 binds to one of the sites and E2 binds to the other two sites. Sp1 is required for the stimulation of P97 and both E2 binding sites are required for full repression of P97. The binding of Sp1 and E2 is mutually exclusive and it appears that Sp1 displacement by E2 is one facet of this repression mechanism.[15]

D. Proteins Interacting with CRE2

There are three sites associated with the ODC promoter that bind the cAMP-responsive transcription factor CREB; these sites are designated CRE1, CRE2 and CRE3.[33] CRE2 is located adjacent to the TATAA box of the promoter and seems to have unique properties relative to basal activity. A clone of Y1 cells that is deficient in PKA activity not only did not respond to elevated cAMP by induction of ODC expression, but exhibited a significant reduction in the basal expression level of ODC message as well.[34] The loss of basal mRNA level resulted from a specific decrease in rate of transcription of the ODC gene. Activity studies of promoter constructs, in which the three CRE sites were either deleted or mutated, showed that CRE2 contributed significantly to the basal ODC expression.[33] In contrast, neither CRE1 nor CRE3 appeared to function in basal expression, although CRE3 plays a major role in regulation by cAMP (see below). The member or members of the CREB/ATF family of transcription factors that mediates basal transcription has not yet been defined. The attributes of CRE2 that provide its unique role in basal transcription have also not been identified, although one suspects that the proximity to the TATAA box may be of importance.

III. IS ODC A PRIMARY RESPONSE GENE?

Exposure of quiescent cells to growth factors and other stimuli leads within minutes to activation of the expression of over 100 genes, which are distinguished both by their kinetics of expression and by the roles of their products in growth and cell cycle progression.[35-38] Among these mitogen-induced genes are the so-called "primary response genes", expression of which is thought to be controlled directly by the mitogen-activated signal transduction pathways. An important diagnostic feature of the primary response genes is that their expression is independent of the induction of other protein factors and thus is insensitive to exposure of the cells to inhibitors of protein synthesis. Of particular note among the primary response genes are those, such as c-*jun*, c-*fos* and c-*myc*, that encode components of transcription factors, which in turn regulate expression of genes that are induced later in progression through the cell cycle. For example, the products of the *jun* and *fos* families of genes dimerize to form the AP-1 family of transcription factors,[39-42] which regulates expression of many genes (termed "secondary response genes" in this essay), including certain metalloproteinases involved in the metabolism of the extracellular matrix.[43,44] In contrast to the primary response genes, the secondary response genes require synthesis of protein factors prior to their induction and thus are sensitive to inhibitors of protein synthesis. It is noteworthy that the protein product of c-*myc* may be important in regulating ODC expression under some conditions, but not others (see Section VII. below).

Is ODC a primary or secondary response gene? The answer to this question is critical in defining the mechanisms of transcriptional regulation of ODC. However, one might anticipate that the answer could be complex, since there seem to be multiple signal transduction pathways coupled to induction of ODC expression.[3] Indeed, on surveying the literature, one finds that different results have been obtained with different cell types when inhibition of protein synthesis was applied as a test (Table 2.1). In constructing Table 2.1, we have excluded from consideration those cells in which inhibition of protein synthesis by itself activated ODC expression, since this complicates the interpretation of effects observed upon cell stimulation.[45-47] The induction of ODC mRNA expression is unaffected by blocking protein synthesis in some cell types, suggesting a primary response, while it is strongly inhibited in others, indicative of a secondary response. Even in very

Table 2.1. Influence of protein synthesis inhibition on ODC mRNA induction by a variety of stimuli in various cell types

Cell Type	Stimulus	Inhibitor Effect	Reference
Fibrosarcoma	TGF-β	NI[a]	107
Pancreatic tumor	βFGF	NI	108
Swiss 3T3 fibroblasts	Serum	PI[b]	10
Pheochromocytoma	NGF	PI	109
Macrophage-like	LPS or TPA	I[c]	51
Macrophage-like	(cAMP + LPS) or (cAMP + TPA)	NI	51
Primary embryo fibroblasts	Serum + TPA	I	50
Transformed embryo fibroblasts	Serum + TPA	NI	50
BALB/c 3T3 fibroblasts	Serum	I	48
Fibroblasts	Insulin or TPA	I	110
SV40-immortalized keratinocytes	EGF	I	111

[a] NI indicates no inhibition due to protein synthesis inhibitor
[b] PI indicates partial inhibition
[c] I indicates complete inhibition

similar cell lines responding to the same stimulus, divergent results have been obtained (Table 2.1): cycloheximide strongly inhibits induction of ODC mRNA by serum in resting BALB/c 3T3 cells,[48] whereas it has only a minor influence with the same stimulus in Swiss 3T3 cells.[49] Cycloheximide inhibits elevation of ODC mRNA by a combination of serum and TPA in primary hamster fibroblasts, whereas chemical transformation of the same cells abolishes the requirement for protein synthesis.[50] Activation of a macrophage-like cell line, RAW264, with either LPS or TPA led to a substantial increase in ODC mRNA that required prior protein synthesis.[51] In contrast, addition of cAMP to either of these stimuli largely abrogated the necessity for protein synthesis, dramatically illustrating the dependence of this phenomenon on the particular signal transduction pathway used.

Based on the apparent direct interaction of the PKA-regulated CREB/ATF family of transcription factors with the promoter region of the ODC gene and the fragmentary results in Table 2.1, it is tempting to suggest that the ODC response to PKA is primary. It is equally tempting to opine that the response to TPA may be secondary in many cell types, but perhaps not in some tumors. Additionally, there is probably at least one more signal transduction pathway linked to regulation of ODC expression yet to be accounted for, which is probably not linked to either protein kinase C (PKC) or PKA.[52] These considerations provide an attractive explanation for the observed differences between the various cell types and stimuli tested. The results obtained would be expected to depend on the mix of signal transduction pathways involved with a particular stimulus in a given cell type.

IV. REGULATION BY cAMP

It has been known for some time that both ODC activity and mRNA levels can be regulated by a variety of effectors that act upon cells to elevate intracellular cAMP and thereby activate the PKA signaling pathway (reference 3 and references therein). The elevation of ODC mRNA levels in response to activation of PKA in both quiescent Swiss 3T3 cells and Y1 adrenal tumor cells was due to an increase in the rate of transcription, as measured by nuclear run-on transcription.[6,34,53] Interestingly, in T lymphocytes and myeloid cells, an increase in cAMP exhibited exactly the opposite effect; not only was there no increase in the rate of transcription, but cAMP inhibited growth factor-stimulated increases in ODC mRNA levels (reference 3 and references therein).

The liganding of specific cell surface receptors leads to activation of adenylate cyclase, mediated by GTP-binding proteins (G-proteins). The resultant elevation of intracellular levels of cAMP in turn activates PKA. Elevated intracellular cAMP levels can result in either an increase or a decrease in gene expression in a particular cell type, depending on the specific gene; cyclic AMP response elements (CREs) have now been identified in many genes. Since there is a wide variety of known ligands that can provide the initial stimulus and there are isoforms and/or families of proteins at every level in this pathway, starting with the cell surface receptor and ending with transcription factors, the opportunity exists for considerable complexity within the pathway.[54,55]

The first CRE to be identified was that of the somatostatin gene, which is an 8bp palindrome (TGACGTCA). This promoter element has been shown to mediate the effects of increased intracellular levels of cAMP in a number of genes and sequence comparisons have derived a consensus sequence in which the TGACG nucleotides are best conserved (reviewed in refs. 54 and 55). A family of transcription factors have been identified,[56-58] the CREB/ATF family, that bind to this consensus sequence or to similar sequences. All members of the family contain the basic region/leucine zipper (bZip) motif, but diverge in other domains. The bZip domain allows for dimerization between bZip proteins and dimerization appears to be a requirement for binding to the CRE. Intrinsic properties of the leucine zipper seem to dictate which members of the bZip family can dimerize with one another. The CRE binding proteins seem to differ functionally from one another and members of the family with either repressor or activator activities have been identified. Two well studied representatives of the CREB/ATF family are CRE-binding protein (CREB), and cAMP-responsive element modulator (CREM). CREB was the first CRE binding protein to be identified. When CREB is phosphorylated by PKA, it is capable of increasing the transcription rate of certain cAMP-responsive genes. Remarkably, through alternative splicing, the CREM gene can encode either a repressor or an activator protein.[54,55]

In the mouse, rat and human ODC genes, there are three putative CREs located around the ODC transcriptional start site at positions -175 (CRE1), -48 (CRE2), and +195 (CRE3).[3] They are identical to the core CRE sequence (TGACG, Fig. 2.1). Three *cis*-acting elements contribute to regulation of ODC transcription by PKA in the adrenal carcinoma cell line, Y1. Inactivation of either CRE2, CRE3 or the GC-box (-136 to -91) results in additive loss of responsiveness of the ODC promoter to PKA.[33] Mutation of CRE2 results in a 5-fold decrease in inducibility and a 2-fold decrease resulted when either CRE3 or the GC-box was deleted or mutated. Interestingly, although CRE1 is highly conserved between species and is capable of binding recombinant CREB, it is not involved in PKA-mediated regulation in the experimental system employed. Electrophoretic mobility shift assays and DNAse I footprint experiments indicate that protein complexes form over CRE2, the TATAA box and the GC-box of the ODC gene. Protein binding at the TATAA box, but not the GC-box, is dependent on an intact CRE2 site. Recombinant CREB, and endogenous CREB or an antigenically related protein in nuclear extracts, bind to wildtype CRE2, but not to CRE2 sequences that have been functionally inactivated by mutation.[33]

These studies suggest a model where a member of the CREB/ATF family binds to CRE2, regulating assembly of a complex over the TATAA box and the ultimate formation of a transcriptional initiation complex. Palvino and coworkers have also delineated a region of cAMP responsiveness in NIH 3T3 cells to the vicinity of CRE2 and described a 70kDa protein in several mouse tissues that bound to this region.[23] The relationship between these observations and those of Abrahamsen and coworkers[33] in a different cell type is not yet clear.

As mentioned earlier, CRE2 appears to be important for basal transcription in addition to cAMP-induced transcription. CRE elements of other genes, including those encoding phosphoenolpyruvate carboxykinase and tyrosine hydroxylase, have also been shown to play this dual role in transcriptional regulation (reference 59 and references therein). Recently, CREB has been shown to contain a protein domain that is required for basal transcription activity.[60] This domain, called the constitutive activation domain (CAD), is distinct from and independent of the kinase inducible domain (KID) that is phosphorylated during cAMP activation of CREB. CAD is believed to interact with multiple elements in the RNA polymerase II initiation complex.[55,60]

V. RESPONSE TO PHORBOL ESTERS AND MODULATION BY RETINOIC ACID

Induction of ODC expression is among the first cellular events to be triggered by tumor-promoting phorbol esters, which are thought to exert their influence through activation of PKC. Early studies of mouse skin carcinogenesis demonstrated rapid elevations of cellular levels of ODC activity and ODC mRNA in response to treatment with the potent tumor promoter, TPA (reviewed in ref. 61). Similar to these observations with epidermis in vivo, activation of PKC with TPA is sufficient to induce growth-associated expression of ODC mRNA in a variety of cell types (reviewed in ref. 3). This elevation of ODC mRNA by TPA is regulated at the transcriptional level in those cell types where it has been examined.[6,62]

Despite the importance of regulation of ODC by phorbol esters, the mechanism has not been defined with precision. In order to define TPA-responsive elements in the ODC gene, promoter activities have typically been studied in cultured cells transfected with plasmids containing a reporter gene driven by promoter regions of interest. Using this approach, luciferase, attached to a fragment of the ODC promoter from -42 to +54, was shown to be strongly induced (ca. 10-fold) by TPA.[63] This area of the promoter contained no recognizable consensus sequences for known sequences responding to phorbol esters. An overlapping region of the promoter (-72 to +130) was shown to regulate a reporter gene when cotransfected with a PKC expression vector,[64] suggesting the likely possibility that TPA may be acting on the promoter through a PKC-responsive element. However, the interpretation of these results is complicated by the existence of elements in the backbones of some plasmids that respond to several stimuli, including TPA (R.-S. Li and G. Spies, unpublished results). Therefore, until a protein binding site within the ODC promoter can be shown to have biological activity through specific mutagenesis, these studies can only be considered preliminary. A prior study was unable to detect a TPA-response element (TRE) associated with the ODC gene.[65] However, in this instance, TPA administration was for only four hours; empirically, we find that at least 12 hours are required after TPA treatment to detect significant increase in expression of the same reporter gene in a transfected cell line.

The initial TRE to be identified was the sequence TGAC/GTCA, which is the binding site for the AP-1 family of transcription factors.[43,66] The AP-1 factors are composed of dimers of the products of the *jun* and *fos* families of proto-oncogenes (reviewed in refs. 37 and 67). There are a few sites resembling AP-1 TRE-like elements in the 5'-flanking region and intron 1 of mammalian ODC genes. Two of them are located within a few base pairs of one another at -1621 and -1636 in the 5'-flanking region of the murine ODC gene. This proximity could promote synergism between the sites, as has been observed in the glutathione transferase P promoter.[68] Despite the enticing existence of these sequences, the evidence is not strong for regulation of the ODC promoter by AP-1. The affinity of these TRE-like sites for the c-Jun/c-Fos heterodimer is not strong.[69] Additionally, although TPA greatly enhances expression of *jun*B and c-*jun* in transformed NIH 3T3 cells, ODC expression is insensitive to the stimulation,[70] suggesting a lack of involvement of these proto-oncogene products in ODC regulation, at least in this cell type. Other known TPA-responsive elements, such as the binding sites for NF-κB and AP-2, have been considered as possible candidates for regulating ODC expression,[3] but these possibilities have not been rigorously tested. At this time, we are still left without conclusive data showing how these important tumor promoters regulate ODC transcription.

Trans-retinoic acid (RA) is one of the most potent retinoids in counteracting the promotion of skin tumors by TPA. A strong correlation has been found between the ability of retinoids to inhibit TPA-induced expression of mouse epidermal ODC and their ability to block tumor promotion.[62,71] RA does not inhibit epidermal ODC expression induced by a complete carcinogenic dose of dimethylbenzanthracene (reviewed in ref. 61).

The profound effects of retinoids on mammalian development and cell proliferation are predominantly mediated by nuclear retinoic acid receptors (RAR), which are themselves transcription factors.[72] Interestingly, the human ODC promoter (-1450 to +810) is inhibited almost 50-fold by RA in Hela cells expressing RAR-α. A truncated ODC promoter (-250 to +514) lacks this strong responsiveness, consistent with specific inhibition by RA.[73] The known modes of action of RAR are either through binding to a steroid hormone-responsive element (HRE) (reviewed in ref. 74) or through physical interaction with transcription factor AP-1.[75,76] There are three good matches (10 or 11 out of 13 base pairs) to the consensus HRE palindrome upstream of the murine ODC promoter at -1135, -670 and -600 (reviewed in ref. 3). However, inconsistent with this simple model, RA did not suppress ODC mRNA levels in the presence of inhibitors of protein synthesis,[71] suggesting that suppression of ODC gene expression may not be a direct effect of RA, but depends on synthesis of an intermediary protein or proteins. Similar results were obtained in testosterone-treated rat sertoli cells,[77] suggesting that liganded receptors for neither RA nor testosterone interact directly with the ODC promoter to exert their inhibitory effects. In considering the relationship between TPA and RA, it should be noted that application of TPA to mouse skin rapidly reduces that levels of the mRNAs encoding RAR-α, RAR-γ and RXR-α.[78]

VI. CROSSTALK BETWEEN PKA AND PKC

The PKA and PKC signal transduction pathways were initially characterized as separate and independent pathways. In recent years data have emerged suggesting that these two systems may be interconnected and that there is extensive crosstalk between the two pathways in a number of experimental systems (refs. 54, 55 and references therein).

There are several lines of evidence suggesting interaction between these two pathways in regulating the ODC gene. In growth-arrested Swiss 3T3 cells, treatment with TPA and the adenylate cyclase activator forskolin leads to a synergistic enhancement of the rate of ODC transcription.[6] Perhaps related, the TPA induction of ODC transcription in Y1 cells is nearly eliminated in cells deficient in PKA activity.[34] In a macrophage-like cell line, RAW 264, cAMP alone had no effect on ODC transcription, but cotreatment with bacterial lipopolysaccharide (LPS), which may act through PKC, and cAMP elevated both the rate of ODC transcription and its mRNA level.[79] Studies of the promoter narrowed the region responsible for the synergistic activation to between nucleotides -63 and +12. CRE2 is located in this region, but when mutations that had previously been shown to inhibit CREB binding were introduced into this site, there was no affect on promoter activity. It should be noted, however, that no protein binding studies were presented to correlate with loss of biological activity of mutated constructs. Although CRE2 is in this region, the binding site for the 70 kDa protein of Palvino and coworkers[23] is also of interest in this context.

There are several known instances of convergence of the PKA and PKC pathways at the level of transcription factors. For example, the nucleotide sequences of the binding sites for the phorbol ester responsive

AP-1 factors (TRE) and the CREB/ATF family differ by only one base pair (TRE = TGAC/GTCA; CRE = TGACGTCA). In specific instances, CRE-binding factors may bind TRE elements or TRE-binding factors may bind CRE elements (discussed in refs. 54 and 55). The complexity of binding site recognition is enriched by the possibility of heterodimerization between members of the AP-1 and CREB/ATF families of proteins. As well, the transcriptional activator protein, AP-2, has also been shown to be responsive to both PKC and PKA.[80] Sites have also been identified in both the proenkephalin gene[81,82] and in the human cytochrome CYP11A and CYP17 genes[83] that appear to be involved in cAMP and TPA dependent transcriptional regulation.

VII. REGULATION BY ONCOGENE PRODUCTS

ODC expression has often been found to be activated in tumor cell lines when compared to normal cells, making the possible regulation of ODC expression by oncogene products of interest. Strong evidence has linked c-*myc* expression to ODC transcription.[84,86] Proto-oncogene c-*myc* encodes an evolutionarily conserved nuclear phosphoprotein that is ubiquitously expressed in somatic cells. Alterations in c-*myc* gene structure and expression by retroviral insertion, amplification and chromosomal translocation are associated with tumorigenesis in different species. The protein product of the c-*myc* gene together with its dimerization partner, Max, binds to the sequence CACGTG.[87,88] This sequence will also bind weakly to c-Myc homodimers, which form at high protein concentrations. The Myc-Max dimer can transactivate a few promoters that contain the CACGTG motif.[88,90,91] Conversely, c-Myc was shown to repress transcription dependent on the initiator element.[92]

The human ODC gene contains three perfect c-Myc-binding motifs, one of which is located at -485 and two others in intron 1 separated by 20 base pairs (Fig. *2*.1). Murine ODC contains only the two motifs in intron 1. Both the endogenous ODC gene[86] and ODC promoter constructs[84,85] are transactivated when cells are transfected with c-Myc expression constructs. Site-directed mutagenesis suggests that all three c-Myc-binding elements in the human gene contribute to activating ODC transcription.[84,85] However, the three elements seem not to be functionally equivalent. For example, CAT transcription driven by human ODC promoter from -787 to + 811, which contains the three c-Myc-binding motifs, is activated by approximately 2-fold in resting cells after stimulation with serum.[84] Mutation of the c-*myc*-binding motif at -470 abolishes the regulation by serum, indicating that this element, but not the others in the human ODC gene, can function as a "serum-responsive element". The basis of the functional difference between the three elements is unclear, as is the reason why this serum-responsive element does not exist in murine ODC gene. To our knowledge, this is the first piece of evidence suggesting that there may be regulatory differences between the human and murine ODC genes.

Regulation of ODC by c-Myc is probably complex. Max does not bind by itself to the CACGTG motif, but Max expression can significantly activate transcription driven by the ODC promoter, and this effect diminished when the c-Myc-binding motif at -480 was mutated.[84] One explanation for these results is that Max depends on dimerization with an endogenous factor, perhaps c-Myc for transcriptional activation. Further complicating the issue, truncated c-Myc, lacking the leucine zipper dimerization domain required for complex formation with Max in vitro and for binding of Myc/Max complexes to the CACGTG motif at -480, still efficiently transactivates the ODC promoter via the CACGTG motifs in intron 1.[85] This suggests that c-Myc is capable of regulating ODC transcription in a Max-independent way. Since proteins containing leucine zippers, specifically Jun and Fos, have been found to interact with helix-loop-helix proteins through their respective dimerization domains,[93,95] it is possible that Myc dimerization with other partners through similar interactions may account for its transactivation of the ODC gene.[85] Taken together, these results suggest that c-Myc may regulate ODC transcription by both Max-dependent and -independent mechanisms and the location of the c-Myc-binding motifs could influence the participation of particular dimer pairs.

Regulation of ODC transcription by c-Myc may be modulated by other factors. Other proteins can bind to the CACGTG element,[96,97] although their effects on ODC promoter activity have not been reported. One of these proteins, called USF, opposes the influence of c-Myc on the activity of the adenovirus-2 major late promoter.[92] Additionally, Mad[98] and MXi1,[99] two members of the family of Myc-related proteins, associate with Max to form heterodimers that are inactive in transcriptional activation. Mad/Max complexes have been shown to exert a strong repressive effect on transactivation by Myc.[100] Over-expression of Max by itself inhibits gene transcription and cell proliferation.[101] It seems clear from these studies that varying ratios of Myc-related proteins could alter regulation of ODC transcription in a cell-type-dependent way. For example, this could explain why concanavalin A did not activate ODC transcription in T cells,[6] although it induces expression of c-myc in the same cells.[102]

Other oncogenes, specifically *ras, fos* and *mos*, can also activate ODC expression, but by less defined mechanisms. ODC expression was shown to be elevated in fibroblasts transformed by Ha-*ras*.[103] Over-expression of v-*mos*, the product of which is a serine/threonine kinase, was also found to activate ODC expression in NIH 3T3 cells.[104] In this case, c-Myc could be involved, since expression of this proto-oncogene product was induced by v-Mos over the same time frame as ODC. Over-expression of c-*fos* dramatically stimulated ODC transcription in a clone of PC12 cells.[69] As discussed above in the context of phorbol ester regulation, there is reason to believe that the AP-1 TRE-like elements of the ODC gene are not involved in transcriptional control of ODC and thus are probably not the target of c-Fos in this case. Elevated levels of c-Fos could be acting via another mode other than dimerization with a member of the Jun family to form AP-1. For example, c-Fos was discovered in a complex formed with nuclear proteins on the TGF-β inhibitory element of the transin gene.[105] c-Fos also was found to interact with CRE-binding proteins,[106] which could be of interest because of the biologically active CREs associated with the ODC promoter.[33]

In summary, regulation of ODC expression by oncogene products will be a fruitful area of research in the future. This area is particularly important because of the deregulation of ODC expression that is often seen in neoplastic transformation.

References

1. Davis RH, Morris DR, Coffino P. Sequestered end products and enzyme regulation—the case of ornithine decarboxylase. Microbiol Rev 1992; 56:280-290.
2. Morris DR. A new perspective on ornithine decarboxylase regulation—prevention of polyamine toxicity is the overriding theme. J Cell Biochem 1991;46:102-105.
3. Abrahamsen MS, Morris DR. Regulation of expression of the ornithine decarboxylase gene by intracellular signal transduction pathways. In: Campisi J, Cunningham DD, Inouye M, Riley M, eds. Perspectives on Cellular Regulation: From Bacteria to Cancer. New York: Wiley-Liss, 1991:107-119.
4. White MW, Oberhauser AK, Kuepfer C, Morris DR. Different early-signaling pathways coupled to transcriptional and post-transcriptional regulation of gene expression during mitogenic activation of T lymphocytes. Mol Cell Biol 1987;7:3004-3007.
5. White MW, Kameji T, Pegg AE, Morris DR. Increased efficiency of translation of ornithine decarboxylase mRNA in mitogen-activated lymphocytes. Eur J Biochem 1987;170:87-92.
6. Abrahamsen MS, Morris DR. Cell Type-specific mechanisms of regulating expression of the ornithine decarboxylase gene after growth stimulation. Mol Cell Biol 1990;10:5525-5528.
7. Auvinen M, Paasinen A, Andersson LC, Holtta E. Ornithine decarboxylase activity is critical for cell transformation. Nature 1992;360:355-358.
8. Moshier JA, Dosescu J, Skunca M, Luk GD. Transformation of NIH/3T3 cells by ornithine decarboxylase overexpression. Cancer Res 1993;53:2618-2622.
9. Shantz LM, Pegg AE. Overproduction of ornithine decarboxylase caused by relief of translational repression is associated with neoplastic transformation. Cancer Res 1994;54:2313-2316.

10. Briggs MR, Kadonaga JT, Bell SP, Tjian R. Purification and biochemical characterization of the promoter-specific transcription factor, Sp1. Science 1986;234:47-52.
11. Li RS, Abrahamsen MS, Johnson RR, Morris DR. Complex interactions at a GC-rich domain regulate cell type-dependent activity of the ornithine decarboxylase promoter. J Biol Chem 1994;269:7941-7949.
12. Pugh BF, Tjian R. Mechanism of transcriptional activation by Sp1: evidence for coactivators. Cell 1990;61:1187-1197.
13. Ackerman SL, Minden AG, Williams GT, et al. Functional significance of an overlapping consensus binding motif for Sp1 and Zif268 in the murine adenosine deaminase gene promoter. Proc Natl Acad Sci USA 1991;88:7523-7527.
14. Bessereau JL, Mendelzon D, LePoupon C, et al. Muscle-specific expression of the acetylcholine receptor alpha-subunit gene requires both positive and negative interactions between myogenic factors, Sp1 and GBF factors. EMBO J 1993;12:443-449.
15. Tan SH, Gloss B, Bernard HU. During negative regulation of the human papillomavirus-16 E6 promoter, the viral E2 protein can displace Sp1 from a proximal promoter element. Nucleic Acids Res 1992;20:251-256.
16. Desai-Yajnik V, Samuels HH. The NF-kappa-B and Sp1 motifs of the human immunodeficiency virus type-1 long terminal repeat function as novel thyroid hormone response elements. Mol Cell Biol 1993; 13:5057-5069.
17. Cornwell MM, Smith DE. SP1 activates the MDR1 promoter through one of two distinct G-rich regions that modulate promoter activity. J Biol Chem 1993;268:19505-19511.
18. Hagen G, Muller S, Beato M, Suske G. Cloning by recognition site screening of two novel GT box binding proteins: a family of Sp1 related genes. Nucleic Acids Res 1992;20:5519-5525.
19. Imataka H, Sogawa K, Yasumoto K, et al. Two regulatory proteins that bind to the basic transcription element (BTE), a GC box sequence in the promoter region of the rat P-4501A1 gene. EMBO J 1992;11:3663-3671.
20. Sogawa K, Imataka H, Yamasaki Y, et al. cDNA cloning and transcriptional properties of a novel GC box-binding protein, BTEB2. Nucleic Acids Res 1993;21:1527-1532.
21. Hagen G, Muller S, Beato M, Suske G. Sp1-mediated transcriptional activation is repressed by Sp3. EMBO J 1994;13:3843-3851.
22. Zhang DE, Hetherington CJ, Tan S, et al. Sp1 is a critical factor for the monocytic specific expression of human CD14. J Biol Chem 1994; 269:11425-11434.
23. Palvino JJ, Eisenberg LM, Janne OA. Protein-DNA interactions in the cAMP responsive promoter region of the murine ornithine decarboxylase gene. Nucleic Acids Res 1991;19:3921-3927.
24. Muller SR, Huff SY, Goode BL, et al. Molecular analysis of the nerve growth factor inducible ornithine decarboxylase gene in PC12 cells. J Neurosci Res 1993;34:304-314.

25. Drummond IA, Madden SL, Rohwer-Nutter P, et al. Repression of the insulin-like growth factor II gene by the Wilms tumor suppressor WT1. Science 1992;257:674-678.
26. Rauscher FJ, Morris JF, Tournay OE, et al. Binding of the Wilms tumor locus zinc finger protein to the Egr-1 consensus sequence. Science 1990;250:1259-1262.

References 27-30 were omitted in proofs.

31. Kitadai Y, Yamazaki H, Yasui W, et al. GC factor represses transcription of several growth factor/receptor genes and causes growth inhibition of human gastric carcinoma cell lines. Cell Growth Differ 1993;4:291-296.
32. Kageyama R, Pastan I. Molecular cloning and characterization of a human DNA binding factor that represses transcription. Cell 1989;59:815-825.
33. Abrahamsen MS, Li RS, Dietrichgoetz W, Morris DR. Multiple DNA elements responsible for transcriptional regulation of the ornithine decarboxylase gene by protein kinase-A. J Biol Chem 1992;267:18866-18873.
34. Clegg CH, Abrahamsen MS, Degen JL, et al. Cyclic AMP-dependent protein kinase controls basal gene activity and steroidogenesis in Y1 adrenal tumor cells. Biochemistry 1992;31:3720-3726.
35. Denhardt DT, Edwards DR, Parfett CLJ. Gene expression during the mammalian cell cycle. Biochim Biophys Acta 1986;865:83-125.
36. Bravo R. Growth factor-responsive genes in fibroblasts. Cell Growth Differ 1990;1:305-309.
37. Herschman HR. Primary response genes induced by growth factors and tumor promoters. Annu Rev Biochem 1991;60:281-319.
38. Muller R, Mumberg D, Lucibello FC. Signals and genes in the control of cell-cycle progression. Biochim Biophys Acta 1993;1155:151-179.
39. Vogt PK, Bos TJ. jun: oncogene and transcription factor. Adv Cancer Res 1990;55:1-35.
40. Distel RJ, Spiegelman BM. Proto-oncogene *c-fos* as a transcription factor. Adv Cancer Res 1990;55:37-55.
41. Ransone LJ, Verma IM. Nuclear proto-oncogenes *FOS* and *JUN*. Ann Rev Cell Biol 1990;6:539-557.
42. Angel P, Karin M. The role of Jun, Fos and the AP-1 complex in cell-proliferation and transformation. Biochimica et Biophysica Acta 1991;1072:129-157.
43. Angel P, Imagawa M, Chiu R, et al. Phorbol ester-inducible genes contain a common cis element recognized by a TPA-modulated trans-acting factor. Cell 1987;49:729-739.
44. Angel P, Allegretto EA, Okino ST, et al. Oncogene *jun* encodes a sequence-specific trans-activator similar to AP-1. Nature 1988;332:166-171.
45. Olson EN, Spizz G. Mitogens and protein synthesis inhibitors induce ornithine decarboxylase gene transcription through separate mechanisms in the BC3H1 muscle cell line. Mol Cell Biol 1986;6:2792-2799.
46. Greenberg ME, Greene LA, Ziff EB. Nerve growth factor and epidermal growth factor induce rapid transient changes in proto-oncogene transcrip-

tion in PC12 cells. J Biol Chem 1985;260:14101-14110.
47. Butler AP, Mar PK, Mcdonald FF, et al. Involvement of protein kinase-C in the regulation of ornithine decarboxylase messenger RNA by phorbol esters in rat hepatoma cells. Exp Cell Res 1991;194:56-61.
48. Katz A, Kahana C. Transcriptional activation of mammalian ornithine decarboxylase during stimulated growth. Mol Cell Biol 1987;7:2641-2643.
49. Stimac E, Morris DR. Messenger RNAs coding for enzymes of polyamine biosynthesis are induced during the G0-G1 transition but not during traverse of the normal G1 phase. J Cell Physiol 1987;133:590-594.
50. Gilmour SK, O'Brien TG. Regulation of ornithine decarboxylase gene expression in normal and transformed hamster embryo fibroblasts following stimulation by 12-O-tetradecanoylphorbol-13-acetate. Carcinogenesis 1989;10:157-162.
51. Shurtleff SA, Mcelwain CM, Taffet SM. Rapid expression of ornithine decarboxylase mRNA in a macrophage-like cell line: cAMP repression of the requirement for prior protein synthesis. J Cell Physiol 1988;134:453-459.
52. Hovis JG, Stumpo DJ, Halsey DL, Blackshear PJ. Effects of mitogens on ornithine decarboxylase activity and messenger RNA levels in normal and protein kinase C-deficient NIH-3T3 fibroblasts. J Biol Chem 1986; 261:10380-10386.
53. Clegg CH, Ran W, Uhler MD, Mcknight GS. A mutation in the catalytic subunit of protein kinase A prevents myristylation but does not inhibit biological activity. J Biol Chem 1989;264:20140-20146.
54. Lalli E, Sassone-Corsi P. Signal transduction and gene regulation: the nuclear response to cAMP. J Biol Chem 1994;269:17359-17362.
55. Delmas V, Molina CA, Lalli E, et al. Complexity and versatility of the transcriptional response to cAMP. Rev Physiol Biochem Pharmacol 1994;124:1-28.
56. Ziff EB. Transcription factors—a new family gathers at the cAMP response site. Trends Gen 1990;6:69-72.
57. Montminy MR, Gonzalez GA, Yamamoto KK. Regulation of cAMP-inducible genes by CREB. Trends Neurosci 1990;13:184-188.
58. Lee KAW, Masson N. Transcriptional regulation by CREB and its relatives. Biochimica et Biophysica Acta 1993;1174:221-233.
59. Xing L, Quinn PG. Three distinct regions within the constitutive activation domain of cAMP regulatory element-binding protein (CREB) are required for transcriptional activation. J Biol Chem 1994;269:28732-28736.
60. Quinn PG. Distinct activation domains within cAMP response element-binding protein (CREB) mediate basal and cAMP-stimulated transcription. J Biol Chem 1993;268:16999-17009.
61. DiGiovanni J. Multistage carcinogenesis in mouse skin. Pharmacol Ther 1992;54:63-128.
62. Verma AK. Inhibition of phorbol ester-induced ornithine decarboxylase gene transcription by retinoic acid: a possible mechanism of antitumor

promoting activity of retinoids. Prog Clin Biol Res 1988;259:245-260.
63. Kim YJ, Pan H, Verma AK. Non-AP-1 tumor promoter 12-O-tetradecanoylphorbol-13-acetate responsive sequences in the human ornithine decarboxylase gene. Mol Carcinogenesis 1994;10:169-179.
64. Tseng CP, Kim YJ, Kumar R, Verma AK. Involvement of protein kinase C in the transcriptional regulation of 12-O-tetradecanoylphorbol-13-acetate-inducible genes modulated by AP-1 or non-AP-1 transacting factors. Carcinogenesis 1994;15:707-711.
65. Van Daalen Wetters T, Brabant M, Coffino P. Regulation of mouse ornithine decarboxylase activity by cell growth, serum and tetradecanoyl phorbol scetate is governed primarily by sequences within the coding region of the gene. Nucleic Acids Res 1989;17:9843-9860.
66. Lee W, Mitchell P, Tjian R. Purified transcription factor AP-1 interacts with TPA-inducible enhancer elements. Cell 1987;49:741-752.
67. Ryseck RP, Bravo R. c-Jun, Jun B, and Jun D differ in their binding affinities to AP- 1 and CRE consensus sequences: effect of Fos proteins. Oncogene 1991;6:533-542.
68. Okuda A, Imagawa M, Sakai M, Muramatsu M. Functional cooperativity between two TPA responsive elements in undifferentiated F9 embryonic stem cells. EMBO J 1990;9:1131-1135.
69. Wrighton C, Busslinger M. Direct transcriptional stimulation of the ornithine decarboxylase gene by fos in PC12 cells but not in fibroblasts. Mol Cell Biol 1993;13:4657-4669.
70. Sistonen L, Holtta E, Makela TP, et al. The cellular response to induction of the p21 c-Ha-ras oncoprotein includes stimulation of *jun* gene expression. EMBO J 1989;8:815-822.
71. Olsen DR, Hickok NJ, Uitto-J. Suppression of ornithine decarboxylase gene expression by retinoids in cultured human keratinocytes. J Invest Dermatol 1990;94:33-36.
72. Nagpal S, Saunders M, Kastner P, et al. Promoter context- and response element-dependent specificity of the transcriptional activation and modulating functions of retinoic acid receptors. Cell 1992;70:1007-1019.
73. Mao Y, Gurr JA, Hickok NJ. Retinoic acid regulates ornithine decarboxylase gene expression at the transcriptional level. Biochemical Journal 1993;295:641-644.
74. Lehmann JM, Zhang X-K, Pfahl M. RARg expression is regulated through a retinoic acid response element embedded in SP1 sites. Mol Cell Biol 1992;12:2976-2985.
75. Schule R, Rangarajan P, Kliewer S, et al. Functional antagonism between oncoprotein C-Jun and the glucocorticoid receptor. Cell 1990;62:1217-1226.
76. Yang-Yen H-F, Zhang X-K, Graupner G, et al. Antagonism between retinoic acid receptors and AP-1: implications for tumor promotion and inflammation. New Biol 1991;3:1206-1219.
77. Weiner KX, Dias-J-A. Protein synthesis is required for testosterone to decrease ornithine decarboxylase messenger RNA levels in rat sertoli cells.

Mol Endocrinol 1990;4:1791-1798.
78. Kumar R, Shoemaker AR, Verma AK. Retinoic acid nuclear receptors and tumor promotion: decreased expression of retinoic acid nuclear receptors by the tumor promoter 12-O-tetradecanoylphorbol-13-acetate. Carcinogenesis 1994;15:701-705.
79. Zheng S, Mcelwain CM, Taffet SM. Regulation of mouse ornithine fecarboxylase gene expression in a macrophage-like cell line—synergistic induction by bacterial lipopolysaccharide and cAMP. Biochem Biophys Res Commun 1991;175:48-54.
80. Imagawa M, Chiu R, Karin M. Transcription factor AP-2 mediates induction by two different signal-transduction pathways: protein kinase C and cAMP. Cell 1987;51:251-260.
81. Comb M, Mermod N, Hyman SE, et al. Proteins bound at adjacent DNA elements act synergistically to regulate human proenkephalin cAMP inducible transcription. EMBO J 1988;7:3793-3805.
82. Comb MJ, Kobierski L, Chu HM, et al. Regulation of opioid gene expression: a model to understand neural plasticity. NIDA Res Monogr 1992;126:98-112.
83. Begeot M, Shetty U, Kilgore M, et al. Regulation of expression of the CYP11A (P450scc) gene in bovine ovarian luteal cells by forskolin and phorbol esters. J Biol Chem 1993;268:17317-17325.
84. Pena A, Reddy CD, Wu SJ, et al. Regulation of human ornithine decarboxylase expression by the cMyc-Max protein complex. J Biol Chem 1993;268:27277-27285.
85. Bello-Fernandez C, Packham G, Cleveland JL. The ornithine decarboxylase gene is a transcriptional target of c-myc. Proc Natl Acad Sci USA 1993;90:7804-7808.
86. Wagner AJ, Meyers C, Laimins LA, Hay N. c-Myc induces the expression and activity of ornithine decarboxylase. Cell Growth Differ 1993;4:879-883.
87. Blackwell TK, Huang J, Ma A, et al. Binding of myc proteins to canonical and noncanonical DNA sequences. Mol Cell Biol 1993;13:5216-5224.
88. Prendergast GC, Lawe D, Ziff EB. Association of Myn, the murine homolog of max, with c-Myc stimulates methylation-sensitive DNA binding and ras cotransformation. Cell 1991;65:395-407.
89. Kerkhoff E, Bister K, Klempnauer KH. Sequence-specific DNA binding by Myc proteins. Proc Natl Acad Sci USA 1991;88:4323-4327.
90. Eilers M, Schirm S, Bishop JM. The Myc protein activates transcription of the alpha-prothymosin gene. EMBO J 1991;10:133-141.
91. Reisman D, Elkind NB, Roy B, Beamon J, Rotter V. c-Myc transactivates the p53 promoter through a required downstream CACGTG motif. Cell Growth Differ 1993;4:57-65.
92. Li LH, Nerlov C, Prendergast G, MacGregor D, Ziff EB. c-Myc represses transcription in vivo by a novel mechanism dependent on the initiator element and Myc box II. EMBO J 1994;13:4070-4079.
93. Bengal E, Ransone L, Scharfmann R, et al. Functional antagonism be-

tween c-Jun and MyoD proteins: a direct physical association. Cell 1992;68:507-519.

94. Li L, Chambard JC, Karin M, Olson EN. Fos and Jun repress transcriptional activation by myogenin and myoD - The amino terminus of Jun can mediate repression. Gen Dev 1992;6:676-689.
95. Blanar MA, Rutter WJ. Interaction cloning: identification of a helix-loop-helix zipper protein that interacts with c-Fos. Science 1992;256:1014-1018.
96. Blackwell TK, Kretzner L, Blackwood EM, et al. Sequence-specific DNA binding by the c-Myc protein. Science 1990;250:1149-1151.
97. Carr CS, Sharp PA. A helix-loop-helix protein related to the immunoglobulin E box-binding proteins. Mol Cell Biol 1990;10:4384-4388.
98. Ayer DE, Kretzner L, Eisenman RN. Mad: a heterodimeric partner for Max that antagonizes Myc transcriptional activity. Cell 1993;72:211-222.
99. Zervos AS, Gyuris J, Brent R. Mxi1, a protein that specifically interacts with Max to bind Myc-Max recognition sites. Cell 1993;72:223-232.
100. Lahoz EG, Xu L, Schreiber-Agus N, DePinho RA. Suppression of Myc, but not E1a, transformation activity by Max-associated proteins, Mad and Mxi1. Proc Natl Acad Sci USA 1994;91:5503-5507.
101. Gu W, Cechova K, Tassi V, Dalla-Favera R. Opposite regulation of gene transcription and cell proliferation by c-Myc and Max. Proc Natl Acad Sci USA 1993;90:2935-2939.
102. Morris DR, Allen ML, Rabinovitch PS, et al. Mitogenic signaling pathways regulating expression of c-myc and ornithine decarboxylase genes in bovine T-lymphocytes. Biochemistry 1988;27:8689-8693.
103. Holtta E, Sistonen L, Alitalo K. The mechanisms of ornithine decarboxylase deregulation in c-Ha-ras oncogene-transformed NIH 3T3 cells. J Biol Chem 1988;263:4500-4507.
104. Jaggi R, Friis R, Groner B. Oncogenes modulate cellular gene expression and repress glucocorticoid regulated gene transcription. J Steroid Biochem 1988;29:457-463.
105. Kerr LD, Miller DB, Matrisian LM. TGF-beta 1 inhibition of transin/stromelysin gene expression is mediated through a Fos binding sequence. Cell 1990;61:267-278.
106. Chatton B, Bocco JL, Goetz J, et al. Jun and Fos heterodimerize with ATFa, a member of the ATF/CREB family and modulate its transcriptional activity. Oncogene 1994;9:375-385.
107. Hurta RA, Greenberg AH, Wright JA. Transforming growth factor beta 1 selectively regulates ornithine decarboxylase gene expression in malignant H-ras transformed fibrosarcoma cell lines. J Cell Physiol 1993;156:272-279.
108. Sarfati P, Seva C, Scemama JL, et al. Effect of basic fibroblast growth factor on ornithine decarboxylase activity and mRNA expression in a pancreatic tumoral cell line (AR4-2J). Pancreas 1992;7:657-663.
109. Feinstein SC, Dana SL, McConologue L, et al. Nerve growth factor rapidly induces ornithine decarboxylase mRNA in PC12 rat pheochrom-

ocytoma cells. Proc Natl Acad Sci USA 1985;82:5761-5765.

110. Manzella JM, Rychlik W, Rhoads RE, et al. Insulin induction of ornithine decarboxylase. Importance of mRNA secondary structure and phosphorylation of eucaryotic initiation factors eIF-4B and eIF-4E. J Biol Chem 1991; 266:2383-2389.

111. Prystowsky JH, Clevenger CV, Zheng ZS. Epidermal growth factor induces ornithine decarboxylase in SV40-immortalized human keratinocytes. Exp Dermatol 1993;2:125-132.

CHAPTER 3

Mammalian S-Adenosylmethionine Decarboxylase Regulation and Processing

Bruce A. Stanley, Ph.D.

I. INTRODUCTION

Mammalian S-adenosylmethionine decarboxylase (SAMDC) is a highly regulated enzyme whose levels can fluctuate many-fold depending on the growth state and intracellular polyamine concentrations of the cell. SAMDC catalyzes the formation of S-adenosyl-(5')-deoxy-(5')-3-methylthiopropylamine, usually referred to as decarboxylated S-adenosylmethionine (dcSAM), which is the aminopropyl donor for the biosynthesis of the polyamines spermidine (Spd) and spermine (Spm). The decarboxylation reaction is one of the rate limiting steps in polyamine biosynthesis and also acts as a crucial branch point for intracellular S-adenosylmethionine (SAM) metabolism. While SAM is the major methyl donor in cells, its decarboxylation commits it to use by the polyamine pathway, since dcSAM is not a substrate for transmethylation pathways. The polyamines are involved in a number of cellular processes, and in particular seem to have a necessary enabling role in rapid cellular growth, since depletion of polyamines with various inhibitors prevents or diminishes the response normally seen after a variety of mitogenic or growth-promoting stimuli, including tumor promoters. In addition, the activity of both SAMDC and the other rate limiting enzyme of the polyamine biosynthetic pathway,

Polyamines: Regulation and Molecular Interaction, edited by Robert Casero. © 1995 R.G. Landes Company.

ornithine decarboxylase (ODC), have been observed to increase in cells stimulated to proliferate, and are greatly decreased in quiescent cells. The other enzymes in the polyamine biosynthetic pathway are spermidine synthase, which catalyzes the addition of the aminopropyl group donated by dcSAM to the diamine putrescine (the product of the ODC-catalyzed reaction) to form Spd, and spermine synthase, which forms Spm by catalyzing the addition of another dcSAM aminopropyl group to existing Spd. There is also an interconversion or back-formation pathway involving an acetylation of Spd or Spm catalyzed by spermidine-spermine-N^1-acetyltransferase (SSAT), followed by an oxidation reaction which is catalyzed by polyamine oxidase. These reactions form Spd from acetylated Spm and putrescine from acetylated Spd. SSAT is induced by high levels of polyamines in cells and may be a way to excrete excess polyamines, since the less-charged acetylated forms are more easily eliminated. The reader is referred to previous reviews of these various pathways.[1-6] One of the consequences of the existence of both a biosynthetic and an interconversion pathway has been that addition of any exogenous polyamine can lead to changes in intracellular pool sizes of the other polyamines as well, thus making assignment of regulatory or metabolic functions to a particular polyamine more difficult. Fortunately specific inhibitors of all the biosynthetic enzymes now exist, so changes in intracellular polyamine pools can be limited to a single polyamine, and functional assignments in regulation are being clarified. As will be discussed below, putrescine, Spd and Spm can all regulate the activity of SAMDC, although in very different ways that comprise both feedback inhibition by Spd and Spm and feedforward stimulation by putrescine.

SAMDC is a cytosolic enzyme which is synthesized as a ~38 kDa proenzyme, and subsequently cleaved in an apparently autocatalytic process to form the ~31 and ~7 kDa subunits of the mature enzyme. The native enzyme is an $\alpha_2\beta_2$ heterotetramer, although larger aggregates can readily form and may be active. Although the smaller subunit was once thought not to remain with the mature enzyme, subsequent experiments have clearly shown that it is part of the active enzyme complex, and in fact contains amino acid residues which play critical roles in both catalytic activity and the regulation of that activity. Unlike the pyridoxal-5'-phosphate cofactor seen with most decarboxylases, SAMDC uses a covalently bound pyruvate molecule as its cofactor, which places it in a very small group of enzymes with pyruvate cofactors.[7] The pyruvate is formed during proenzyme processing from an internal serine residue located at the carboxyl side of the proenzyme cleavage site, i.e., the pyruvate is the amino-terminal moiety of the cleaved 31 kDa subunit of the mature enzyme. The primary amino acid sequence of the enzyme is highly conserved among all mammalian species for which sequence data are available, and in addition the 5' untranslated region (5'-UTR) of the mRNA is highly conserved among all mammalian species. In addition to the highly conserved nature of the SAMDC 5'-UTR, the mRNA is among a relatively rare group of mRNAs with 5'-UTRs longer than 200 bases, a group whose members include a disproportionate number of proto-oncogenes, growth and transcription factors, and other regulatory proteins.[8] Both of these facts suggest a role of the 5'-UTR in control of expression of the enzyme. Southern blots from rat tissues are consistent with the existence of more than one copy of the gene,[9] and recent work demonstrates the existence of four copies of the rat gene, two functional genes on chromosome 20, a pseudogene on chromosome X, and an uncharacterized gene on chromosome 3.[10] A functional human SAMDC sequence has been mapped to chromosome 6 and an intronless pseudogene to chromosome X,[11] and the identified genomic sequences of the human and (both functional) rat genes reveals potential alternate polyadenylation signals, whose use would account for the two SAMDC mRNA sizes commonly observed on Northern blots (a third polyadenylation site in the human transcript does not seem to be used to any measurable extent).[10,12]

In addition to the highly conserved sequences of the known mammalian SAMDCs, there are also regions of primary amino acid sequence similarity, particularly around residues (such as the proenzyme cleavage site) which are critical for function, in such widely separated eukaryotes as yeast, trypanosomes, *Acanthamoeba* and sweet potato (see Fig. 3.1). In contrast there is essentially no similarity with the *E. coli* SAMDC other than a somewhat similar primary sequence near the cysteine residues which appear to be located in the active sites of each of the enzymes. The subunit sizes (12.4 and 18 kDa) and the native structure of the *E. coli* enzyme $(\alpha\beta)_4$ are also very different,[13,14] and in contrast to the human enzyme, the smaller β-subunit of the *E. coli* enzyme is apparently not crucial for catalytic activity, since activity was not affected after total tryptic digest of the subunit.[14] All of this suggests convergent rather than divergent evolution of the eukaryotic vs. prokaryotic enzyme.

The intracellular levels of the both SAMDC and ODC are rapidly altered in response to changes in the extra- or intracellular environment. For example, serum refeeding of cells in culture, exposure to various hormones or mitogenic stimuli, exposure to calcium ionophores or changes in intracellular polyamine levels all lead to changes in SAMDC activity.[15-17] There is less evidence for changes in SAMDC activity during normal cell cycling, although an increase is associated with G_0 to G_1 transition.[18] The cellular controls involved in these changes include changes in mRNA accumulation, changes in the translational efficiency of the mRNA, changes in the rate of proenzyme processing, changes in the K_m of the enzyme for the substrate SAM, and changes in the half-life of the enzyme. The changes in enzyme activity occurring in response to decreased or increased intracellular polyamines make sense

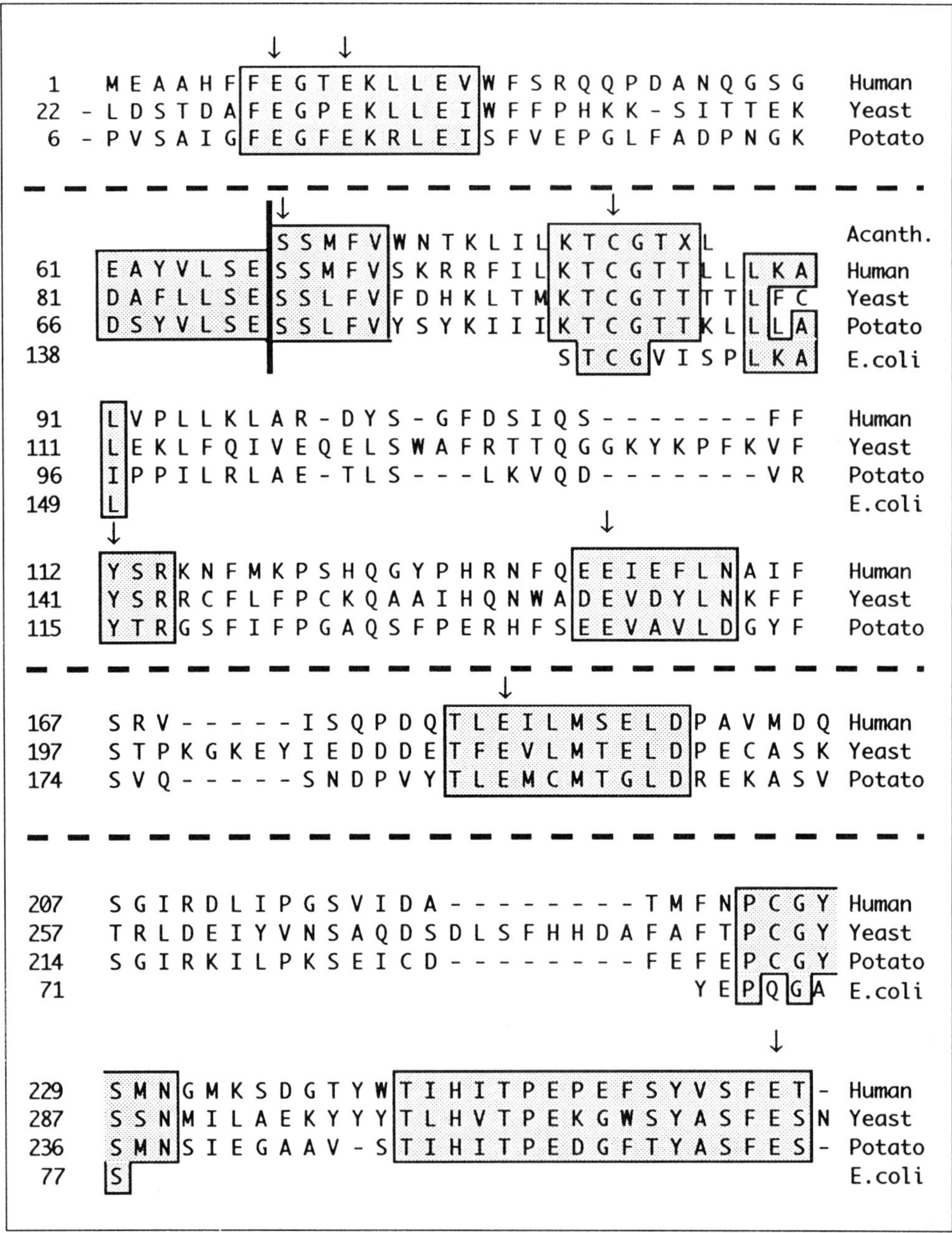

*Fig. 3.1. Primary amino acid sequence of human, yeast, and potato SAMDCs. Amino acid sequences are deduced from the cDNA sequence of human, yeast (*S. cerevisiae*), potato (*S. tuberosum*), and* E. coli *SAMDCs. The other mammalian sequences are >90% homologous to the human sequence and are not shown. The* Acanthamoeba castellani *(Neff) sequence is from direct peptide sequencing, so the residue numbering relative to the proenzyme is unknown. The site of proenzyme processing/pyruvate cofactor formation for the human and yeast enzyme is indicated by the heavy line between human sequence glutamate 67 and serine 68 (which is converted to pyruvate during the cleavage reaction). Regions of primary sequence homology are shown in the shaded boxes and mutations which have changed processing or activity kinetics in the human enzyme are indicated by arrows.*

in terms of regulation, i.e., when intracellular polyamine levels are decreased, enzyme activity goes up and when intracellular polyamine levels increase, enzyme activity goes down. However, it is not a simple one-to-one correspondence, and after depletion of intracellular polyamines with various inhibitors, exogenous polyamines can reduce SAMDC and ODC activities to unperturbed levels while only partially restoring intracellular polyamine levels. This suggests that exogenous polyamines may enter a different compartment than that occupied by steady-state intracellular polyamines, complicating interpretation of replacement experiments. It has been estimated that most polyamines, due to their cationic nature, are bound to anions such as RNA, DNA, phospholipids and ATP in cells,[19] while the small fraction of free polyamines may be the metabolically active fraction that regulates the biosynthetic enzyme levels. If exogenous polyamines enter preferentially into the free pool inside cells, this could account for their more efficient regulation of the enzyme levels. In any case, it is clear that changes in polyamine levels profoundly affect SAMDC activity, as do changes in proliferative state.[4,17,20-24] As will become clearer below, it is also apparent that Spd and Spm exercise their major control in somewhat independent ways.

II. TRANSCRIPTIONAL CONTROL OF SAMDC

Most treatments which increase SAMDC activity in cells act at least partially through increases in SAMDC mRNA,[9,25] and this has been assumed to represent changes in transcription rate (changes in SAMDC mRNA half-life, which could also contribute to the mRNA accumulation, have been measured much less frequently than the mRNA levels). Treatment of rats with difluoromethylornithine (DFMO), a specific enzyme-activated inhibitor of ODC, decreased polyamine levels in prostate, and increased the SAMDC mRNA levels about 7-fold.[25] Treatment of Swiss 3T3 cells with either a spermidine synthase inhibitor (S-adenosyl-1,8-diamino-3-thiooctane, AdoDato) or a spermine synthase inhibitor (S-methyl-5'-thiomethyladenosine, $AdoS^{+}(CH_3)_2$) depleted polyamine levels and caused a 2-fold increase in SAMDC mRNA.[9,15] Similarly, stimulation of growth-arrested 3T3 cells led to SAMDC mRNA increases.[18] Pharmacological doses of testosterone also greatly increased SAMDC activity and mRNA content in epithelial cells of prostate and seminal vesicles from castrated rats or mice, with a parallel increase seen in both the larger and smaller SAMDC mRNA species,[26] and a similar parallel change in the abundance of both mRNA species is also seen, for example, upon serum stimulation of SV-3T3 cells.[27] Interestingly, kidney SAMDC enzyme and mRNA levels did not change in response to testosterone, while ODC enzyme and mRNA levels increased in the kidney as well as in the prostate and seminal vesicles.[26] This led the authors to hypothesize that either the kidney cells were lacking a *trans*-acting factor that allowed testosterone responsiveness of the SAMDC gene, or that the kidney (but not the prostate or seminal vesicle) expresses a protein that prevents the androgen receptor from interacting with the SAMDC but not the ODC gene. There has also been a report of testosterone regulation of liver SAMDC, but in this case administration of testosterone to female rats lowered liver SAMDC activity to about half its normal level in the female rat, bringing it down to the level observed in males. Estrogen administration to rats of either sex had no effect, and it is not known what changes in mRNA level accompany the activity changes.[28]

The genomic sequences of human SAMDC contains DNA elements for binding of transcription factors AP-1, AP-2, CREB and SP-1, as well as several steroid receptor sites between position -3,158 and the transcription start site. An intron-exon boundary is located immediately in front of the codons for the processing site of the SAMDC proenzyme, suggesting that the two subunits may have evolved as separate protein domains.[12] The 5'-flanking sequence of the initially isolated rat genomic sequence[29] diverges from the human gene at -65 and is therefore missing many of the transcriptional control elements seen in the human gene. As pointed out by Maric et al,[12] it is somewhat difficult to rationalize such a large discrepancy between the regulatory sites of the two genes, since SAMDC is considered to be a housekeeping gene expressed in essentially all tissues, albeit a highly regulated one. A second rat genomic clone with a different 5'-flanking sequence starting at -63 has since been reported and this second gene shows several promoter elements in common with the human gene.[10] Another interesting but unexplored aspect of the sequences of the rat and human genes is the possible significance of the 100 codon in-frame downstream ORF found just past the termination codon of the protein coding regions.[9]

Of the polyamines, Spd seems to have the predominant effect on SAMDC mRNA accumulation. As is the case in most cell lines studied, when serum-starved SV-3T3 cells are refed serum, a large (7- to 8-fold) increase in SAMDC activity is observed 4 to 6 hours later. Inclusion of 5 or 50 μM Spd in the refeeding medium inhibited the serum-induced SAMDC activity increases by 45-53%, and all of this decrease in enzyme activity could be accounted for by a parallel 51-55% decrease in SAMDC mRNA accumulation. In contrast, inclusion of 5 or 50 μM Spm had a much greater overall inhibitory effect (83-88%) on serum-induced SAMDC increases, but without any change in mRNA levels.[27] If COS-7 cells were depleted of putrescine and Spd by treatment with the specific ODC inhibitor difluoromethylornithine (DFMO), Spm levels remained unchanged, while endogenous SAMDC mRNA levels increased 6- to 9-fold and SAMDC activity increased even more (25- to 32-fold). In contrast, depletion of Spm in the COS-7 cells (and concurrent augmentation of Spd levels) with a specific inhibitor of spermine synthase (n-butyl-1,3-diaminopropane, or BDAP[30,31]) had no effect on SAMDC mRNA levels. Addition of Spm along with the BDAP prevented the 18-fold increase in SAMDC activity seen with the BDAP alone, showing that the BDAP effect was due specifically to decreased Spm.[27] The lack of mRNA level change indicates that the increased SAMDC activity must be the result of either decreased degradation of SAMDC protein, or increased translational efficiency of SAMDC mRNA engendered by lowered Spm concentrations. Since BDAP had no effect on levels of SAMDC activity expressed by a plasmid (pSAMh1) which lacked the first 72 bases of the full-length 5'-UTR, stabilization of the SAMDC protein is unlikely to have contributed under those experimental conditions, which implies that a Spm-responsive translational control element is located in the first 72 bases of the 5'-UTR. To avoid complications of interpretation due to potential polyamine contributions to SAMDC protein stability changes, COS-7 or CHO cells depleted of either Spd or of Spm as above were also transfected with chimeric plasmids linking portions of the human SAMDC promoter region to a chloramphenicol acetyltransferase (CAT) reporter gene. Supplying exogenous Spd decreased both CAT mRNA and protein amounts, while Spm had little or no effect on mRNA levels. Since the smallest SAMDC promoter fragment used (-172 to +112, with the normal SAMDC transcription start site as +1) was fully responsive to inhibition by Spd, this indicates that a Spd-responsive element is probably contained in this region.[27] These experiments also provide evidence consistent with a transcriptional effect of Spd mediated through the SAMDC promoter region, since Spd levels did not alter CAT expression from a plasmid driven by a Rous sarcoma virus promoter in the same cells; however, it is also formally possible that the Spd-responsive signal in the -173/+112 SAMDC region could destabilize both the SAMDC and the CAT mRNA in these cells, since either decreased transcription or increased mRNA degradation in response to Spd could account for the observed decreases in CAT mRNA and protein. It is evident in any case that Spd, but not Spm, decreased accumulation of both endogenous SAMDC mRNA levels and CAT mRNA driven by SAMDC promoter regions, supporting the idea that Spd is the more important polyamine in regulation of SAMDC mRNA levels.

III. TRANSLATIONAL CONTROL OF SAMDC EXPRESSION

Translational control of SAMDC expression has been implied by numerous experiments in which changes in enzyme activity or amount could only be partially accounted for by changes in mRNA amount and enzyme half-life (reviewed in ref. 32). Various treatments with inhibitors of the polyamine biosynthetic enzymes cause decreased polyamine levels and apparently lead to more efficient translation of SAMDC mRNA. For example, depletion of spermine in SV-3T3 cells by the spermine synthase inhibitor S-methyl-5'-thiomethyladenosine ($AdoS^+(CH_3)_2$) caused up to 30-fold, dose-dependent increase in SAMDC activity (and antigenic enzyme). This was accompanied by a 6-fold stabilization of the enzyme and a 2.8-fold increase in SAMDC mRNA, which accounts for only 55% of the increase in SAMDC enzyme amount observed.[15] In another cell line (Ehrlich ascites tumor cells), depletion of polyamines by growth in 5 mM DFMO led to a 5.2-fold increase in SAMDC synthesis rate (measured by incorporation of [^{35}S]-methionine into immunoprecipitable protein during a 25 min pulse labeling), but only a 2-fold increase in mRNA level was observed, insufficient to account for the increase in synthesis rate. SAMDC activity per cell was also increased 5.8-fold with only a 20% increase in SAMDC half-life and both of these measures are consistent with a strong translational component of the observed changes in SAMDC expression.[20] In a third cell line, addition of the specific SAMDC inhibitor 5'-{[(Z)-4-amino-2-butenyl]methylamino}-5'-deoxyadenosine ([(Z)-AbeAdo]) completely inhibited SAMDC activity and depleted both Spd and Spm in L1210 cells. Although the presence of the inhibitor prevented measurement of SAMDC activity, the amount of antigenic SAMDC protein doubled in these cells, and the synthesis rate of the enzyme was increased 40- to 50-fold, again as judged by changes in [^{35}S]-methionine incorporation into immunoprecipitable protein during a short labeling period. Adding Spd at 1 μM or Spd at 2-5 μM to the growth media could prevent these increases. The difference between the increase in the synthesis rate and the actual increase in SAMDC accumulation was postulated to have occurred through increased degradation of the enzyme. We have shown that [(Z)-AbeAdo] inactivates SAMDC through transamination of the pyruvate cofactor, and such transaminations (which occur naturally as well as by the actions of an inhibitor) may target inactivated SAMDC for more rapid degradation.[33] In the above experiments, the mRNA level for SAMDC only went up 2-fold in response to the [(Z)-AbeAdo], not nearly enough to account for the 40- to 50-fold increase in synthesis rate, particularly if the enzyme degradation rate was actually increased by the inhibitor, indicating a large translational component.

In addition to the polyamine-mediated translational control implied above, there are also indications that increases in SAMDC caused by growth stimulation have a translational component. Mach et al showed that during mitogen activation of bovine lymphocytes, an 8- to 10-fold increase in SAMDC activity was only accompanied by a 4-fold increase in SAMDC mRNA.[34] In further investigating this implied translational control, they examined polysome profiles after sucrose density gradient separation, and showed that the average number of ribosomes associated with SAMDC mRNA in mitogen-stimulated lymphocytes was 2.7, whereas the unstimulated cell mRNA was in a fraction of monosomes (1.4 ribosomes per mRNA on average).[34] This directly demonstrated a mitogen-induced change in SAMDC mRNA translational efficiency. Treatment of lymphocytes with low doses of cycloheximide, which only slows elongation rather than completely stopping synthesis,[35,36] moved all of the SAMDC mRNA into polysomes of about 10 ribosomes, about the expected number for a messenger RNA the size of SAMDC.[32] Both the cycloheximide and the mitogen-stimulated movement of SAMDC mRNA into larger polysomes are consistent with a mRNA which is being translationally controlled at the level of initiation.

By gel filtration of a rabbit reticulocyte lysate to remove endogenous polyamines in a cell-free translation system, Kameji and Pegg were able to demonstrate a direct effect of the polyamines, particularly Spm, on translational efficiency of both SAMDC and ODC mRNA contained in rat prostate poly(A)$^+$ RNA.[37] Adding back spermine to only 80 µM in the filtered lysate could directly inhibit the translation of mRNAs for both SAMDC (about 75% inhibition) and ODC (about 25% inhibition), as measured by [^{35}S]-methionine incorporation into immunoprecipitable protein. The translation of albumin mRNA and total protein synthesis in the lysate were not decreased at 80 µM Spm, although they were decreased at higher concentrations. At 10-fold higher concentrations, Spd could also specifically inhibit SAMDC and ODC mRNA translation, about 85% and 55% respectively, while albumin and total protein synthesis remained at the same levels as in the filtered lysate.[37] Although this clearly indicates that increasing polyamine concentrations can inhibit translation of these two messages (and eventually total protein synthesis), low levels of polyamines appeared to be stimulatory to both specific mRNA translation and to overall protein synthesis, and the rate of total protein synthesis continued to rise as polyamine concentrations were raised to levels that were clearly detrimental to translation of the specific SAMDC and ODC mRNAs. This is consistent with other reports of polyamine roles in reducing premature termination of proteins, in increasing elongation rates during protein synthesis, and in increasing overall protein synthetic rates in various cell-free translation systems.[38-47] The stimulatory effect of low concentrations of the polyamines was accentuated at suboptimal Mg^{++} concentrations, indicating that some of the general stimulatory effects may be due to the cationic nature of the polyamines as opposed to a specific function requiring the exact charge separation of the polyamines. It is also possible that the concentration of polyamines required for optimal translation of many different mRNAs may vary in a sequence-dependent manner, as indicated in several previous reports.[38,41,48] The optimal polyamine concentration for ODC and SAMDC translation certainly occurs at the low end of any such range of polyamine concentrations, since overall protein synthesis directed by prostate poly(A)$^+$ RNAs was not inhibited at all by levels which inhibited synthesis of SAMDC and ODC. This inhibitory effect has also been demonstrated in reticulocyte lysates by *bis*-ethyl analogs of the two polyamines. In assays essentially identical to the above, 80 μM N^1,N^{12}-*bis*(ethyl)spermine (BESpm) inhibited SAMDC mRNA translation as well as did Spm, and with a similar lack of effect on either albumin or total protein synthesis. N^1,N^9-*bis*(ethyl)spermidine (BESpd) was slightly less effective even at 10-fold higher concentrations. The Spm and Spd analogs thus have relative efficacies similar to those of the parent Spm and Spd molecules.[49] In whole cells, methylated Spm and putrescine analogs which could not be converted into other polyamines were used to study ODC and SAMDC mRNA translation rates. In Ehrlich ascites tumor cells, 5,8-dimethylspermine (but not 1,4-dimethylputrescine) decreased SAMDC synthesis rates (as measured by ^{35}S-methionine incorporation), while *both* analogs decreased ODC synthesis rates.[50] The authors concluded that Spm is more important in regulating SAMDC translation based on the effect of the Spm analog; however, it is somewhat incongruous that the "ineffective" 1,4-dimethylputrescine treatment actually *raised* intracellular Spm levels (while depleting putrescine and Spd). By using a specific inhibitor of ODC (DFMO) vs. a specific inhibitor of spermine synthase (BDAP), Shantz et al were able to selectively deplete either putrescine and Spd with DFMO, or Spm with BDAP, in HT29, CHO, and COS-7 cells. Although either treatment increased the amount of endogenous SAMDC activity (and antigenic SAMDC as measured by Western blots from the COS-7 cell extracts), depletion of COS-7 cell Spd led to an increase in SAMDC mRNA, while depletion of Spm did not. The increase in SAMDC activity with Spd depletion could be accounted for by the increase in mRNA and the increased half-life of the protein, while the Spm depletion results were consistent with translational regulation. Taken together, the above experiments indicate that growth stimulation or decreases in polyamine concentration can increase the translational efficiency of SAMDC mRNA, and that Spm is the most potent of the polyamines at decreasing SAMDC mRNA translational efficiency. More recent experiments have begun to shed light on the mechanisms through which this control is exerted.

Based on comparison with conclusions from many studies of general mRNA features contributing to translational control,[51-53] the long 5'-UTR of SAMDC mRNA has two features which might contribute to its translational control. The first is a stable predicted secondary structure with a ΔG of -69 kcal/mole,[54] as calculated by the FOLD algorithm.[55] In the ribosome-scanning model of protein synthesis initiation, the 40S ribosomal subunit binds at the 5' capped structure of mRNAs and then scans the 5'-UTR until it reaches an AUG initiator codon.[52,56] Stable RNA structures are predicted to inhibit initiation by impeding ribosome scanning between the 7-methylguanosine cap (5'-cap) found on mRNAs and the AUG start codon and stem-loop structures with ΔGs of -50 to -60 kcal/mole greatly decrease initiation at downstream AUG codons both in vitro and in vivo.[57-59] The second potentially important feature of the SAMDC 5'-UTR is a short upstream open reading frame (uORF) located approximately 11-14 bases from the cap-site in both bovine and human SAMDC mRNAs. This uORF encodes the hexapeptide MAGDIS, and is located about 300 bases upstream from the major ORF encoding the SAMDC protein. Since eukaryotes do not appear to utilize bicistronic mRNAs often or well,[60] such uORFs can also have profound effects on translation of the downstream reading frames, the magnitude of the effect being influenced by (1) the distance of the uORF from the cap structure (uORFs within 12 bases of the cap are "skipped" about 50% of the time, so that initiation occurs at the downstream ORF, while increasing the distance to just 20 nucleotides leads to 100% initiation at the uORF);[61,62] (2) the relative strengths of the context of the start codon in the uORF vs. downstream ORF (the bases surrounding a start codon, particularly but not limited to the +4 and -3 positions, make a particular AUG codon a stronger or weaker initiator codon);[63,64] (3) the presence or absence of secondary structure downstream from the uORF (stable secondary structure, optimally at about 12-15 bases downstream from a "weak" initiator, increases the strength of that codon as an initiator).[57,65] Perhaps the best known example of translational control by uORFs is in expression of the yeast gene GCN4.[66-68]

Recent experiments from several laboratories have determined that the strong, predicted secondary structure of the 5'-UTR is of less importance than the uORF in control of SAMDC mRNA translation, in contrast with translational control of ODC expression, where the secondary structure in the 5'-UTR may be of greater importance than the ODC uORF.[69-72] Suzuki et al made chimeric constructs from mouse SAMDC cDNAs and showed that the polyamine effects on cell-free translation of RNA produced from these constructs were dependent on the presence of the 5'-most 100 bases of the original mouse cDNA containing the uORF from the mouse sequence. This argues against control by any secondary structure, since the "minus 100" RNA in their chimeric constructs still had a relatively stable predicted structure of ΔG = -58 kcal/mol, yet translated more efficiently and were without polyamine responsiveness.[73] Interestingly, this was true of both the *stimulatory* effects at low polyamine concentration and the *inhibitory* effects at higher concentrations.

A series of experiments in which the secondary structure of either bovine or human SAMDC 5'-UTRs was systematically removed showed only minor effects on translational efficiency of such deletion constructs, while any construct in which the uORF was removed greatly increased translation. This was true as well (with minor quantitative differences) when the wild-type or altered SAMDC 5'-UTRs were linked to reporter genes whose amount or activity was then monitored, thus showing that the control exerted by the 5'-UTR could be transferred to expression of an independent protein. These effects have been demonstrated in several cell types, with some as yet unresolved cell-specific differences. Shantz et al[54] made a series of expression constructs which contained the full length 5'-UTR of human SAMDC, plus the entire coding region and all of the 3'-UTR which was contained in our original clone of human SAMDC, pSAMh1.[9] The alterations to the 5'-UTR depicted in Figure 3.2 were made, as well as chimeric plasmids containing the SAMDC 5'-UTR upstream of a luciferase reporter gene coding region and the various constructs were transfected into COS-7 cells. In some experiments, the transfected cells were treated with the spermine synthase inhibitor BDAP[30,31] (50 μM), which low-

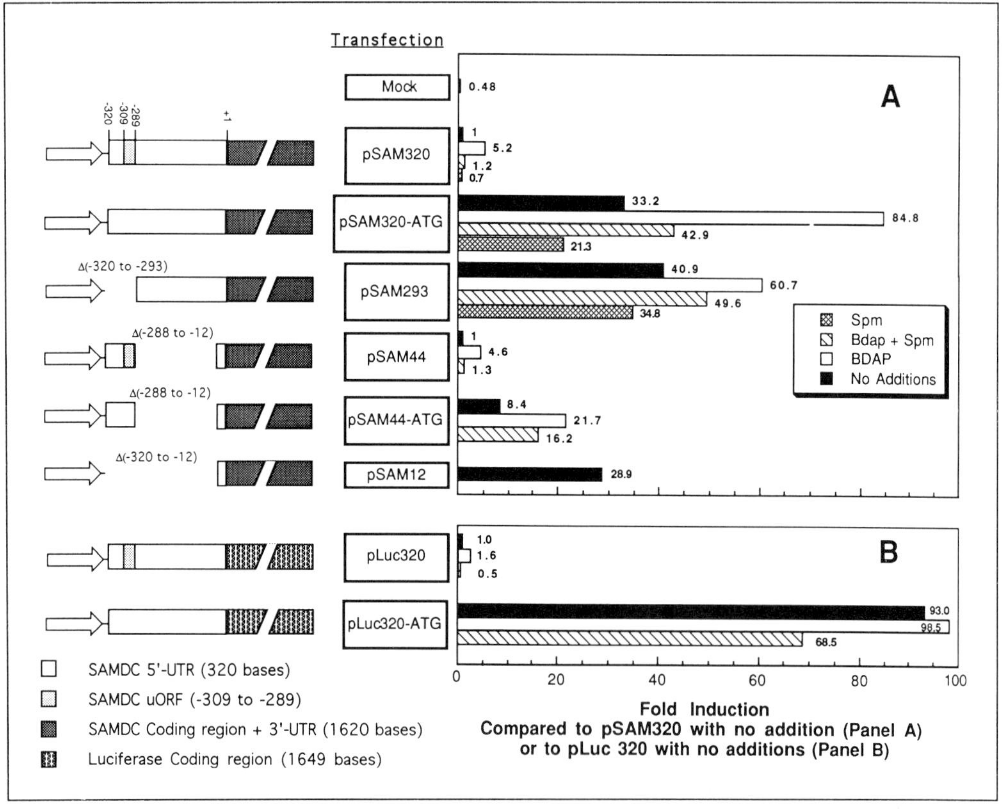

Fig. 3.2. Effect of alterations to the SAMDC 5'-UTR on translational efficiency. COS-7 cells were transfected with pEUK-C1 vectors (Clontech) containing human SAMDC cDNA just after the SV40 late promoter transcription start site. Where indicated, BDAP (50 μM), a specific inhibitor of spermine synthase,[31] was used to deplete cellular Spm levels. It also tripled Spd levels and depleted putrescine levels. Addition of 10 μM Spm along with BDAP where indicated returned Spd and Spm levels to control values, with no change in putrescine levels. Values are normalized to levels of SAMDC activity or luciferase protein after transfection with full-length SAMDC 5'-UTR constructs in the absence of BDAP or exogenous Spm. These values were 0.16 to 0.30 nmol CO_2 released/30 min/mg protein for the SAMDC plasmid, and 20.3 pg luciferase for the luciferase chimeras. Data are recalculated from Table 3.1, Fig. 3.3, and Fig. 3.4 of Shantz et al.

ered intracellular Spm levels to 10-20% of untreated, transfected cells (from 20-25 nmol/mg protein to 3-7 nmol/mg protein in BDAP-treated cells), lowered putrescine to essentially undetectable levels (less than 5% of control values), while simultaneously raising Spd levels to 250-300% of controls. Administration of 10 μM Spm along with the BDAP restored Spd levels to control values, while putrescine remained depressed and Spm was elevated to 200-240% of controls. As determined by slot blots, transfected levels of RNA were the same among the different constructs. As summarized in Figure 3.2, any construct which contained the intact uORF had much lower expression of either SAMDC or the luciferase reporter gene, while deletion of almost the entire region between the uORF and the downstream start codon had very little effect (p44SAM compared to p320SAM). Elimination of the uORF by simple alteration of the start codon AUG to CGA had the same magnitude effect as did any complete deletions of the uORF (as did subsequent mutation to CUG, L. Shantz, personal communication). In the absence of the uORF, the intervening sequence -288 to -12 seemed to have, if anything, a positive effect on translation, as seen in the much smaller increase noted upon alteration of the uORF in the pSAM44-ATG vs. pSAM44 (8.3-fold) compared to the pSAM320-ATG vs. pSAM320 (18.6-fold). This is consistent with work with other cDNAs which has indicated that increasing the distance between the cap and the ORF between 17 and 80 bases with random, unstructured synthetic leaders causes a proportional increase in the relative strength of translation from the ORF,[51,62,74] although it is unusual that a highly structured 5'-UTR such as that of SAMDC would have a positive effect.

Morris and colleagues have also extensively studied the translational control of SAMDC in several other cell lines by measuring the location of both native and chimeric SAMDC mRNA in polysome profiles after sucrose gradient centrifugation.[32] They first showed that in resting bovine T-lymphocytes, endogenous SAMDC mRNA was found primarily in monosomes, while mitogenic stimulation of the cells led to a broad redistribution of the mRNA into medium-sized polysomes.[34] This regulation is in contrast to that of other translationally-controlled mRNAs for ODC and for the ribosomal protein L32, where the primary location of the mRNAs in quiescent cells was bimodal, in ribonucleoprotein fractions (mRNP) containing dissociated ribosomes and in larger polysomes, with movement into polysome fractions upon stimulation.[17,75-77] They then examined this translational control phenomenon in detail using a chimeric 5'-UTR construct where 27 bases identical to the human genomic SAMDC cDNA region containing the transcription start site were spliced onto the 5' end of a bovine SAMDC cDNA containing a 310 base 5'-UTR. (Thus, part of the uORF at -316 to -296, with numbering relative to the start site of the major ORF, consisted of bases known to be identical in both the human and bovine clones (-310 to -296), while the 5' most end is identical to the human sequence. The known bovine 5'-UTR sequence shows 94% identity with the human sequence, with 100% identity from -310 to -277 of the bovine cDNA). The translational efficiency of this construct, either attached to the native bovine SAMDC coding region or in a chimeric construct with a human growth hormone (hGH) cDNA, was studied after expressing the constructs in several cell types. In various nonlymphoid cell types, including Swiss 3T3 fibroblasts, Y1 adrenal carcinoma cells, NIH 3T3 cells, BHK cells, primary human foreskin fibroblasts and HeLa cells, SAMDC or growth hormone mRNA was found in a broad distribution in sucrose gradient fractions corresponding to small- to medium-sized polyribosomes (1-10 ribosomes). In contrast, the SAMDC mRNA in two cell lines of T-cell origin, Jurkat and EL-4 cells, distributed in monosomes and disomes.[17,78,79] This led to a conclusion that there was tissue-specific control of SAMDC mRNA expression by translational control in cells of lymphoid origin but not in other cell types of nonlymphoid origin,[78] which contradicts the results of Shantz et al showing translational control through the SAMDC 5'-UTR in COS-7 cells, which are of nonlymphoid origin. This conflict will be discussed further below.

To look for the mechanism by which the strong translational control observed in the lymphoid Jurkat cells was implemented, a series of alterations were made to both the uORF and the sequences between the uORF and the downstream hGH ORF (the intercistronic region) in a chimera containing the human/bovine SAMDC 5'-UTR described above and the reporter gene human growth hormone (hGH). Most, but not all, alterations to the uORF of the SAMDC 5'-UTR:hGH chimeras moved the mRNA into polysome fractions, unlike the intact uORF:hGH constructs whose mRNA was found, like endogenous SAMDC mRNA, primarily in monosomes.[80] Four different deletions spanning the 295 bases of 5'-UTR between the uORF and the +1 major ORF had no effect on the inhibited translation observed in the Jurkat cells,[79] in agreement with the results in COS-7 cells indicating that the translational control resides in the uORF, not in the strong predicted secondary structure of the 5'-UTR. Also quite striking is the well-supported conclusion that the regulation resides in the expressed amino acids of the hexapeptide, but not the ribonucleic acid sequence encoding them and that it is mandatory that the last three amino acids of the hexapeptide MAGDIS occur right before the stop codon. As seen in Table 3.1 (data taken from Tables II and III in reference 80), mutations altering any of the DIS residues diminished the translational repression observed, while six alterations in the nucleotide sequence at wobble positions (so that the altered sequence still coded for DIS) had no effect. Alterations of the second or third codons (A2→E or G3→A) were also without effect. Two different 3' extensions of the uORF by 21 nucleotides so that DIS no longer immediately preceded the stop codon also eliminated the regulatory effect of the uORF,[80] while deletion of the nucleotides immediately after the uORF had no effect.[79] This strongly indicates that the regulation occurs through the actual peptide and its juxtaposition to the stop codon and not through any characteristics of the nucleotide sequence (other than what they encode). The peptide apparently only acts in *cis*, since cotransfection of a wild-type SAMDC 5'-UTR/human growth hormone chimera had no effect on ribosomal loading of a SAMDC mRNA containing an altered (MAGRIS, nonsuppressing) 5'-UTR, even when the wild type mRNA accumulated to 10-fold higher levels. It is not known what the expressed level of hexapeptide is under any of these conditions. Extensions at the 5' end of the uORF of 9 or 21 nucleic acids (three or seven amino acid residues) also eliminated the regulation by the uORF.[80]

It is also possible that MxxDIS is the actual peptide motif involved in this regulation and that amino-terminal extensions

Table 3.1. Effects of mutations of the uORF of the SAMDC 5'-UTR

Nucleic acid sequence of uORF							Amino acid sequence encoded by uORF	Polysome size
ATG	GCC	GGC	GAC	ATT	AGC	TAG	MAGDIS *	Monosomes
---	AG-	---	---	---	---	---	-E---- *	Monosomes
---	---	-C-	---	---	---	---	--A--- *	Monosomes
---	---	---	CG-	---	---	---	---R-- *	5-10
---	---	---	- G	---	---	---	---E-- *	5-10
---	---	---	---	GCC	---	---	----A- *	2-4
---	---	---	---	---	GCT	---	-----A *	5-10
---	--A	--G	--T	--C	TC--	---	------ *	Monosomes
---	GAC	AGC	ATT	GGC	GCC	---	-DSIGA *	5-10
							------LRSLYSL*	5-10
							------ASAGRIS*	5-10
							------ASA*	5-10
							MAGRISAS------ *	5-10
							MSAS ------ *	5-10

The data shown are combined from Tables II and III in reference 80. Jurkat cells were transfected with chimeric plasmids containing either the wild type SAMDC uORF (encoding the hexapeptide MAGDIS), a modified uORF which encoded the indicated peptides, or a modified uORF in which the codons had been altered but still coded for the wild-type hexapeptide MAGDIS. In the constructs where the uORF is lengthened, the distance between the 5'-cap site and the initiator AUG codon remained constant (approximately 11-14 nucleotides). The average size polysomes in which each mRNA was recovered is shown in the right hand column. The asterisk (*) indicates the stop codon; dashes (-) indicate no change from the wild type sequences in line 1. The nucleotides coding for the 5'- and 3' uORF extensions are not shown for clarity of presentation.

containing MxxDIS at the carboxy terminal might be tolerated. Although this possibility has not been eliminated by the experiments to date, the parsimonious explanation of the above results is that translation in *cis* of a uORF of specific length and carboxy-terminal amino acid sequence (DIS) is crucial for the translational repression observed in Jurkat cells. Although Shantz et al concluded that the Spm-mediated translational control exerted through the SAMDC 5'-UTR in COS-7 cells involved the uORF region but not necessarily translation of the uORF (based on increased SAMDC activity from the pSAM320-ATG and pSAM44-ATG constructs when Spm was depleted by BDAP, see Fig. 3.2), they noted that in those experiments Spm depletion also stabilized the enzyme itself.[54] Their results are therefore consistent with a Spm effect on both translation and stability of SAMDC, i.e., the increase seen with the pSAM320-ATG and pSAM44-ATG constructs and BDAP-depleted Spm (~2.5-fold and ~2.6-fold respectively over that of the same constructs in the presence of normal Spm levels) might represent stabilization of the enzyme rather than continued Spm-mediated translational control in the absence of the uORF—note in particular that the SAMDC activity increase upon BDAP-induced Spm depletion is much greater (~5.2-fold and ~4.6-fold respectively) when the otherwise identical but ATG-containing constructs pSAM320 or pSAM44 were used, as one would expect if Spm were exerting an additional translational effect through the uORF. Their results with the chimeric pLuc320 and pLuc320-ATG constructs are also consistent with this view, since Spm presumably does not stabilize the luciferase protein. With these chimeric constructs, increased levels of luciferase following BDAP-induced Spm depletion occurred only with the pLuc320 construct, which contained the ATG start codon of the uORF, and not with the pLuc320-ATG construct which lacked the ATG (Fig. 3.2). However, adding Spm with the BDAP (which produced intracellular Spm levels about 2-fold higher than control) did decrease luciferase production somewhat in both constructs and further experiments are clearly needed to precisely define the mechanism of Spm regulation of SAMDC mRNA translation. Evidence to support an additional locus of possible Spm effects comes from demonstration of Spm-induced in vitro binding of a 39 kDa cytosolic protein to a labeled RNA containing the -50 to +22 region of the mouse SAMDC sequence (but not the equivalent uORF).[81] This could represent another level of control of SAMDC translation by Spm and since the +1 to +22 portion of this highly conserved region is missing in the pLuc320 constructs used by Shantz et al, Spm might no longer be able to act through this mechanism on these constructs.

A final argument against major contributions of 5'-UTR secondary structure to translational control of SAMDC is found in work by Shantz and Pegg, where ODC and SAMDC constructs, as well as chimeric constructs containing the 5'-UTR of ODC or SAMDC and a luciferase reporter gene, were expressed in either NIH 3T3 cells or the NIH 3T3 cell line 3T3 pMV7-4E (P2 cells). The P2 cells over-express eukaryotic initiation factor eIF-4E, the α-subunit of eIF-4F, also called the cap-binding protein or CPB-II. This initiation factor is the least abundant initiation factor in cells, and in addition to cap-binding, eIF-4E has been reported to improve the translation of mRNAs containing secondary structure in their 5'-UTRs.[82] As mentioned earlier, highly structured, long 5'-UTRs seem to occur disproportionately in oncogene and growth control related mRNAs and over-expression of eIF-4E in NIH 3T3 or Rat 2 fibroblasts causes a transformed phenotype, apparently through a *ras*-dependent pathway.[83-85] Expression of ODC or a chimeric luciferase cDNA containing the 5'-UTR of ODC mRNA was greatly increased in the P2 cells compared to the NIH 3T3 cells expressing lower amounts of eIF-4E, but expression of SAMDC or a chimeric luciferase cDNA containing the 5'-UTR of SAMDC mRNA was equal (and poor) in the two cell types. Removal of the uORF of the SAMDC 5'-UTR greatly increased the expression of the luciferase chimera in both cell types, but there was still no difference in expression resulting from the increased eIF-4E in the P2 cells. Thus, while over-expression of eIF-4E significantly increased translation from the ODC 5'-UTR constructs, whose secondary structure has previously been shown to be important for control of ODC mRNA translation,[70] no effects were seen on translation from the SAMDC 5'-UTR constructs, consistent with the idea that regulation of SAMDC mRNA translation resides in the uORF rather than in the 5'-UTR secondary structure.

Having defined initiation at the uORF as the probable mechanism in the strong translational repression of SAMDC seen in the lymphoid Jurkat cells, a logical next step was to define the means by which nonlymphoid cells, expressing the same constructs, escape from that control and more efficiently express the downstream ORF. Relief of inhibition can be envisaged to occur in several ways and any models developed have to account for the repression, the cell-specific differences observed and the relief of the translational repression in growth stimulated and/or Spm-depleted cells. Although both the uORF and the major ORF of SAMDC mRNA are in strong contexts for initiation according to the rules developed by Kozak,[63,64,86] the proximity of the uORF to the 5' cap site (11-14 nucleotides) should make initiation at this upstream site relatively inefficient, so that translation at the downstream, major ORF could occur through "leaky scanning" wherein once initiation at the first uORF did not occur, the ribosome would continue scanning and thus initiate at the second strong AUG start codon, the major ORF. Cellular conditions which increased the leakiness of the upstream initiation would thus lead to more efficient translation of the downstream ORF, accompanied by increased appearance of the mRNA in polyribosomes. That the 11-14 nucleotide leader length of the SAMDC 5'-UTR was involved in the relative leakiness of the uORF in HeLa cells (a nonlymphoid cell line where SAMDC 5'-UTR suppression was normally minimal) was directly demonstrated by increasing the length of the 5'-UTR leader by an additional 30 nucleotides. The translation of the growth hormone chimeras containing the extended leader sequence was decreased by 5-fold, and the 5'-extended mRNA was found in the monosome fraction, unlike wild-type length mRNA, which was found in a broad peak associated with 4-6 ribosomes. The inclusion of the 30 extra bases also decreased the already low expression of the growth hormone chimera in Jurkat cells another 2.5-fold compared to a chimera containing the wild-type 5'-UTR.[79] The effect of leader length on efficiency of initiation at start codons close to the 5'-cap has been postulated to relate to the time that a scanning ribosome needs to accumulate all the initiation factors needed for efficient initiation—a longer leader may give the ribosome more time to amass whatever factors are needed before encountering a start codon, and if those factors are not in place when the first AUG is reached, the ribosome simply continues scanning. The amount of time needed could also change depending on the effective concentrations of limiting factors in the cell, and a possible explanation of the leaky scanning in HeLa cells could be relatively lower concentrations of some limiting initiation factor. Conversely, in the inhibited Jurkat cells, growth stimulus or Spm-depletion which led to lower levels or decreased affinity of a limiting factor could change the relative frequencies of initiation at the uORF vs. downstream major ORF, thus relieving the translational repression under those conditions. As pointed out by Morris and colleagues, in the case where 90% of the ribosomes are initiating at the uORF in Jurkat cells, a relatively small decrease in initiation at the uORF would lead to a large increase in initiation at the major ORF, e.g., a decrease from 90% initiation at the uORF to 80% is only a ~11% change, but the amount of initiation at the major ORF, going from 10% up to 20%, would double.[79] A second possibility for altering the relative rates of initiation at a "leaky" initiation site is through secondary structure slightly downstream from the initiator codon. Such structures seem to make a scanning ribosome pause and are most efficient when placed 14 nucleotides downstream from the AUG codon, which corresponds to the distance (12-15 nucleotides) from the codon recognition site to the leading edge of the ribosome. Such structures can greatly increase the efficiency of initiation at weaker context or even non-AUG start codons.[65] Once initiation has occurred and the full 80S ribosome is assembled, elongation is possible because the 80S ribosome can read through structures that the 40S preinitiation complex cannot. In the SAMDC 5'-UTR-hGH chimeras expressed in HeLa cells, addition of an inverted repeat (ΔG = -61 kcal/mol) 21 bases downstream from the upstream AUG codon caused translational suppression not seen with the parent SAMDC 5'-UTR[79] and increased initiation at the uORF is one possible mechanism; however, the inverted repeat had previously been demonstrated to block 40S ribosome scanning[87] and downstream initiation may have been reduced simply by suppressed scanning through the intercistronic 5'-UTR, rather than through increased initiation at the uORF. If such secondary structures were involved in determining the different relative efficiencies of initiation at the uORF in Jurkat vs. HeLa cells, this would have to be a regulatable process such that mitogenic stimulation of the Jurkat cells could lead to a decrease in the secondary structure. Their experiments where alteration or elimination of the nucleotides downstream from the start codon of the uORF had little effect on translational repression in the Jurkat cells argues against this possibility, however.[78,80]

Cap-independent translation, another possible explanation of the relatively efficient translation observed in the nonlymphoid cells examined by Morris and colleagues, occurs where the initial contact between the ribosome and an mRNA occurs not at the 5'-cap, but rather at an internal ribosome entry site (IRES). Examples of IRESs are fairly common in viral mRNAs (e.g., cytomegalovirus, adenovirus or poliovirus) and limited examples are known in eukaryotic mRNAs as well. To test whether an IRES was involved in the relatively efficient translation of the downstream ORF of SAMDC in HeLa cells, Ruan et al first eliminated portions of the intercistronic region between the uORF and the downstream ORF as described above. If an IRES had been involved in translation of the downstream ORF, elimination of the region containing the IRES should have greatly reduced the expression of the growth hormone reporter gene. Although the overlapping deletions tested covered all of the intercistronic region, none of the deletions caused a decrease in growth hormone expression.[79] The second line of evidence against an IRES is seen in the experiment mentioned in the previous paragraph where insertion of an inverted repeat just 3' of the uORF caused decreased expression of the growth hormone chimera in HeLa cells. Since the entire wild type intercistronic region still existed just past the inverted repeat, any IRES other than one directly after the uORF should have been intact and available for ribosome entry. The decreased translation caused by the inverted repeat suggests that translation of the downstream ORF requires scanning through the area where the repeat was inserted, consistent with initial ribosome binding at the 5'-cap. The final experiment which argues against the existence of an IRES in SAMDC mRNA involved infection of HeLa cells with poliovirus, which inhibits cap-dependent cellular mRNA translation through a proteolytic cleavage of the p220 subunit of eIF-4F,[88,89] thus commandeering cellular ribosomes for translation of its own IRES-containing RNA. Before poliovirus infection, 90% of the endogenous SAMDC mRNA in HeLa cells was associated with polysome-containing fractions of sucrose gradients, while 4 hours after infection less than 10% was in polysome fractions. This is in contrast to the IRES-containing cellular mRNA for the heavy chain immunoglobulin binding protein,[90] which remained in polysomes 4 hours after poliovirus infection. Taken together, the above evidence against internal ribosome entry in SAMDC mRNA is quite convincing.

A fourth possibility is that translation of the downstream reading frame occurs through re-initiation after translation of the uORF. This is known to occur with eukaryotic mRNAs, e.g., regulated reinitiation is the basis for the well-studied GCN4 control system in yeast.[66] Longer intercistronic distances tend to encourage reinitiation, presumably because the occasional ribosomal subunit which continues to scan after protein synthesis termination at a uORF needs to reacquire initiation factors necessary for a second initiation at a downstream ORF. Short uORFs also lend themselves to more frequent reinitiation. In the case of SAMDC mRNA, which has a very long intercistronic region and a very short uORF, the limited synthesis seen in quiescent Jurkat cells would thus result from infrequent continuation of scanning after termination of synthesis from the uORF, and conditions in the HeLa cells, where translation of SAMDC mRNA is much more efficient, would somehow increase the frequency of continued scanning after translation of the uORF. Although the increased rather than decreased expression of the major ORF seen upon shortening the bovine SAMDC intercistronic regions argues against this mechanism somewhat, the more compelling argument against reinitiation is demonstrated by additional experiments by Ruan et al. If leaky scanning is the mechanism by which the downstream ORF in SAMDC (or in the chimeric SAMDC 5'-UTR/hGH constructs) gets translated, then overlapping the uORF into the downstream ORF should have little or no effect on translation of the downstream ORF. In contrast, if re-initiation is the mechanism involved, overlapping the uORF should eliminate initiation at the downstream ORF, since the ribosome would already be past the start codon for the second ORF when terminating synthesis from the first. After transfecting a construct that had all but 35 nucleotides of the intercistronic region of SAMDC 5'-UTR removed, they found translational efficiency (growth hormone protein divided by growth hormone chimera mRNA amount) in HeLa cells very slightly higher than that of the parent construct containing the full length SAMDC 5'-UTR. After alteration of the uORF stop codon, the now-extended uORF overlapped the out of frame growth hormone ORF by 101 nucleotides. Transfected HeLa or Jurkat cells expressed slightly more growth hormone protein per growth hormone mRNA than when transfected with the nonextended uORF construct of the same size and the mRNA distribution in polysomes in both HeLa and Jurkat cells was almost identical to that of the parent, full-length UTR construct.[79] Considering their earlier results in which any 3' extension of the uORF completely abolished translational control by the SAMDC 5'-UTR and moved the mRNA from monosomes to polysomes of size 5-10,[80] it is somewhat surprising that the growth hormone expression and polysome distribution with this extended uORF construct was not changed much more in the Jurkat cells. However, the much shorter overall length of the 5'-UTR in this extended uORF construct (~66 nucleotides compared to ~327 in the parent construct) may account for this lack of effect, since shorter leader sequences in other constructs have led to decreased translation relative to longer ones.[62,74,91] It is clear in any case that the overlapping uORF did not *decrease* translation in either the Jurkat or the HeLa cells and this is incompatible with the idea of reinitiation contributing significantly to translation of the downstream ORF in SAMDC.

This leaves us with a model for translational control of SAMDC expression in which ribosomes scanning from the 5'-cap initiate *either* at the uORF *or* at the downstream ORF, but not both. Control of translation of the downstream ORF is thus achieved through modulating the frequency or "leakiness" at the uORF. One unusual result requiring further clarification is the appearance of lymphoid cell SAMDC mRNA in the monosome rather than in mRNP fractions of sucrose gradients. If translational control in these cells is indeed occurring through more frequent initiation at the uORF, one would expect the ribosome to very rapidly synthesize the hexapeptide, terminate, and release the mRNA. Thus, the mRNA would tend to spend most of its time in mRNP fractions. The mRNA location in monosome fractions of gradients obviously says that this is not occurring. A very rapid cycling of the mRNA from ribosome to ribosome could account for this, and the strict length and amino acid residue requirements in the uORF encoded hexapeptide could mean that the hexapeptide somehow served as the molecular signal to speed up cap recognition by the next 40S ribosome; however, the crucial residues appear to be the carboxyterminal DIS residues of the hexapeptide and they would still be protected by the translating ribosome at the moment of termination. A more attractive hypothesis is that release of the mRNA from the ribosome is inhibited by some interaction of the DIS (or MxxDIS) sequence with parts of the ribosome. Termination has already been shown to be much slower than elongation and may rival initiation as a rate limiting step in protein synthesis.[92] When a stop codon is reached, a termination or eukaryotic release factor (eRF) moves into the ribosome A site. This leads to hydrolysis of the tRNA-peptide bond at the ribosome P-site, releasing the nascent peptide and normally causing the dissociation of the now-uncharged tRNA and the 60S and 40S ribosomal subunits.[93] This process is poorly understood, but the binding of an eRF at the A-site is clearly different than binding a new tRNA. In theory, the newly synthesized DIS residues attached to the $tRNA_{Ser}$ in the P-site might interact with the ribosomal machinery in such a way that attachment of the eRF at the A-site is inhibited, but attachment of another charged tRNA molecule is not. Thus, DIS sequences not located next to the stop codon would have no effect, but would inhibit termination when immediately preceding a stop codon. If this motif is active in such a manner, one would expect the DIS to appear rarely right before stop codons, and an initial analysis of the NCBI nonredundant protein database (which included 140,534 sequences and over 40 million individual residues the day of the analysis) suggests that this motif is under-represented at the position immediately preceding termination codons (expected occurrences based on the overall frequency of appearance of each residue = 29.25, observed occurrences 11). Another implication of this hypothesis is that any peptide ending in DIS should be inhibitory, yet the 5'-extensions of the SAMDC uORF all eliminated the translational control exhibited by the MAGDIS uORF, in spite of the fact that the DIS was still in the same position relative to the stop codon; as mentioned earlier, the possibility that the active motif might be MxxDIS has not yet been fully tested and this might explain the lack of translational inhibition with the 5'-extended uORFs tested so far. It is intriguing to note that DIS or DVS tripeptides occur in the same internal position in the few eukaryotic release factors (eRFs) whose sequences are known. DIS occurs in the human eRF protein HuTB3 and in the *X. laevis* protein XlCll, while DVS occurs in the *S. cerevisae* and *A. thaliana* proteins.[94] Thus the "DIS" motif in the inhibiting hexapeptide might act through competitively binding at a ribosomal site where the eRF "D(I/V)S" residues would normally dock. Mutation of I→A in the SAMDC "MAGDIS" uORF caused a partial but not total release of translational inhibition, so that the structural specificity for the inhibitory effect at that position is less stringent than at the "D" and "S" residues ("MAGDAS" uORF mRNA was found with an average of two to four ribosomes attached, see Table 3.1). If the above model is correct, then a similar tolerance might be exhibited in the eRF sequence, explaining the appearance of the conservative I↔V substitution among the different species.

Another example of a translationally controlled mRNA which is found in monosomes rather than in mRNP fractions is the cytomegalovirus gp48 (gpUL4) mRNA, and here again translational inhibition by its longer (22 amino acid) uORF seems to be dependent on the exact peptide produced. Missense mutations of the residues at the carboxy end of the protein released the translational block, while silent mutations caused no release.[95] Like the SAMDC uORF, the gp48 peptide only acted in *cis*. A third example of control by sequence-dependent uORFs (reviewed in reference 67) is the arginine-regulated yeast gene CPA1,[96] which is similarly controlled by a 25 codon uORF. Insertion of this uORF into heterologous 5'-UTRs confers arginine control of translation to the heterologous construct.[97] Analysis of seven yeast mutants selected for constitutive CPA1 expression showed that single missense or nonsense mutations of this uORF were responsible for loss of control.[97,98] Unfortunately for the speculative theory advanced above, the amino acid sequences of these two *cis*-acting uORFs (MQPLVLSAKKLSSLLTCKYIPP and MFSLSNSQYTCQDYISDHIWKTSSH) bear no obvious resemblance to each other or to the MAGDIS peptide, nor have any homologies to termination factors been noted so far. However, they might inhibit in a similar way by binding either to other ribosome sites or to the eRF and blocking the interaction necessary to cleave the nascent peptide and release the ribosomal subunits. Inhibition of CPA1 involves CPAR (CPA81), an unlinked gene product, as well as arginine, so the important interactions of the nascent uORF peptide in inhibition of CPA1 may be with this protein, rather than with an eRF or a ribosome.[98] Potential differences in structural features of the regulatory molecules with which these three uORF-encoded peptides interact might explain the lack of obvious similarity at the primary sequence level without negating the possibility that all three are acting in ultimately similar ways to inhibit termination.

Resolving the apparent conflict as to whether SAMDC mRNA is translationally controlled in nonlymphoid tissues as well as lymphoid tissues may require expression of SAMDC or 5'-UTR:reporter gene chimeras from the exact same vector in the different cells, since different promoters may initiate transcription at different sites. The SV40 late promoter used in some vectors, for example, is known to use variable transcription start sites, and thus might be better avoided for experiments where leader length is an issue. In any case, very stringent determination of the actual transcription start sites from whatever vectors are used will be required. The importance of the sequence length before the SAMDC uORF in controlling frequency of initiation at the SAMDC uORF, and hence the level of translational control exhibited in different cells, has been clearly demonstrated above, since addition of 30 nucleotides conferred much greater control of SAMDC mRNA translation in the HeLa cells. It is not known if shorter additions to the SAMDC 5'-UTR could have the same effect, but similar experiments with other systems have demonstrated that increases in leader length between 12 and 80 nucleotides showed a proportional increase in translational efficiency.[62,74,91] Given the obvious Spm contribution to the translational control exhibited in the COS-7 cells, it is also possible that free Spm concentrations might be higher in the COS-7 cells than in other nonlymphoid tissues—COS-7 cells have, for example, about 33% higher Spm than do CHO cells.[27] Regardless of the explanation, the difference between lymphoid and nonlymphoid translational control of SAMDC is a quantitative rather than a qualitative difference, since alteration of the upstream AUG codon in the SAMDC 5'-UTR:hGH chimera doubled hGH production even in the "noncontrolled" HeLa cells.[79] Although the effect was much more dramatic in the Jurkat cells, it is still clear that the uORF repressed SAMDC mRNA translation in these nonlymphoid derived cells as well as in those of lymphoid origin, albeit to a much smaller degree.

IV. REGULATION OF RATE OF PROENZYME PROCESSING

As shown by both site-specific mutagenesis and by peptide sequence of the purified human SAMDC expressed in *E. coli*, proenzyme processing to a mature, catalytically active enzyme occurs between residues Glu 67 and Ser 68, as indicated in Figure 3.1. The Ser 68 residue is converted to pyruvate during the serinolysis reaction of cleavage, and Edman degradation of the purified subunit yields no sequence unless the blocking pyruvate is first reductively aminated to form an alanine residue; sequencing after reductive amination yielded the sequence ASMFVSKRRFILxTCG, which corresponds exactly to the predicted amino acid sequence starting at Ser 68 (with the exception of the Ser→Pyruvate→Ala in the first position, and a poorly detected lysine residue marked by the x).[99] This processing, which produces a catalytically active mature enzyme, occurs quite readily whether the enzyme is expressed in mammalian cells, in cell-free lysates, or in *E. coli*, and the apparently normal processing in *E. coli* suggests strongly that no other eukaryotic proteins are necessary for the processing to occur—the highly dissimilar primary structure of the *E. coli* SAMDC enzyme, even at the processing site,[9,13,99,100] makes the existence of any *E. coli* "SAMDC processing proteins" that could substitute for potential eukaryotic homologues highly unlikely. This evidence is consistent with autocatalytic SAMDC proenzyme processing, and the autocatalytic nature of the proenzyme cleavage is further supported by analogy to the serinolysis reaction in another pyruvate-containing enzyme, histidine decarboxylase (HisDC) of *Lactobacillus* 30A. HisDC is the best studied example of the few known pyruvate-containing enzymes, and there is no reason to believe that an identical type of cleavage is not occurring in all of the pyruvoyl-enzymes.[7,101,102] The extensive studies of HisDC from several groups, following the pioneering studies of Snell and colleagues,[101] indicate that the mechanism of HisDC proenzyme cleavage and pyruvate formation occurs through serinolysis involving a nucleophilic attack of the hydroxyl group of the serine 82 residue on the nitrogen of the preceding peptide bond. Although mutational studies in the HisDC proenzyme have indicated that some residues surrounding the cleavage site may be important (i.e., substitution of the Ser 81 residue with Ala inhibited cleavage almost totally at physiological salt concentrations and at high salt slow cleavage occurred between residues 80 and 81, rather than at the normal cleavage site between residues 81 and 82), there is no consensus seen even at the residue immediately before the serinolysis site in any of the known cleavage sites in other enzymes. These sequences are shown in Table 3.2 (with the cleavage occurring before the underlined S residue in each case).

Cleavage of the HisDC proenzyme occurs even after covalent attachment of the proenzyme to a column, extensive washing with 9M urea and renaturation,[111] or after the enzyme is released from pure protein crystals (from which the three dimensional structure was determined by X-ray diffraction).[112] This unequivocal evidence of processing of pure HisDC lends support to the idea of autocatalytic cleavage of the SAMDC proenzyme. Other studies of HisDC indicate that the autocatalytically cleaved Ser-Ser bond is highly sensitive to cleavage by chemical means as well. Hydroxylaminolysis under mild conditions cleaved that specific bond, while no other peptide bonds were cleaved, including two other Ser-Ser bonds within the same protein.[112] This cleavage occurs in the native enzyme, but not the denatured proenzyme, reinforcing the idea that secondary and tertiary structure-induced strains on that particular bond contribute importantly to its cleavage. Since the amino acid residues in a protein which lead to a particular conformation are often widely distributed in the primary amino acid sequence, it is not surprising that local primary sequence in the different cleaved proteins might not show any obvious similarity. In the human SAMDC enzyme, where cleavage occurs before residue Ser 68, mutations of S66A, E67Q, or S69A had no effect on proenzyme

Table 3.2. Amino acid sequences of various pyruvate-containing enzymes

Proenzyme	Cleavage site	Reference
SAMDC (human)	A Y V L S E **S** S M F V S	99
(rat, presumed site)	A Y V L S E **S** S M F V S	9, 99
(potato, presumed site)	S Y V L S E **S** S L F V Y	103
(*S. cerevisiae*)	A F L L S E **S** S L F V F	104
(*E. coli*)	V A H L D K **S** H I C V H	13, 14
HisDC (*Lactobacillus 30a*)	N M T T A S **S** F T G V Q	105, 106
(*L. buchneri*)	N M T T A S **S** F S G V G	107
(*C. perfringens*)	N M L T A S **S** F C G V A	107, 108
Aspartate DC (*E. coli*)	D L H Y E G **S** C A I N Q	109
Phosphatidylserine DC (*E. coli*)	G R F K L G **S** T V I N L	110

The internal serine residue which is converted to pyruvate during proenzyme processing is underlined in each case. Although members of the same enzyme family show similarities, there is very little similarity between enzyme families at the primary sequence level.

processing, while distal mutations such as E11K, Y112A, or L259Stop completely prevented processing, again indicating the importance of overall folding (and probably induced strain on the E67-S68 bond) rather than local primary sequence in facilitating the processing reaction. One piece of evidence that might argue against the proposed autocatalytic nature of the processing reaction is the much greater appearance of proenzyme when the human SAMDC is expressed in *E. coli*, even when grown in the presence of exogenous putrescine (which stimulates the processing reaction, see below). However, the antigenic 38 kDa proenzyme band seen on denaturing SDS-PAGE gels does not process to the 31 and 7 kDa subunits of the mature enzyme even upon extended incubation of the purified enzyme in the presence of putrescine. A small fraction (<10%) of "unprocessable" proenzyme is sometimes seen with in vitro synthesized SAMDC as well, and probably represents (in both cases) protein which has misfolded into a stable, nonprocessable form. A fraction of the HisDC expressed in *E. coli* also accumulates as a very slowly processing proenzyme, and the evidence in that case also points to misfolding rather than the absence of an activating protein.[7,113,114] Since the amounts of unprocessed proenzyme found in vivo in mammalian cells are very low (often undetectable), the possibility still exists that there is a chaperone protein in whole mammalian cells that prevents this misfolding, and its absence in *E. coli* or in lysates accounts for the unprocessed fraction observed; however, this would not change the idea that the processing reaction is autocatalytic. An alternate explanation of the low appearance of unprocessed proenzyme in mammalian cells compared to in *E. coli* might be that any misfolded proenzyme is more rapidly degraded in mammalian cells, given the rapid turnover of even apparently correctly folded SAMDC protein.

As indicated above, in spite of the autocatalytic nature of the proenzyme cleavage, the presence of putrescine stimulates the rate of proenzyme processing 3- to 5-fold. In cells with a normal concentration of putrescine, this process is very fast, and particularly in view of the fact that the processing is likely to continue during cell or tissue harvest procedures (where putrescine is typically included in the harvest buffer), it is not surprising that evidence for the proenzyme was originally difficult to find. However, depletion of cellular putrescine with the ODC enzyme-activated inhibitor difluoromethylornithine (DFMO) raised the detectable amount of 38 kDa proenzyme in rat prostate from about 4% of the total in untreated rats to 25% in the treated animals; conversely, MGBG treatment of rats led to a large accumulation of putrescine, and proenzyme bands were undetectable.[15] Pharmacological treatment to alter putrescine levels can thus clearly influence the rate of proenzyme processing, and combined with the activation of mature enzyme catalytic activity (discussed below), it seems likely that intracellular putrescine levels can modulate the amount of active

SAMDC in cells, although the extent to which fluctuations of putrescine level within the normal physiological range might regulate SAMDC has never been directly studied.

The amino acid residues through which putrescine stimulates processing of the SAMDC proenzyme have been partially defined by site-specific mutagenesis experiments. Unlike the very rapid processing reactions in *Lactobacillus* HisDC or even the potato (*S. tuberosum*) SAMDC enzymes (unpublished observations), the processing reaction in the wild type human SAMDC enzyme occurs slowly enough to follow the disappearance of the [^{35}S]methionine-labeled 38 kDa proenzyme band into 31 kDa and 7 kDa bands of the mature enzyme. In studying the activation effects of putrescine, which is positively charged at physiological pH, we assumed that putrescine interactions with the SAMDC protein would occur through acidic residues (Glu (D) and Asp (E)). Although an abundance of such residues are contained in the primary sequence of all of the mammalian enzymes, we were able to narrow our search to those residues which were also found in the putrescine-stimulated yeast SAMDC enzyme,[104] and further reduce our initial choices by comparison of these conserved residues to the sequence of the human spermidine synthase protein,[115] which uses putrescine as a substrate. Our reasoning was that this putrescine-utilizing enzyme might have a "putrescine-binding" motif similar to the one we were looking for in SAMDC. Through somewhat arbitrary criteria, we were able to identify three weakly conserved regions containing Glu residues in similar positions relative to hydrophobic and hydrophilic residues in the three enzymes. Glu→Gln mutations were made in the conserved positions to minimize steric changes while removing the negative charge found in the wild-type residues. Proenzyme processing was followed over time in the presence or absence of putrescine, and after separation of the radiolabeled products on SDS/PAGE gels, both the processed and unprocessed bands were quantitated by laser densitometry of exposed films. This initial screen showed that Glu 11 was crucial for putrescine stimulation of proenzyme processing, as the processing rate of the Glu 11→Gln mutant was identical in the presence or absence of added putrescine, and in both cases the rate was equal to that of the wild-type enzyme in the *absence* of putrescine (Fig. 3.3).[116] For an effectively zero order reaction of A→B (proenzyme→processed enzyme), the rate constant k can be calculated by fitting the data to the formula $B_{(t)}=B_{(0)} + A_{(0)}*(1-e^{-kt})$, as is shown in Figure 3.3. By comparing the calculated rate constants for the wild type human enzyme in the presence and absence of putrescine, it is seen in Figure 3.3 that the presence of putrescine stimulated the processing over 5-fold. Additional mutations of conserved acidic residues not found in the spermine synthase comparison revealed that Glu 178 and Glu 256 were also essential for putrescine stimulation of proenzyme processing,[117] and mutation of any one of these three glutamate residues resulted in an equal rate of processing in the presence or absence of putrescine (Fig. 3.3). Mutations E8Q, E15Q, E61Q, E67Q, E132Q, E135Q, E183Q, D185N, E247Q, and E249Q had no effect on putrescine stimulation of the processing rate, while as noted earlier, mutation E11K produced a proenzyme that did not process at all.[116,117] Since putrescine has two positive charges, each putrescine molecule will probably interact with two acidic residues. The identification of three rather than two residues involved in this stimulation suggests that another acidic residue remains to be identified, and that the binding of at least two putrescine molecules is necessary for activation to occur. It is possible however that one site for putrescine binding could occur between the same residue of two dimerized proenzyme molecules. Since the mature na-

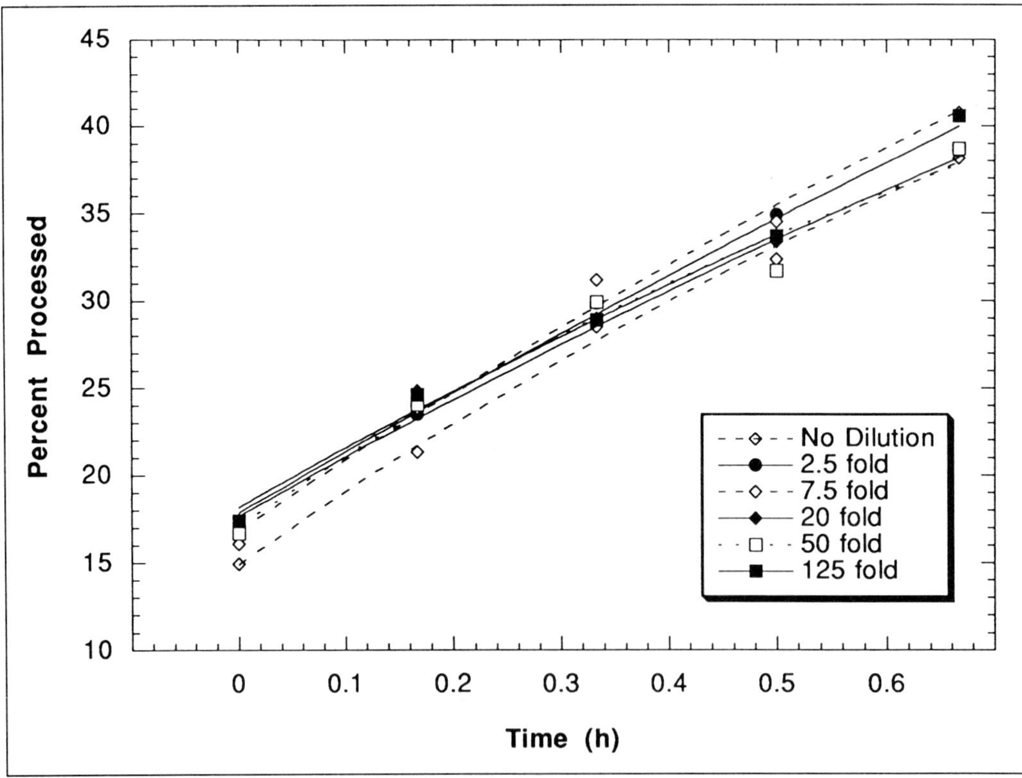

Fig. 3.3. Time course of wild type vs. mutant SAMDC processing +/– putrescine. Rabbit reticulocyte lysate cell-free translation reactions containing [^{35}S]methionine were incubated 30 min at 30° C, and translation reactions were halted by the addition of cycloheximide to 100 µM and 15 µg RNase A. Aliquots were incubated at 30° C with (filled symbols) or without (open symbols) 0.2 mM putrescine for the indicated times, denatured, and electrophoresed on 12.5% PAGE gels. Autoradiograms were exposed at –70° C and bands were quantified by laser densitometry of the developed films and corrected with equal-sized background bands from the same lane. Percent processed was calculated at each time point by dividing the density of the 31 kDa band by the combined density of the 31 kDa and 38 kDa bands. The curves and the rate constants (k) were calculated by fitting the data to the fromula $B_{(t)} = B_{(0)} + A_{(0)}*(1-e^{-kt})$ using the Levenberg-Marquardt algorithm.

tive enzyme is an $\alpha_2\beta_2$ heterotetramer, processing could either occur in preformed proenzyme dimers, forming the tetramer after cleavage of both proenzymes, or processing could occur in the monomers which then joined together to form the tetramer. If the processing had to occur in a preformed proenzyme dimer, however, one would expect the rate of processing to increase with increased concentration of proenzyme, and this does not seem to be the case. As seen in Fig. 3.4, when in vitro synthesized, radiolabeled proenzyme was immediately diluted between 1- and 125-fold, the subsequent rates of processing

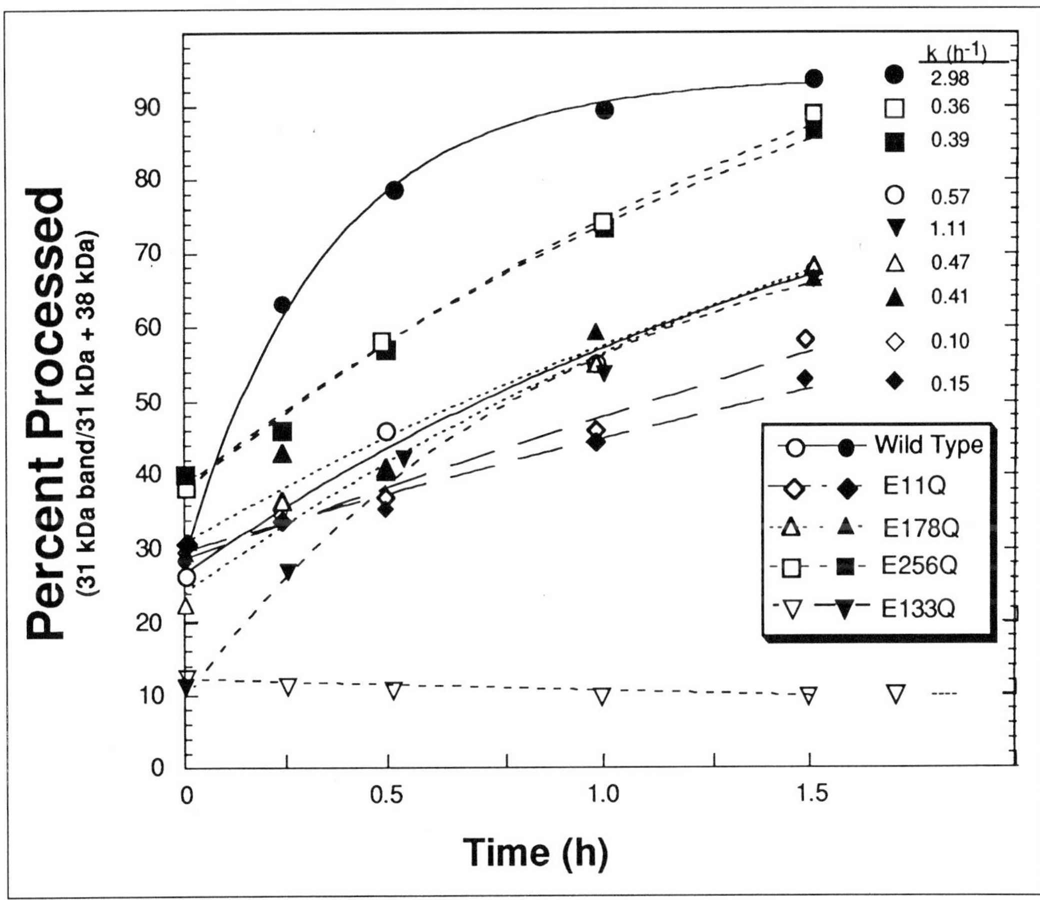

Fig. 3.4. Effect of dilution on wild type SAMDC processing. Coupled transcription/translation (TnT®, Promega) reactions containing [^{35}S]methionine were incubated 30 min at 30°C and halted by the addition of cycloheximide to 100 μM and RNase A to 0.1 μg/μl. The dilutions indicated were made by aliquoting the synthesis reaction into 50 mM sodium phosphate, pH 6.8, 1.25 mM DTT, incubating at 37° for the indicated times and separating on 12.5% PAGE gels. Bands were quantified on a Molecular Dynamics PhosphorImager and corrected with equal sized background bands from the same lane. Percent processed was calculated at each time point by dividing the density of the 31 kDa band by the combined density of the 31 kDa and 38 kDa bands.

at each of the concentrations of proenzyme were indistinguishable. We do, however, have indirect evidence that unprocessed proenzyme can at least form a dimer (really a proenzyme/αβ trimer) with already processed SAMDC. If one prevents processing (for example, by mutation at the Ser 68 cleavage site), this unprocessed mutant enzyme will not bind at all to an MGBG affinity column, while wild type and mutant forms which process will bind quite readily (this is one of the major purification steps used in preparing pure SAMDC).[33,118] As mentioned earlier, when human SAMDC is expressed in *E. coli*, unprocessed proenzyme is bound to the MGBG column along with the processed enzyme. Since the unprocessed enzyme cannot bind to the column by itself, it must be retained through interactions with processed enzyme which is bound. While not certain, it seems likely that the intermolecular interactions involved in forming these trimers will be quite similar to the intermolecular interactions between the subunits of the mature enzyme heterotetramers, and makes it likely that unprocessed proenzyme molecules can dimerize. It is also possible that proenzyme association before processing occurs might inhibit the processing reaction, and that the higher levels of SAMDC production in the heterologous *E. coli* expression system might encourage such nonproductive association, thus leading to higher yields of unprocessed proenzyme; however, the dilution experiments cited above are consistent with proenzyme concentration having little to do with processing rate, at least in the range of concentrations tested.

Substitution of any one of three Glu residues eliminates putrescine activation, which indicates that the binding of at least two putrescine molecules is necessary to achieve an activated conformation of the proenzyme, with a single putrescine being insufficient. The processing rates of the mutants are very close to that of the wild type enzyme in the absence of putrescine, which permits distinction between two possible activating interactions of putrescine. If putrescine were activating by *allowing* protein regions containing two negatively charged residues to come closer together in a more favorable conformation, i.e., by simple charge shielding, removal of a negative charge in the mutant should remove the charge repulsion, and the rate of processing should be higher than the wild type in the absence of putrescine. However, if putrescine is activating by virtue of *forcing* or pulling two or more regions of the protein into a more favorable conformation, removal of the negatively charged residue would make that impossible, and this seems to be the case. The fact that neither monovalent nor divalent cations activate processing of mammalian SAMDC is also consistent with a "forced conformation" mechanism of putrescine activation.

Although residue Glu 133 is clearly not directly involved in putrescine activation of proenzyme processing (processing of a E133Q mutant is still stimulated by putrescine, Fig. 3.3), its absence in the E133Q mutant confers an absolute requirement for putrescine.[117] This result is consistent with coordination of Glu 133 with a positively charged residue in the wild type enzyme, an interaction which holds the proenzyme in a basal conformation allowing processing to occur without the presence of putrescine. The binding of two or more molecules of putrescine would then force the enzyme into a still more favorable conformation. In the absence of the Glu 133 residue, processing depends totally on the conformation-forcing interactions of the putrescine molecules.

V. ACTIVATION OF MATURE ENZYME

In addition to putrescine stimulation of the processing reaction, putrescine stimulates the catalytic activity of the mature processed enzyme 3- to 10-fold (depending on pH). This effect is achieved at least partially by decreasing the enzyme K_m for the substrate SAM. Although it has often been stated (correctly) that the stimulation of catalytic activity by putrescine is greater at lower pH, the *maximal* putrescine-stimulated activity of the purified enzyme occurs between pH 6.8 and 7.2, and falls off at lower or higher pH. In contrast, the nonputrescine stimulated activity continued to rise steadily through at least the tested range of 6.5 through 7.5, thus accounting for the greater stimulation at lower pH (unpublished observations). In addition, the amount of putrescine needed to stimulate semi-purified rat enzyme activity is higher at lower pH, i.e., the rate constant of activation (K_a) is higher at lower pH.[119]

Mutation of Glu 11 (or Glu 8) to Gln produced an inactive enzyme, so the possible contributions of these residues to wild-type putrescine stimulation of activity cannot be assessed by mutational analysis. However, mutation of either Glu 178 or Glu 256 to Gln produced a catalytically active enzyme that was not stimulated by putrescine, mimicking the effects of these mutations on the processing reaction. SAM analog studies have previously shown that any inhibitor lacking a positive charge at the position equivalent to the sulfonium ion in the SAM substrate was ineffective.[120,121] This suggests that the positive charge of the substrate interacts with an acidic residue at the active site of the enzyme. Glu 8 and Glu 11 are thus prime candidates for such acidic residues, and one or both of them are likely to be found in the active site. Antibodies raised to the purified mature enzyme had already been shown to recognize the smaller subunit, demonstrating its continued presence in the mature enzyme, and the above Glu 8/11 results show that the small subunit also contains residues critical for enzyme activity. In separate mutant experiments, Cys 82 of the larger β-subunit was also shown to be critical for catalytic activity.[116] This residue is located in one of the only regions showing any similarity between the *E. coli* and eukaryotic enzymes (Fig. 3.1), and the equivalent residue Cys 140 in the *E. coli* enzyme has been demonstrated to be alkylated by a substrate-derived acrolein-like molecule, suggesting its location in the active site.[122] Either the glutamates or the cysteine residue might serve as the proton donor needed during pyruvate-dependent decarboxylation reactions.[7]

While mutation of Glu 178 or Glu 256 eliminated putrescine stimulation of catalytic activity, mutations E15Q, E61Q, E67Q, E132Q, E135Q, E183Q, D185N, E247Q, and E249Q had no effect on catalytic activity.[116,117] In close agreement with previous reports on enzyme purified directly from mammalian sources,[119,121] kinetic studies on purified mutant vs. wild-type SAMDC showed that the *E. coli* expressed wild type enzyme had a K_m for the SAM substrate of 312 μM in the absence of putrescine and 78 μM with saturating putrescine.[117] Although the E256Q mutant had reduced catalytic activity, the K_m calculated for both it and the fully active E178Q mutant was in the range of 46-114 μM, regardless of the presence or absence of putrescine (Table 3.3). This suggests that the absence of the negative charges at these residues obviates the need for putrescine to lower the K_m, i.e., the enzyme without putrescine is in a fully-activated conformation (at least in the E178Q mutant). The further implication of this observation is that, although the putrescine stimulation of processing and of catalytic activity seem to occur through the same residues, different mechanisms may be involved. The putrescine stimulation of processing seems to occur through forcing a particular proenzyme conformation, and the negative charges of the involved Glu residues are necessary for this forced conformation to occur. In contrast, the putrescine activation of catalytic activity seems to involve a permissive effect—the negative charges of the glutamate residues prevent the fully active conformation from occurring, but their removal or interaction with putrescine allows the enzyme to fold into the conformation which lowers the K_m for substrate. Kinetic analysis of the E133Q mutant, which

Table 3.3. Kinetic properties of purified wild type and mutant AdoMetDC

Mutation	k_{cat} (sec^{-1})	k_{cat}/K_m ($M^{-1}sec^{-1}$)	K_m SAM (μM)	K_m SAM (μM) in absence of putrescine
None	0.481	6167	78	312
E133Q	0.002	0.5	3600	ND
E178Q	0.383	8333	46	61
E256Q	0.036	312	114	70

ND= not determined

Wild type and mutant SAMDCs were purified as described in reference 117 and K_m values determined from double reciprocal plots of SAM concentration and initial reaction velocities. Except where noted, the assays were carried out in the presence of 0.2 mM putrescine and varying concentrations of SAM from 10 μM to 3 mM. The k_{cat} values were determined as described by Fersht.[123]

had an absolute putrescine requirement for processing, showed greatly reduced catalytic activity (lowered k_{cat} and greatly elevated K_m). However, the putrescine effect on stimulation of catalytic activity of the processed enzyme was identical to that observed with the wild type enzyme.[117]

The inactivity of the E11Q mutant makes it impossible so far to determine if the wild type E11 residue is involved in the putrescine activation of catalytic activity in the same way that it is obviously involved in putrescine stimulation of processing. This also prevents any firm conclusions as to whether one vs. two putrescine molecules per SAMDC αβ unit are involved in stimulation of catalytic activity. However, kinetic analyses performed with putrescine and a series of 2-substituted analogs were consistent with the involvement of two or more independent but equivalent binding sites for putrescine, and with competition between activator putrescine and substrate SAM at the sites.[119] Glu 11 is likely to interact with the sulfonium group of the SAM substrate based on the inactivity of the E11Q mutant, and is known to interact with putrescine from the processing experiment data. It is probable that it is involved with putrescine activation of catalytic activity as well, and may be the site through which the competitive substrate/activator kinetics are produced.

VI. POSTTRANSLATIONAL MODIFICATIONS AND INACTIVATION OF ENZYME

It has been generally accepted that changes in SAMDC activity are caused by changes in antigenic amount of protein, i.e., that there are no posttranslational modifications that either activate or inactivate the enzyme in any regulatable manner. However, there are some relatively unexplored suggestions of posttranslational modifications whose presence or absence may still contribute to enzyme regulation, for example by determining the wide range of intracellular half-lives observed under different conditions. Measurements of SAMDC activity vs. enzyme amount in rat liver or prostate showed a very close correlation; however, in the presence of cycloheximide the enzyme activity still disappeared faster than did the antigenic protein, only slightly so in the liver but three times faster in total prostate protein.[124] This inactivation was not accompanied by any obvious change in the size of the protein, which suggests that either a modification, a small proteolytic clip or a denaturation of the protein may inactivate the enzyme, perhaps committing the altered protein to rapid degradation. Another suggestion of such modified forms of the enzyme is seen in prostate regional comparisons in AXC rats, where 1.7-2.0 times as much antibody was required to immunotitrate one unit of SAMDC activity from dorsolateral vs. ventral prostate extracts. This result could obtain either if there was a larger pool of inactive enzyme in the dorsolateral prostate or if the affinity of the antibody for the dorsolateral enzyme was lower. In either case, a posttranslational modification could account for the result (as could a simple misfolding of the protein which prevented catalytic activity but retained antigenicity). Evidence from yeast and *E. coli* SAMDC studies has suggested that a large fraction of the enzyme (up to 25-50%) from these organisms is inactive, apparently due to a transamination of the substrate SAM amine group to the carbonyl group of the pyruvate cofactor, converting it to alanine and inactivating the enzyme.[100,104] This inactivation probably occurs at some rate in the mammalian enzyme as well, and it has been suggested that the rapid turnover of SAMDC in the various organisms is necessary to replace this pool of inactive enzyme.[100]

With the overwhelming number of phosphorylated proteins already known, a frequently heard phrase among phosphorylation researchers is "if there is no evidence of phosphorylation of your protein, you haven't looked hard enough yet." The possibility that regulation might occur through changes in phosphorylation state of SAMDC is made more interesting by the observation that polyamines stimulate the activity of casein kinase II (CKII),[125-128] that CKII activity has been reported to be decreased by inhibitors of ODC activity, and that a 32 kDa protein was among the three proteins whose level of phosphorylation was observed to increase in prostate tumor cells in the presence of exogenous Spm.[129] Nonetheless, in spite of the presence of numerous potential casein kinase II and other consensus phosphorylation sites (as described in the PROSITE database[130,131]), there is to date no reported evidence of phosphorylation of SAMDC. No phosphorylation of immunoprecipitable SAMDC could be detected in insulin stimulated or unstimulated mouse embryo 3T3-L1 preadipocytes (unpublished observations), nor was any evidence found of phosphorylation in human cytomegalovirus-infected (HCMV, strain AD169) human diploid embryonic lung (MRC5) cells (E.L. White, personal communication), in spite of the postinfection appearance of SAMDC of greatly altered kinetics and salt tolerance (the completely sequenced AD169 strain contains no obvious evidence of a virally-encoded SAMDC enzyme).[132]

There is also evidence that Arg 294 (numbering from the human sequence) may be modified in vivo, since the amino acid sequence from that region of the purified bovine protein showed an unidentifiable residue in the position predicted from the cDNA sequence to be an arginine residue.[34]

Two apparently different forms of SAMDC were purified from rat liver compared to rat psoas muscle, with differences observed in isoelectric point (pI), extent of activation by putrescine, K_m for substrate SAM, and MGBG affinity,[133] suggesting a possible difference in posttranslational modification. Another unexplored possible explanation is that such altered forms arise from alternate splicing. In our original cloning of the rat SAMDC cDNA, a clone was isolated that was missing a 21 bp region. This region was found in both the human clone and in subsequently isolated rat clones,[134] including rat genomic clones.[10,29] Although our initial interpretation was that this was a cloning artifact, Maric et al have recently noted that in both the human and the rat gene, this 21 nucleotide sequence forms the beginning of exon 4 and has an AG dinucleotide at the 3'-end, so that mRNAs missing that region might be produced by alternate splicing.[12] The difference in the calculated pIs for the ±21 nucleotide encoded proteins is 0.356 units, which is very close to the difference previously observed between the psoas muscle (pI=5.3) and liver (pI=5.7) enzymes.[133]

VII. REGULATION THROUGH CHANGES IN DEGRADATION RATE

Many stimuli which cause changes in intracellular SAMDC activity do so at least partially by causing changes in the half-life of the enzyme.[3,15,17,135-137] Growth stimuli which cause increases in the activity of both SAMDC and ODC are often accompanied by stabilization of the enzyme. In contrast, when polyamine levels are high, there is generally a more rapid degradation of SAMDC and ODC, and the half-lives of the two proteins are among the fastest known. What signals targets proteins such as SAMDC have for rapid degradation, while other cytosolic proteins have biological half-lives of days, is an area of ongoing research. Two "unifying theories" of such signals have been proposed and SAMDC contains both these purported signals. Based on the degradation rates of a series of cleaved fusion proteins whose only difference was their amino terminal residue, Varshavsky and colleagues proposed the "N-end rule", that the primary determinant of proteolytic rate was the N-terminal amino acid of a protein. This N-terminal residue then acts as a signal to the recognition proteins of the ubiquitination system. Ubiquitination of the protein serves as a tag which leads to delivery of the tagged protein to the 26S proteasome, where the protein is degraded and ubiquitin is recycled.[138-141] In spite of the evidence of such a signal (with slight variations) in bacteria, yeast, and mammalian cells, it is difficult to fully reconcile this idea with the fact that a great number of intracellular proteins are modified at their N-terminal residues (and all of them are of course initially translated with N-terminal methionine, which is a stabilizing residue according to the N-end rules proposed). This means that the signals in peptides or nascent peptide chains which govern the uncovering of destabilizing penultimate (or other interior) amino acid residues would ultimately govern intracellular degradation rates, and the very residues which the N-end rule shows as destabilizing are the same penultimate residues that discourage aminopeptidase removal of methionine residues, for example, or promote acetylation.[142,143] Residues two through seven of the SAMDC proenzyme are destabilizing according to the N-end rules; the penultimate Glu 2 residue is considered to be a secondary signal in that posttranslational arginylation of Glu (or Asp) residues by transfer from a $tRNA_{Arg}$ molecule must precede rapid degradation.[144,145] Preliminary studies of SAMDC degradation in the same rabbit reticulocyte system used to study degradation of other proteins, including ODC, suggests that such a reaction may be involved in SAMDC degradation, since pretreatment of the degrading lysate with RNase A, which degrades tRNAs, greatly slowed the rate of SAMDC degradation, as seen in Fig. 3.5. Since the rabbit reticulocyte lysate did not show decreased SAMDC degradation in the presence of MGBG, which is known to stabilize it in vivo, lysates from COS-7, BHK and L1210 cells were also prepared, all of which rapidly degraded ODC, while SAMDC was not degraded (manuscript in preparation). This suggests that either the proteases involved in ODC and SAMDC deg-

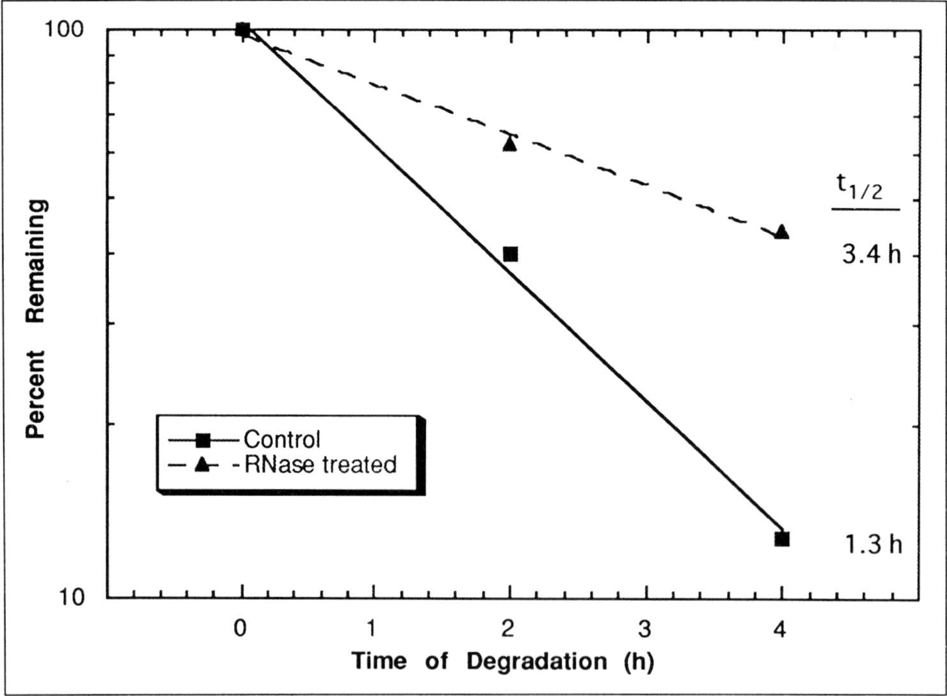

Fig. 3.5. Effect of RNase pretreatment on wild type SAMDC degradation. Rabbit reticulocyte lysate cell-free translation reactions containing [^{35}S]methionine were incubated 90 min at 30°C in the presence of in vitro synthesized RNA for wild-type SAMDC and translation reactions were halted by the addition of cycloheximide to 100 μM. Equal aliquots containing approximately 50,000 cpms were diluted into a crude (nonhemin containing) rabbit reticulocyte lysate with an ATP regenerating system ± RNase A at 1 μg/10μl final volume, and incubated at 37°C for the indicated times, immunoprecipitated with anti-SAMDC antibody and protein A, and separated on SDS-PAGE gels. Radioactivity remaining in the 31 + 37 kDa bands of SAMDC was quantitated from the dried gels with a Betagen Betascope 603 detector. The curves and the rate constants of degradation (k) were calculated by fitting the data to the formula $A_{(t)}=A_{(0)}*(e^{-kt})$, with the half life = ln(2)/k.

radation are different, or that some factor specific for SAMDC targeting to the same 26S proteasome responsible for ODC[146,147] and ubiquitin-dependent[148] degradation was absent in these lysates.

A second proposed signal is that of PEST regions. In comparing the sequence of rapidly vs. slowly degraded proteins, Rogers et al noted the presence in rapidly degraded proteins of regions dense in proline (P), glutamic acid (E), serine (S) and threonine (T) residues, bordered by basic amino acids, while slowly degraded proteins did not have this feature.[149] Since one would expect proteolytic targeting signals to be contained in surface regions, clusters of hydrophilic residues such as Glu, Ser and Thr residues could provide the surface region necessary, while the required Pro residues might form kinks in the surface-exposed regions which would be recognized by proteases or targeting proteins, for example. Both SAMDC and ODC contain PEST regions and truncation or alteration of portions of the C-terminal PEST region of ODC stabilize the enzyme.[150-152] A crucial signal in that region was Cys 441, which was necessary for ODC interactions with Antizyme, a nonproteolytic targeting protein whose induction and interaction with ODC delivers the ODC molecule to the 26S proteasome for degradation.[153] In spite of the differences noted above in lysate degradation of SAMDC vs. ODC, the observed coordinated changes in SAMDC and ODC half-lives in vivo make it tempting to speculate that Antizyme, which has also been shown to affect PA transport,[154] might also regulate SAMDC degradation in some way. However, anti-antizyme antibodies (kindly provided by Dr. S. Matsufuji and S.-I. Hayashi) which coprecipitate labeled ODC in the presence of Antizyme do not coprecipitate labeled SAMDC; conversely, anti-ODC Ab will coprecipitate a labeled Antizyme chimeric construct (pZ1N),[155-157] while anti-SAMDC antibodies will not (data not shown). These results, combined with the differences in the lysates noted above, make it unlikely that SAMDC degradation involves Antizyme.

Various reports have indicated that polyamines cause more rapid degradation of SAMDC in vivo. Depletion of polyamine pools using the specific ODC inhibitor difluoromethylornithine (DFMO) stabilizes SAMDC in vivo, and re-addition of polyamines to DFMO-inhibited cells again destabilizes the enzyme,[17] but the DFMO treatment induces SAMDC and causes accumulation of decarboxylated S-adenosylmethionine (dcSAM) which may in itself stabilize SAMDC. Inhibition of spermidine and spermine synthase to decrease Spd and Spm pools respectively, has also been shown to stabilize SAMDC, but again there is a concomitant accumulation of dcSAM, and many of the inhibitors are themselves nonmetabolizable SAM analogs which may stabilize SAMDC. Putrescine has also been reported to stabilize the enzyme during protein purification and possibly in vivo.[158] Because of the interconversion of the polyamines, and the confounding dcSAM pool changes, a specific role for Spd vs. Spm in SAMDC degradation has never been clearly defined, although the evidence is all strongly consistent with Spd and/or Spm destabilization of SAMDC. Reversal of stabilizing effects of a SAMDC inhibitor by Spm but not Spd has been reported,[23] but Spm may simply compete more successfully than Spd for an enzyme binding site of the stabilizing inhibitor, rather than acting as a primary destabilizer. Both nonspecific (e.g., methylglyoxal*bis*(guanylhydrazone), MGBG) and specific small molecular weight inhibitors of SAMDC have been shown to greatly stabilize the enzyme in vivo, indicating either the presence of destabilizing residues in the vicinity of the active site, or that activity of the enzyme itself causes more rapid turnover.[23,135,159] That binding of all the different inhibitors causes a common change in SAMDC folding which hides a proteolytic signal not near the active site is less likely, but cannot be ruled out. For the *E. coli* enzyme, which like the mammalian enzyme contains a covalently-bound pyruvate cofactor, a substrate (SAM)-dependent transamination of pyruvate to alanine has been demonstrated to occur, inactivating the enzyme.[100] In mammalian cells, depletion of substrate SAM pools by a methionine analog inhibitor of SAM synthase has resulted in a 5-fold increase in SAMDC half-life (65→310 min), and this effect seems to precede a decrease in cellular Spd and Spm levels. However, this effect cannot be completely separated from possible interactions of the methionine analog with SAMDC, since with removal of the analog SAMDC half-life reverted to 70 min with no obvious changes in SAM or Spd/Spm pools.[136] Depletion of SAM pools by a different inhibitor had no effect on SAMDC half-life in carcinoma and lymphoma cells, while depletion of methionine pools doubled the half-life of SAMDC.[160] There is evidence that one enzyme-activated SAMDC inhibitor [(Z)-AbeAdo] destabilizes the enzyme,[161] unlike many of the other inhibitors which seem to stabilize it. We have shown that (Z)-AbeAdo inactivates SAMDC by transamination of the pyruvate cofactor to form alanine, while the stabilizing inhibitor MZHPA (5'-deoxy-5'[(3-hydrazinopropyl)methylamino]adenosine) inactivated the enzyme by transfer of a hydrazone group to the pyruvate cofactor, suggesting that the transamination but not necessarily other modifications of the pyruvate cofactor of SAMDC destabilizes the enzyme.[33] All of the above evidence indirectly suggests that substrate-dependent modifications at the active site of SAMDC may be involved in destabilization of the enzyme in vivo, but there is no direct evidence specifically linking the substrate to destabilization of SAMDC. Other as-yet-undefined posttranslational modifications may also contribute to the changes observed in SAMDC half-life in different conditions, but the very limited evidence accumulated so far does not permit choosing among posttranslational modifications, alterations in potential targeting protein levels or affinities and alterations in a SAMDC-specific proteolytic activity or affinity.

VIII. SUMMARY

In summary, mammalian SAMDC is highly regulated in cells, and these differences in regulation are particularly evident in rapid vs. slow growth and high vs. low polyamine conditions. Evidence for regulation at every possible level has been accumulated—mRNA accumulation, translation, processing, activation of the mature enzyme and regulation of the enzyme half-life, with suggestions of possible regulation at the level of splice variations and posttranslational modification as well. Recent evidence supports the idea that Spd has greater effects on accumulation of SAMDC mRNA while Spm has greater effects on translational control, a process which involves a small upstream open reading frame whose encoded peptide is crucial. Both Spd and Spm seem to destabilize the enzyme as well, while the diamine putrescine acts positively both by activating proenzyme processing and mature enzyme catalytic activity and by stabilizing the enzyme. Recent results suggest that combined alteration of regulation of ODC and SAMDC may be more efficacious than inhibition of either enzyme alone in preventing growth of some tumor cells,[162,163] and inhibition of parasite SAMDC has shown potential for treating infections.[164] A clearer and more detailed understanding of SAMDC structure and regulation in mammalian cells will aid in designing better inhibitors and modifiers of enzyme regulation, new inhibitors which work differentially on infectious agents, and will also satisfy the innate curiosity which drives our shared research efforts.

ACKNOWLEDGMENTS

The author wishes to thank H. Recipon of NCBI who provided the amino acid residue occurrence data, and L.M. Shantz, E.L. White, and D. Stickle for helpful discussions during the preparation of this manuscript, as well as his wife Anne, who single-handedly trained their puppy while he worked.

REFERENCES

1. Williams-Ashman HG and Pegg AE. Aminopropyl group transfers. In: D.R. Morris & L.J. Marton, ed. Polyamines in Biology and Medicine. New York: Marcel Dekker, 1981:43-74.
2. Tabor CW and Tabor H. Polyamines. Annu Rev Biochem 1984;53:749-790.
3. Pegg AE. Recent advances in the biochemistry of polyamines in eukaryotes. Biochem J 1986; 234:249-262.
4. Pegg AE, Kameji T, Shirahata A, et al. Regulation of mammalian S-adenosylmethionine decarboxylase. Adv Enzyme Regul 1988;27:43-55.
5. Seiler N. Polyamine Metabolism. Digestion 1990;46:319-330.
6. Seiler N, Sarhan S, Mamont P, et al. Some biological consequences of S-adenosylmethionine decarboxylase inhibition by MDL 73811. Life Chem Reports 1991;9:151-162.
7. van Poelje PD and Snell EE. Pyruvol-dependent enzymes. Ann Rev Biochem 1990;59:29-59.

8. Kozak M. An analysis of 5'-noncoding sequences from 699 vertebrate messenger RNAs. [Review]. Nucleic Acids Res 1987;15:8125-48.
9. Pajunen A, Crozat A, Jänne OA, et al. Structure and regulation of S-adenosylmethionine decarboxylase. J Biol Chem 1988;263:17040-17049.
10. Pulkka A, Ihalainen R, Suorsa A, et al. Structures and chromosomal localizations of two rat genes encoding S-adenosylmethionine decarboxylase. Genomics 1993;16:342-9.
11. Radford DM, Eddy R, Haley L, et al. Gene sequences coding for S-adenosylmethionine decarboxylase are present on human chromosome 6 and the X and are not amplified in colon neoplasia. Cytogenet Cell Genet 1988;49:285-8.
12. Maric SC, Crozat A and Janne, OA. Structure and organization of the human S-adenosylmethionine decarboxylase gene. J Biol Chem 1992; 267:18915-23.
13. Tabor CW and Tabor H. The speEspeD operon of *Escherichia coli.* Formation and processing of a proenzyme form of S-adenosylmethionine decarboxylase. J Biol Chem 1987;262:16037-40.
14. Anton DL and Kutny R. Escherichia coli S-adenosylmethionine decarboxylase. Subunit structure, reductive amination, and NH2-terminal sequences. J Biol Chem 1987;262:2817-22.
15. Pegg AE, Wechter R, and Pajunen, A. Increase in S-adenosylmethionine decarboxylase activity in SV-3T3 cells treated with S-methyl-5'-methylthioadenosine. Biochem J 1987;244:49-54.
16. Otani S, Matsui-Yuasa I, Mimura-Shimazu Y, et al. Synergistic stimulation of S-adenosylmethionine decarboxylase activity by Ca^{2+} ionophore A23187, cholera toxin and 1-oleoyl-2-acetylglycerol. Eur J Biochem 1988;171:509-13.
17. White MW, Degnin C, Hill J, et al. Specific regulation by endogenous polyamines of translational initiation of S-adenosylmethionine decarboxylase mRNA in Swiss 3T3 fibroblasts. Biochem J 1990;268:657-60.
18. Stimac E and Morris DR. Messenger RNAs coding for enzymes of polyamine biosynthesis are induced during the G_0-G_1 transition but not during traverse of the normal G_1 phase. J Cell Physiol 1987;133:590-4.
19. Watanabe S, Kusama-Eguchi K, Kobayashi H, et al. Estimation of polyamine binding to macromolecules and ATP in bovine lymphocytes and rat liver. J Biol Chem 1991;266:20803-20809.
20. Persson L, Stjernborg L, Holm I, et al. Polyamine-mediated control of mammalian S-adenosyl-L-methionine decarboxylase expression: effects on the content and translational efficiency of the mRNA. Biochem Biophys Res Commun 1989;160:1196-202.
21. Persson L, Khomutov AR, and Khomutov RM. Feedback regulation of S-adenosylmethionine decarboxylase synthesis. Biochem J 1989;257:929-31.
22. Holm I, Persson L, Pegg AE, et al. Effects of S-adenosyl-1,8-diamino-3-thio-octane and S-methyl-5'-methylthioadenosine on polyamine synthesis in Ehrlich ascites-tumour cells. Biochem J 1989;261:205-10.

23. Autelli R, Stjernborg L, Khomutov AR, et al. Regulation of S-adenosylmethionine decarboxylase in L1210 leukemia cells. Studies using an irreversible inhibitor of the enzyme. Eur J Biochem 1991;196:551-6.
24. Pegg AE. S-adenosylmethionine decarboxylase: a brief review. [Review]. Cell Biochem Funct 1984;2:11-5.
25. Shirahata A and Pegg, AE. Increased content of mRNA for a precursor of S-adenosylmethionine decarboxylase in rat prostate after treatment with 2-difluoromethylornithine. J Biol Chem 1986;261:13833-7.
26. Crozat A, Palvimo JJ, Julkunen M, et al. Comparison of androgen regulation of ornithine decarboxylase and S-adenosylmethionine decarboxylase gene expression in rodent kidney and accessory sex organs. Endocrinology 1992;130:1131-44.
27. Shantz LM, Holm I, Jänne OA, et al. Regulation of S-adenosylmethionine decarboxylase activity by alterations in the intracellular polyamine content. Biochem J 1992;288:511-8.
28. Wainfan E, Brody H, Relyea NM, et al. Comparisons of liver transfer RNA methyltransferase and adenosylmethionine decarboxylase activities of male and female rats. Cancer Res 1980;40:620-4.
29. Pulkka A, Ihalainen R, Aatsinki J, et al. Structure and organization of the gene encoding rat S-adenosylmethionine decarboxylase. Febs Lett 1991;291:289-95.
30. Pegg AE and Coward JK. Effect of N-(n-butyl)-1,3-diaminopropane on polyamine metabolism, cell growth and sensitivity to chloroethylating agents. Biochem Pharmacol 1993;46:717-724.
31. Baillon JG, Kolb M, and Mamont PS. Inhibition of mammalian spermine synthase by N-alkylated-1,3-diaminopropane derivatives in vitro and in cultured hepatoma cells. Eur J Biochem 1989;179:17-21.
32. Morris DR and White MW. Growth-regulation of the cellular levels and expression of the mRNA molecules coding for ornithine decarboxylase and S-adenosylmethionine decarboxylase. [Review]. Adv Exp Med Biol 1988;250:241-52.
33. Shantz LM, Stanley BA, Secrist J, III, et al. Purification of human S-adenosylmethionine decarboxylase expressed in *Escherichia coli* and use of this protein to investigate the mechanism of inhibition by the irreversible inhibitors, 5'-deoxy-5'-[(3-hydrazinopropyl)methylamino]adenosine and 5'-([(Z)-4-amino-2-butenyl]methylamino)-5'-deoxyadenosine. Biochemistry 1992;31:6848-55.
34. Mach M, White MW, Neubauer M, et al. Isolation of a cDNA clone encoding S-adenosylmethionine decarboxylase. Expression of the gene in mitogen-activated lymphocytes. J Biol Chem 1986;261:11697-703.
35. Lodish HF and Desalu O. Regulation of synthesis of non-globin proteins in cell-free extracts of rabbit reticulocytes. J Biol Chem 1973;248:3520-3527.
36. White MW, Kameji T, Pegg AE, et al. Increased efficiency of translation of ornithine decarboxylase mRNA in mitogen-activated lymphocytes. Eur J Biochem 1987;170:87-92.

37. Kameji T and Pegg AE. Inhibition of translation of mRNAs for ornithine decarboxylase and S-adenosylmethionine decarboxylase by polyamines. J Biol Chem 1987;262:2427-30.
38. Atkins JF, Lewis JB, Anderson CW, et al. Enhanced differential synthesis of proteins in a mammalian cell-free system by addition of polyamines. J Biol Chem 1975;250:5688-95.
39. Abraham AK, Olsnes S, and Pihl A. Fidelity of protein synthesis in vitro is increased in the presence of spermidine. Febs Lett 1979;101:93-6.
40. Abraham AK and Pihl A. Variable rate of polypeptide chain elongation in vitro. Effect of spermidine. Eur J Biochem 1980;106:257-62.
41. Abraham AK. Effect of polyamines on the fidelity of macromolecular synthesis. Med Biol 1981;59:368-73.
42. Sidransky H, Verney E, and Murty CN. Effect of spermine on hepatic polyribosomes and protein synthesis in rats. Biochem Med 1982;27:68-81.
43. Igarashi K, Hashimoto S, Miyake A, et al. Increases of fidelity of polypeptide synthesis by spermidine in eukaryotic cell-free systems. Eur J Biochem 1982;128:597-604.
44. Takemoto T, Nagamatsu Y, and Oka T. The study of spermidine-stimulated polypeptide synthesis in cell-free translation of mRNA from lactating mouse mammary gland. Biochim Biophys Acta 1983;740:73-9.
45. Holtta E and Hovi T. Polyamine depletion results in impairment of polyribosome formation and protein synthesis before onset of DNA synthesis in mitogen-activated human lymphocytes. Eur J Biochem 1985;152:229-37.
46. Ogasawara T, Ito K, and Igarashi K. Effect of polyamines on globin synthesis in a rabbit reticulocyte polyamine-free protein synthetic system. J Biochem 1989;105:164-7.
47. Uzawa T, Yamagishi A, Ueda T, et al. Effects of polyamines on a continuous cell-free protein synthesis system of an extreme thermophile, Thermus thermophilus. J Biochem 1993;114:732-4.
48. Ballas SK, Smith ED, and Marton LJ. Effect of polyamines on hemoglobin and globin chain synthesis. Biochem Biophys Res Commun 1985;129:546-52.
49. Porter CW, Pegg AE, Ganis B, et al. Combined regulation of ornithine and S-adenosylmethionine decarboxylases by spermine and the spermine analog N1 N12-bis(ethyl)spermine. Biochem J 1990;268:207-12.
50. Holm I, Persson L, Heby O, et al. Feedback regulation of polyamine synthesis in Ehrlich ascites tumor cells. Analysis using nonmetabolizable derivatives of putrescine and spermine. Biochem Biophys Acta 1988; 972:239-48.
51. Kozak M. Structural features in eukaryotic mRNAs that modulate the initiation of translation. [Review]. J Biol Chem 1991;266:19867-70.
52. Kozak M. Regulation of translation in eukaryotic systems. [Review]. Annu Rev Cell Biol 1992;8:197-225.
53. Kozak M. A consideration of alternative models for the initiation of translation in eukaryotes. [Review]. Crit Rev Biochem Mol Biol 1992;27:385-402.

54. Shantz LM, Viswanath R, and Pegg, AE. Role of the 5'-untranslated region of mRNA in the synthesis of S-adenosylmethionine decarboxylase and its regulation by spermine. Biochem J 1994;302:765-72.
55. Zuker M and Stiegler P. Optimal computer folding of large RNA sequences using thermodynamics and auxiliary information. Nucleic Acids Res 1981;9:133-48.
56. Kozak M. The scanning model for translation: an update. [Review]. J Cell Biol 1989;108:229-41.
57. Kozak M. Influences of mRNA secondary structure on initiation by eukaryotic ribosomes. Proc Natl Acad Sci USA 1986;83:2850-4.
58. Kozak M. Circumstances and mechanisms of inhibition of translation by secondary structure in eucaryotic mRNAs. Mol Cell Biol 1989;9:5134-42.
59. Guan KL and Weiner H. Influence of the 5'-end region of aldehyde dehydrogenase mRNA on translational efficiency: Potential secondary structure inhibition of translation in vitro. J Biol Chem 1989;264:17764-17769.
60. Kozak M. Bifunctional messenger RNAs in eukaryotes. Cell 1986;47:481-3.
61. Sedman SA, Gelembiuk GW, and Mertz JE. Translation initiation at a downstream AUG occurs with increased efficiency when the upstream AUG is located very close to the 5' cap. J Virol 1990;64:453-7.
62. Kozak M. A short leader sequence impairs the fidelity of initiation by eukaryotic ribosomes. Gene Expr 1991;1:111-5.
63. Kozak M. At least six nucleotides preceding the AUG initiator codon enhance translation in mammalian cells. J Mol Biol 1987;196:947-50.
64. Kozak M. Point mutations define a sequence flanking the AUG initiator codon that modulates translation by eukaryotic ribosomes. Cell 1986;44:283-92.
65. Kozak M. Downstream secondary structure facilitates recognition of initiator codons by eukaryotic ribosomes. Proc Natl Acad Sci USA 1990;87:8301-5.
66. Abastado JP, Miller PF, Jackson BM, et al. Suppression of ribosomal reinitiation at upstream open reading frames in amino acid-starved cells forms the basis for GCN4 translational control. Mol Cell Biol 1991; 11:486-96.
67. Geballe AP and Morris DR. Initiation codons within 5'-leaders of mRNAs as regulators of translation. [Review]. Trends Biochem Sci 1994;19:159-64.
68. Hinnebusch AG. Translational control of GCN4: An in vivo barometer of initiation-factor activity [Review]. Trends in Biochemical Sciences 1994;19:409-414.
69. Grens A and Scheffler IE. The 5'- and 3'-untranslated regions of ornithine decarboxylase mRNA affect the translational efficiency. J Biol Chem 1990;265:11810-11816.
70. Manzella JM, Rychlik W, Rhoads RE, et al. Insulin induction of ornithine decarboxylase. Importance of mRNA secondary structure and phosphorylation of eucaryotic initiation factors eIF-4B and eIF-4E. J Biol Chem 1991;266:2383-9.
71. Kashiwagi K, Ito K, and Igarashi K. Spermidine regulation of ornithine

decarboxylase synthesis by a GC-rich sequence of the 5'-untranslated region. Biochem Biophys Res Comm 1991;178:815-822.

72. Van Steeg H, Van Oostrom CTM, Hodemaekers HM, et al. The translation in vitro of rat ornithine decarboxylase mRNA is blocked by its 5' untranslated region in a polyamine-independent way. Biochem J 1991;274:521-526.
73. Suzuki T, Kashiwagi K, and Igarashi K. Polyamine regulation of S-adenosylmethionine decarboxylase synthesis through the 5'-untranslated region of its mRNA. Biochem Biophys Res Commun 1993;192:627-34.
74. Kozak M. Effects of long 5' leader sequences on initiation by eukaryotic ribosomes in vitro. Gene Expr 1991;1:117-25.
75. White MW, Oberhauser AK, Kuepfer CA, et al. Different early-signaling pathways coupled to transcriptional and posttranscriptional regulation of gene expression during mitogenic activation of T lymphocytes. Mol Cell Biol 1987;7:3004-7.
76. Kaspar RL, Rychlik W, White MW, et al. Simultaneous cytoplasmic redistribution of ribosomal protein L32 mRNA and phosphorylation of eukaryotic initiation factor 4E after mitogenic stimulation of Swiss 3T3 cells. J Biol Chem 1990;265:3619-22.
77. Kaspar RL, Kakegawa T, Cranston H, et al. A regulatory cis element and a specific binding factor involved in the mitogenic control of murine ribosomal protein L32 translation. J Biol Chem 1992;267:508-14.
78. Hill JR and Morris, DR. Cell-specific translation of S-adenosylmethionine decarboxylase mRNA. Regulation by the 5' transcript leader. J Biol Chem 1992;267:21886-93.
79. Ruan H, Hill JR, Fatemie-Nainie S, et al. Cell-specific translational regulation of S-adenosylmethionine decarboxylase mRNA. Influence of the structure of the 5' transcript leader on regulation by the upstream open reading frame. J Biol Chem 1994;269:17905-10.
80. Hill JR and Morris DR. Cell-specific translational regulation of S-adenosylmethionine decarboxylase mRNA. Dependence on translation and coding capacity of the cis-acting upstream open reading frame. J Biol Chem 1993;268:726-31.
81. Waris T, Ihalainen R, Keranen MR, et al. Molecular cloning of the mouse S-adenosylmethionine decarboxylase cDNA: specific protein binding to the conserved region of the mRNA 5'-untranslated region. Biochem Biophys Res Commun 1992;189:424-9.
82. Koromilas AE, Lazaris-Karatzas A, and Sonenberg N. mRNAs containing extensive secondary structure in their 5' non-coding region translate efficiently in cells overexpressing initiation factor eIF-4E [published erratum appears in EMBO J 1992;11(13):5138]. EMBO J 1992;11:4153-8.
83. Lazaris-Karatzas A, Montine KS, and Sonenberg N. Malignant transformation by a eukaryotic initiation factor subunit that binds to mRNA 5' cap. Nature 1990;345:544-7.
84. Lazaris-Karatzas A, Smith MR, Frederickson RM, et al. Ras mediates translation initiation factor 4E-induced malignant transformation. Genes Dev

1992;6:1631-42.

85. Lazaris-Karatzas A and Sonenberg N. The mRNA 5' cap-binding protein, eIF-4E, cooperates with v-myc or E1A in the transformation of primary rodent fibroblasts. Mol Cell Biol 1992;12:1234-8.
86. Kozak M. Point mutations close to the AUG initiator codon affect the efficiency of translation of rat preproinsulin in vivo. Nature 1984;308:241-6.
87. Kozak M. Context effects and inefficient initiation at non-AUG codons in eucaryotic cell-free translation systems. Mol Cell Biol 1989;9:5073-80.
88. Krausslich HG, Nicklin MJ, Toyoda H, et al. Poliovirus proteinase 2A induces cleavage of eucaryotic initiation factor 4F polypeptide p220. J Virol 1987;61:2711-8.
89. Wyckoff EE, Croall DE, and Ehrenfeld E. The p220 component of eukaryotic initiation factor 4F is a substrate for multiple calcium-dependent enzymes. Biochemistry 1990;29:10055-61.
90. Macejak DG and Sarnow P. Internal initiation of translation mediated by the 5' leader of a cellular mRNA [see comments]. Nature 1991;353:90-4.
91. Kozak M. Leader length and secondary structure modulate mRNA function under conditions of stress. Mol Cell Biol 1988;8:2737-44.
92. Wolin SL and Walter P. Ribosome pausing and stacking during translation of a eukaryotic mRNA. Embo J 1988;7:3559-69.
93. Stansfield I and Tuite MF. Polypeptide chain termination in *Saccharomyces cerevisiae.* [Review]. Curr Genet 1994;25:385-95.
94. Frolova L, Le Goff X, Rasmussen HH, et al. A highly conserved eukaryotic protein family possessing properties of polypeptide chain release factor. Nature 1994;372:701-703.
95. Degnin CR, Schleiss MR, Cao J, et al. Translational inhibition mediated by a short upstream open reading frame in the human cytomegalovirus gpUL4 (gp48) transcript. J Virol 1993;67:5514-21.
96. Kinney DM and Lusty CJ. Arginine restriction induced by delta-N-(phosphonacetyl)-L-ornithine signals increased expression of HIS3, TRP5, CPA1, and CPA2 in *Saccharomyces cerevisiae.* Mol Cell Biol 1989;9:4882-4888.
97. Delbecq P, Werner M, Feller A, et al. A segment of mRNA encoding the leader peptide of the CPA1 gene confers repression by arginine on a heterologous yeast gene transcript. Mol Cell Biol 1994;14:2378-90.
98. Werner M, Feller A, Messenguy F, et al. The leader peptide of yeast gene CPA1 is essential for the translational repression of its expression. Cell 1987;49:805-13.
99. Stanley BA, Pegg AE, and Holm I. Site of pyruvate formation and processing of mammalian S-adenosylmethionine decarboxylase proenzyme. J Biol Chem 1989;264:21073-21079.
100. Anton DL and Kutny R. Structural and mechanistic properties of E. coli adenosylmethionine decarboxylase. [Review]. Adv Exp Med Biol 1988;250:81-9.
101. Snell E. Pyruvate-containing enzymes. TIBS 1977;2:131-135.
102. Li QX and Dowhan W. Studies on the mechanism of formation of the

pyruvate prosthetic group of phosphatidylserine decarboxylase from *Escherichia coli.* J Biol Chem 1990;265:4111-4115.

103. Taylor MA, Mad Arif SA, Kumar A, et al. Expression and sequence analysis of cDNAs induced during the early stages of tuberisation in different organs of the potato plant (*Solanum tuberosum* L.). Plant Mol Biol 1992;20:641-651.
104. Kashiwagi K, Taneja SK, Liu TY, et al. Spermidine biosynthesis in Saccharomyces cerevisiae. Biosynthesis and processing of a proenzyme form of S-adenosylmethionine decarboxylase. J Biol Chem 1990;265:22321-8.
105. Huynh QK, Vaaler GL, Recsei PA, et al. Histidine decarboxylase of *Lactobacillus* 30a. Sequences of the cyanogen bromide peptides from the alpha chain. J Biol Chem 1984;259:2826-32.
106. Vanderslice P, Copeland WC, and Robertus JD. Cloning and nucleotide sequence of wild type and a mutant histidine decarboxylase from *Lactobacillus* 30a. J Biol Chem 1986;261:15186-91.
107. Huynh QK and Snell EE. Pyruvoyl-dependent histidine decarboxylases. Preparation and amino acid sequences of the beta chains of histidine decarboxylase from *Clostridium perfringens* and *Lactobacillus buchneri.* J Biol Chem 1985;260:2798-803.
108. van Poelje PD and Snell EE. Cloning, sequencing, expression, and site-directed mutagenesis of the gene from *Clostridium perfringens* encoding pyruvoyl-dependent histidine decarboxylase. Biochemistry 1990;29:132-9.
109. Smith RC. The structure and molecular characterization of L-aspartate a-decarboxylase 1988, MIT.
110. Li QX and Dowhan W. Structural characterization of *Escherichia coli* phosphatidylserine decarboxylase. J Biol Chem 1988;263:11516-22.
111. Snell EE, Recsei PA, and Misono H. Histidine decarboxylase from Lactobacillus30a: Nature of conversion of proenzyme to active enzyme. In: S. Shaltiel, ed. Metabolic Interconversion of Enzymes. NY: Springer-Verlag, 1976:213-219.
112. Recsei PA and Snell EE. Bacterial prohistidine decarboxylase: Kinetics of conversion to the active enzyme. In: H. Holzer, ed. Metabolic Interconversion of Enzymes. Berlin: Springer-Verlag, 1980:335-343.
113. Copeland WC, Vanderslice P, and Robertus JD. Expression and characterization of *Lactobacillus* 30a histidine decarboxylase in *Escherichia coli*. Protein Eng 1987;1:419-23.
114. van Poelje PD. Kinetics and mechanism of activation of prohistidine decarboxylase. 1988; Univ. Texas, Austin.
115. Wahlfors J, Alhonen L, Kauppinen L, et al. Human spermidine synthase: Cloning and primary structure. DNA and Cell Biol 1990;9:103-110.
116. Stanley BA and Pegg AE. Amino acid residues necessary for putrescine stimulation of human S-adenosylmethionine decarboxylase proenzyme processing and catalytic activity. J Biol Chem 1991;266:18502-18506.
117. Stanley BA, Shantz LM, and Pegg AE. Expression of mammalian S-adenosylmethionine decarboxylase in *Escherichia coli*: Determination of sites for putrescine activation of activity and processing. J Biol Chem

1994;269:7901-7907.
118. Pegg AE and Pösö H. S-Adenosylmethionine decarboxylase (rat liver). Methods Enzymol. 1983;94:234-239.
119. Dezeure F, Gerhart F, and Seiler N. Activation of rat liver S-adenosylmethionine decarboxylase by putrescine and 2-substituted 1,4-butanediamines. Int J Biochem 1989;21:889-99.
120. Pankaskie M and Abdel-Monem MM. Inhibitors of polyamine biosynthesis 8: Irreversible inhibition of mammalian S-adenosyl-L-methionine decarboxylase by substrate analogs. J Med Chem 1980;23:121-127.
121. Kolb M, Danzin C, Barth J, et al. Synthesis and biochemical properties of chemically stable product analogs of the reaction catalyzed by S-adenosyl-L-methionine decarboxylase. J Med Chem 1982;25:550-6.
122. Diaz E and Anton DL. Alkylation of an active-site cysteinyl residue during substrate-dependent inactivation of *Escherichia coli* S-adenosylmethionine decarboxylase. Biochemistry 1991;30:4078-81.
123. Fersht A. The basic equations of enzyme kinetics. In: A. Fersht, ed. Enzyme Structure and Mechanism. New York: W. H. Freeman and Co., 1985:98-120.
124. Shirahata, A and Pegg AE. Regulation of S-adenosylmethionine decarboxylase activity in rat liver and prostate. J Biol Chem 1985;260:9583-8.
125. DeBenedette M and Snow EC. Induction and regulation of casein kinase II during B lymphocyte activation. J Immunol 1991;147:2839-2845.
126. Lamprecht SA, Schwartz B, Avigdor A, et al. Growth-related enzyme activities in crypt compartments during rat colon carcinogenesis. Anticancer Res 1990;10:773-778.
127. Filhol O, Cochet C, Delagoutte T, et al. Polyamine binding activity of casein kinase II. Biochem Biophys Res Comm 1991;180:945-952.
128. Chaudhry PS, Nanez R, and Casillas ER. Purification and characterization of polyamine-stimulated protein kinase (casein kinase II) from bovine spermatozoa. Arch Biochem Biophys 1991;288:337-342.
129. Levasseur S, Poleck T, Shaw M, et al. The effects of polyamine antimetabolites on polyamine-responsive casein kinase activity. Life Sci 1987;41:1679-83.
130. Fuchs R. Predicting protein function: a versatile tool for the Apple Macintosh. Comput Appl Biosci 1994;10:171-8.
131. Bairoch A and Bucher P. PROSITE: recent developments. Nucleic Acids Res 1994;22:3583-9.
132. White EL, Arnett G, Secrist J, et al. Characterization of S-adenosylmethionine decarboxylase induced by human cytomegalovirus infection. Virus Res 1994;31:255-63.
133. Poso H and Pegg AE. Comparison of S-adenosylmethionine decarboxylases from rat liver and muscle. Biochemistry 1982;21:3116-22.
134. Pulkka A, Keranen MR, Salmela A, et al. Nucleotide sequence of rat S-adenosylmethionine decarboxylase cDNA. Comparison with an intronless rat pseudogene. Gene 1990;86:193-9.
135. Madhubala R, Secrist JA, and Pegg AE. Effect of inhibitors of S-

adenosylmethionine decarboxylase on the contents of ornithine decarboxylase and S-adenosylmethionine decarboxylase in L1210 cells. Biochem J 1988;254:45-50.
136. Kramer DL, Sufrin JR, and Porter CW. Modulation of polyamine-biosynthetic activity by S-adenosylmethionine depletion. Biochem J 1988;249:581-586.
137. Hyvönen T and Eloranta TO. Regulation of S-adenosyl-L-methionine decarboxylase by 1-aminooxy-3-aminopropane: enzyme kinetics and effects on the enzyme activity in cultured cells. J Biochem 1990;107:339-342.
138. Baker RT and Varshavsky A. Inhibition of the N-end rule pathway in living cells. Proc Natl Acad Sci USA 1991;88:1090-1094.
139. Bachmair A and Varshavsky A. The degradation signal in a short-lived protein. Cell 1989;56:1019-1032.
140. Gonda DK, Bachmair A, Wünning I, et al. Universality and structure of the N-end rule. J Biol Chem 1989;264:16700-16712.
141. Tobias JW, Shrader TE, Rocap G, et al. The N-end rule in bacteria. Science 1991;254:1374-1377.
142. Arfin SM and Bradshaw RA. Cotranslational processing and protein turnover in eukaryotic cells. Biochemistry 1988;27:7979-7984.
143. Yamada R and Bradshaw RA. Rat liver polysome N-alpha-acetyltransferase: substrate specificity. Biochemistry 1991;30:1017-1021.
144. Ferber S and Ciechanover A. Role of arginine-tRNA in protein degradation by the ubiquitin pathway. Nature 1987;326:808-811.
145. Ciechanover A, Ferber S, Ganoth D, et al. Purification and characterization of arginyl-tRNA-protein transferase from rabbit reticulocytes. Its involvement in post-translational modification and degradation of acidic NH2 termini substrates of the ubiquitin pathway. J Biol Chem 1988;263:11155-11167.
146. Murakami Y, Matsufuji S, Kameji T, et al. Ornithine decarboxylase is degraded by the 26S proteasome without ubiquitination. Nature 1992;360:597-9.
147. Murakami Y, Matsufuji S, Tanaka K, et al. Involvement of the proteasome and antizyme in ornithine decarboxylase degradation by a reticulocyte lysate. Biochem J 1993;295:305-8.
148. Kanayama HO, Tamura T, Ugai S, et al. Demonstration that a human 26S proteolytic complex consists of a proteasome and multiple associated protein components and hydrolyzes ATP and ubiquitin-ligated proteins by closely linked mechanisms. Eur J Biochem 1992;262:567-578.
149. Rogers S, Wells R, and Rechsteiner M. Amino acid sequences common to rapidly degraded proteins: the PEST hypothesis. Science 1986;234:364-368.
150. Ghoda L, van Daalen Wetters T, Macrae M, et al. Prevention of rapid intracellular degradation of ODC by a carboxyl-terminal truncation. Science 1989;243:1493-1495.
151. Rosenberg-Hasson Y, Bercovich Z, and Kahana C. Characterization of sequences involved in mediating degradation of ornithine decarboxylase in cells and in reticulocyte lysate. Eur J Biochem 1991;196:647-651.

152. Lu L, Stanley BA, and Pegg AE. Identification of residues in ornithine decarboxylase essential for enzymatic activity and for rapid protein turnover. Biochem J 1991;277:671-675.
153. Miyazaki Y, Matsufuji S, Murakami Y, et al. Single amino-acid replacement is responsible for the stabilization of ornithine decarboxylase in HMOA cells. Eur J Biochem 1993;214:837-44.
154. Mitchell JL, Judd GG, Bareyal-Leyser A, et al. Feedback repression of polyamine transport is mediated by antizyme in mammalian tissue-culture cells. Biochem J 1994;299:19-22.
155. Matsufuji S, Miyazaki Y, Kanamoto R, et al. Analyses of ornithine decarboxylase antizyme mRNA with a cDNA cloned from rat liver. J. Biochem. 1990;108:365-371.
156. Miyazaki Y, Matsufuji S, and Hayashi S. Cloning and characterization of a rat gene encoding ornithine decarboxylase antizyme. Gene 1992;113:191-7.
157. Li X and Coffino P. Distinct domains of antizyme required for binding and proteolysis of ornithine decarboxylase. Mol Cell Biol 1994;14:87-92.
158. Wang JY, Viar MJ, McCormack SA, et al. Effect of putrescine on S-adenosylmethionine decarboxylase in a small intestinal crypt cell line. Am J Physiol 1992;263:G494-501.
159. Pegg AE. Investigation of the turnover of rat liver S-adenosylmethionine decarboxylase using a specific antibody. J Biol Chem 1979;254:3249-3253.
160. Tisdale MJ. Effect of methionine deprivation on S-adenosylmethionine decarboxylase of tumour cells. Biochem Biophys Acta 1981;675:366-72.
161. Stjernborg L, Heby O, Mamont P, et al. Polyamine-mediated regulation of S-adenosylmethionine decarboxylase expression in mammalian cells. Studies using 5'-([(Z)-4-amino-2-butenyl]methylamino)-5'-deoxyadenosine, a suicide inhibitor of the enzyme. Eur J Biochem 1993;214:671-6.
162. Porter CW and Bergeron RJ. Enzyme regulation as an approach to interference with polyamine biosynthesis—an alternative to enzyme inhibition. Adv Enzyme Regul 1988;27:57-79.
163. Casero R Jr, Ervin SJ, Celano P, et al. Differential response to treatment with the bis(ethyl)polyamine analogs between human small cell lung carcinoma and undifferentiated large cell lung carcinoma in culture. Cancer Res 1989;49:639-43.
164. Bacchi CJ, Nathan HC, Yarlett N, et al. Cure of murine *Trypanosoma brucei rhodesiense* infections with an S-adenosylmethionine decarboxylase inhibitor. Antimicrob Agents Chemother 1992;36:2736-40.

CHAPTER 4

Regulation of Spermidine/Spermine N^1-Acetyltransferase

Lei Xiao, Robert A. Casero, Jr.

I. INTRODUCTION

The polyamines spermidine and spermine, and their diamine precursor putrescine, are naturally occurring polycationic components of all eukaryotic cells and are required for cell growth and differentiation.[60,61,81] At physiological pH, polyamines are fully protonated. These cationic molecules are capable of noncovalent interaction with nucleic acids and their ability to affect DNA and chromatin conformation has been well characterized.[1,20,21,52,57,72] It has been found that polyamines are involved in the transition of B- to A- and Z-forms of DNA,[51,84] inducing bending in specific DNA sequences[20] and aid in the condensation and aggregation of DNA.[42] Depletion of polyamines was found to be associated with altered chromatin structure.[233,80] An increasing body of data has revealed the importance of polyamines in the regulation of expression of certain growth-related genes such as *c-myc*, *c-fos*, Egr-1, and ornithine decarboxylase (ODC).[14-16,74] Data from recent studies demonstrate that polyamines also play a role in protein synthesis,[34] membrane stability, neurotransmission, and intrinsic rectification in ion channels.[25,48] However, the roles of the polyamines in specific biochemical processes, especially in vivo, are less well described.

The intracellular concentration of polyamines is controlled by the concerted action of highly regulated rate-limiting enzymatic steps, including the biosynthetic enzymes ODC and S-adenosylmethionine decarboxylase (AdoMetDC), and the catabolic enzyme spermidine/

Polyamines: Regulation and Molecular Interaction, edited by Robert Casero. © 1995 R.G. Landes Company.

spermine N^1-acetyltransferase (SSAT). The intracellular levels of polyamines can be additionally modified by the uptake and efflux mechanisms. Cloning of the ODC and AdoMetDC genes and the development of highly specific inhibitors of polyamine biosynthesis have greatly facilitated the study of polyamine function. The ongoing studies of the regulation of the two key enzymes ODC and AdoMetDC at transcriptional, translational, and posttranslational levels in various systems continue to provide important information for understanding mechanisms in control of the intracellular polyamines (see chapters 2 and 3). The recent cloning of the SSAT gene and cDNA has also allowed the in-depth studies of its regulation. Here, we will focus on recent findings on the regulation of SSAT gene expression and the mechanisms underlying its expression in different systems. Further, we will discuss the significance of SSAT expression in response to antineoplastic polyamine analogs in various human solid tumor models. Finally, the potentially important physiological functions of SSAT in cellular responses to various environmental stimuli including carcinogens and stress will be discussed.

II. THE POLYAMINE METABOLIC PATHWAY: A VERY BRIEF OVERVIEW

The polyamine biosynthetic pathway has been well studied and reviewed.[39,61] Therefore, only a very brief overview will be included here for perspective. Two rate-limiting enzymes ODC and AdoMetDC control the biosynthesis of polyamines. The first step of biosynthesis is the decarboxylation of ornithine by ODC-forming putrescine. Spermidine and spermine are then formed by the sequential transfer of aminopropyl groups to each nitrogen of the putrescine molecule catalyzed by spermidine and spermine synthases, respectively. The aminopropyl donor (decarboxylase S-adenosylmethionine) for these reactions is provided by the action of AdoMetDC (Fig. 4.1). The biosynthesis of polyamines is known to be negatively regulated by intracellular polyamines themselves.

The catabolism and excretion of polyamines is controlled by a two-step enzymatic process. The first and rate-limiting step in this pathway is the acetyl-transfer from acetylcoenzyme A to either spermidine or spermine which is controlled by the highly inducible enzyme, spermidine/spermine N^1-acetyltransferase (SSAT). The N^1-acetyl-products are substrates for the FAD-dependent enzyme, polyamine oxidase (PAO), which completes the conversion of spermidine to putrescine, or spermine to spermidine, depending on the starting compounds (Fig. 4.1). The acetylated polyamines can then either be recycled through the biosynthetic pathway or be excreted from the cell. The activity of SSAT is capable of producing a rapid decline in intracellular polyamine levels, thus coupled with a highly regulated biosynthetic pathway, it provides cells with the ability to finely attenuate their changing requirements for intracellular polyamines.

III. THE SPERMIDINE/SPERMINE N[1]-ACETYLTRANSFERASE PROTEIN

The SSAT protein is normally present at very low levels and is induced by a number of factors including various toxic agents, hormones, growth factors, the natural polyamines, their structural analogs, and environmental stresses.[10,76] In addition to being rapidly induced, the SSAT protein also has a very short half-life of less than 30 min.[53,65] The rapid turn-over of SSAT allows the protein level to be changed very rapidly in response to stimuli. Unlike many other rapidly degraded proteins, including ODC, c-Myc, p53, c-Fos, etc., SSAT

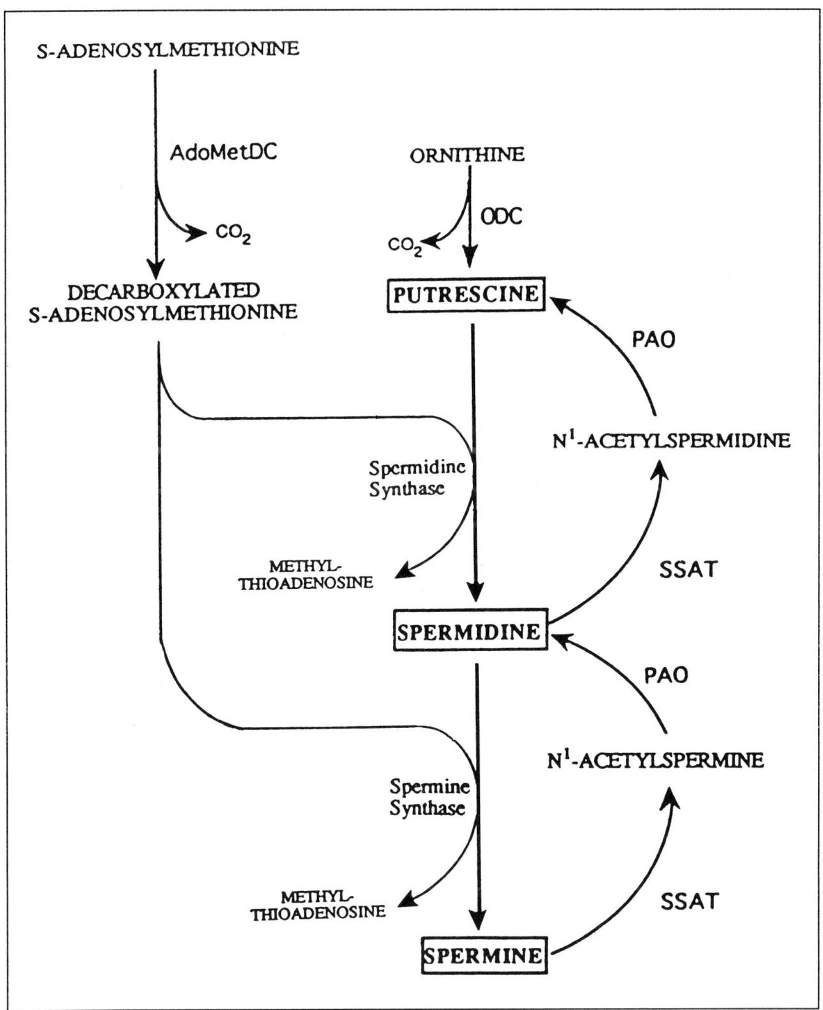

Fig. 4.1. The polyamine metabolic pathway.

does not contain what have been termed "PEST" sequences. PEST sequences (regions rich in the amino acids Proline, glutamatE, Serine, and Threonine) appear to act as signal sequences marking proteins for rapid degradation.[73] Currently, the mechanism by which SSAT protein is degraded is not known.

The SSAT protein has been purified to homogeneity from rat, human and chicken tissues.[7,46,65,79] The native protein appears to be a multimeric protein existing as a dimer or tetramer. The complete amino acid sequences of human, hamster, and mouse SSAT have been deduced from the nucleotide sequences and much of the human sequence has been confirmed by microsequencing.[7] The human SSAT protein is 171 amino acids in length and is highly conserved among the mammalian species. There are only eight amino acid differences between the human and the hamster SSAT, and six between the human and the mouse sequences. To understand the overall regulation of SSAT activity, it will be necessary to determine the location of the catalytic site in SSAT and the amino acid residues in SSAT critical for the rapid degradation. However, since sequence homology is so high among the mammalian species sequenced to date, conserved region analysis has not been helpful in this regard. Therefore, the determination of the homologies of the SSAT in lower eukaryotic species such as yeast and *Drosophila* would be useful to provide more information on such features of the SSAT protein required for enzyme activity and regulation. Unfortunately, the recently cloned *E. coli* spermidine acetyltransferase (SAT) has no similarity with the human SSAT at either the nucleotide or amino acid levels.[28]

The intracellular concentration of polyamines exerts positive feedback regulation on SSAT activity.[77] It was found that SSAT activity is up-regulated by increases in intracellular polyamines after exposure to exogenous spermidine or spermine. Additionally, if polyamines were depleted by inhibitors of polyamine biosynthesis, SSAT was also lowered. More interestingly, the observed changes in SSAT activity are accompanied by parallel changes in steady-state levels of the mRNA, suggesting that the polyamine-mediated expression of the SSAT gene is regulated at the level of transcription and/or stability of the mRNA. It is important to note that increases in SSAT activity are a result of increases in protein and not due to activation of a pre-existing protein.

IV. CLONING AND CHARACTERIZATION OF THE SPERMIDINE/SPERMINE N^1-ACETYLTRANSFERASE GENE

The molecular cloning of the human and subsequently, mouse and hamster cDNAs coding for SSAT[12,26,63] has allowed the initiation of detailed studies into the regulation of SSAT expression at the molecular level. The cloning of the SSAT cDNA was greatly facilitated by the unusual response of a limited number of cell lines that "superinduce" SSAT protein in response to treatment with polyamine analogs.[9,69] Induced SSAT protein from one such cell line, NCI-H157 was purified and partially sequenced.[7] The partial amino acid sequences were then used to produce degenerate oligomers for library screening.[12]

With these tools now in hand, several laboratories are investigating the molecular events regulating the expression of SSAT. The following sections provide an overview of what is currently known about the physical nature and regulation of expression of the SSAT gene.

A. SSAT mRNA

Human cells contain a 1.3 to 1.4 kb of SSAT mRNA whose size difference appears to result from a variable length poly(A)$^+$ tail.[12,88] The SSAT mRNA is present in a variety of human tissues. High levels of mRNA are expressed in lung, liver, kidney, and placenta. There

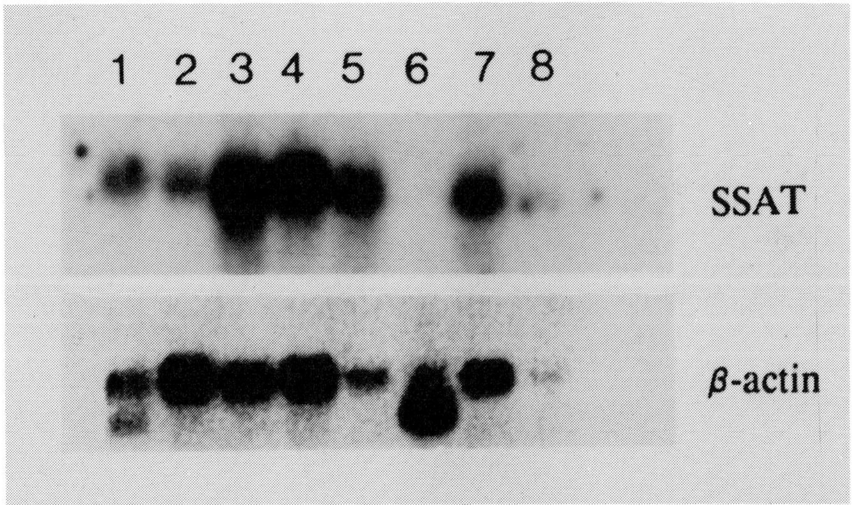

Fig. 4.2. Tissue distribution of the human SSAT mRNA. A Northern blot containing 2 ug of poly (A)+ RNAs from human heart (1), brain (2), placenta (3), lung (4), liver (5), muscle (6), kidney (7), and pancreas (8) was hybridized to the ^{32}P-labeled full-length human SSAT cDNA. The same blot was rehybridized to a human β-actin cDNA probe as indicated.

is no detectable SSAT mRNA in human muscle (Fig. 4.2). Like human cells, mouse and hamster also express SSAT mRNAs with a 513-base open reading frame (ORF) that encodes a 171 amino acid peptide.[26,63]

As indicated above, the human SSAT cDNA was found to be highly homologous to that of mouse and hamster with respect to the length and nucleotide content. The nucleotide sequence of the coding region of the human SSAT cDNA has about 91% homology with the hamster and mouse cDNA. Remarkably, the 5' untranslated region is even more similar with greater than 96% identity within -120 bp upstream from the translational initiation codon. The unusually high homology of this noncoding region strongly suggests a regulatory role for the 5' leader. Both human and mouse mRNAs contain long leader sequences >180 nucleotides in length. Analyses of mammalian SSAT mRNAs have revealed that the leader is GC-rich, particularly in the most 5' portion. The high GC-content could result in the formation of secondary structures possessing regulatory activity.[26,87] However, further study to determine the precise function of this unusual and highly conserved area is necessary. By comparison, the 3' untranslated regions of the various mammalian cDNAs demonstrate considerably less homology. The strong interspecies homology in both the coding and the 5' untranslated region may be functional in that mRNA stability, translational control, and specific polyamine interactions may occur in these regions.

In addition to facilitating the cloning procedure, the availability of cell systems that responded to induction stimuli with several thousand-fold induction of the specific SSAT protein has provided the opportunity to examine the molecular detail of SSAT induction. The marked increase in SSAT activity following exposure to certain polyamine analogs such as N^1, N^{12}-*bis*(ethyl)spermine (BESpm) and N^1, N^{11}-*bis*(ethyl)norspermine (BENSpm) (Fig. 4.3) in specific cell types is partly due to a substantial increase in the amount of SSAT mRNA.[9,12,27,62,68] However, nuclear run-on transcription analyses indicate that the observed increases in SSAT gene transcription are not sufficient to account for the polyamine analog-induced accumulation of the steady-state message RNA. Apparently, changes in posttranscriptional regulation, such as mRNA processing and/or turnover, are also involved in the drug mediated response. Additionally, the induction of SSAT activity in response to a variety of stresses in both rodent and human cells seem to be solely controlled by posttranscription mechanisms.[30,58] It is apparent that SSAT can be regulated at multiple levels that may be cell type- and stimuli-specific.

B. The SSAT Gene

Southern blot analysis of genomic DNA under stringent hybridization conditions has demonstrated that the haploid human genome seems to contain only one copy of the SSAT gene, which was localized to the short arm of the human X-chromosome at the p22.1 locus.[87] Unlike the human genome, the mouse genome seems to have two separate SSAT gene copies.[26] However, it has not been determined if both copies of the mouse gene are expressed.

The structure of human and mouse SSAT genes, like their cDNAs, are very similar. Both contain six exons and five introns. The first exon contains the 5' untranslated region as well as the translational start site. Exon 6, the largest exon, contains the stop codon and the entire 3' untranslated region. More interestingly, the human and mouse SSAT gene both appear to be under the control of a TATA-less promoter. The area 5' to the transcription start site contains several GC-rich regions, consensus sequences of a cAMP responsive element, and several putative binding sites for transcription factors, such as SP1, C/EBP, and AP2. These putative *cis*-elements along with *trans*-acting factors may facilitate the tissue-specific expression of SSAT gene in response to a variety of stimuli and could have functional significance in the general control of SSAT expression.

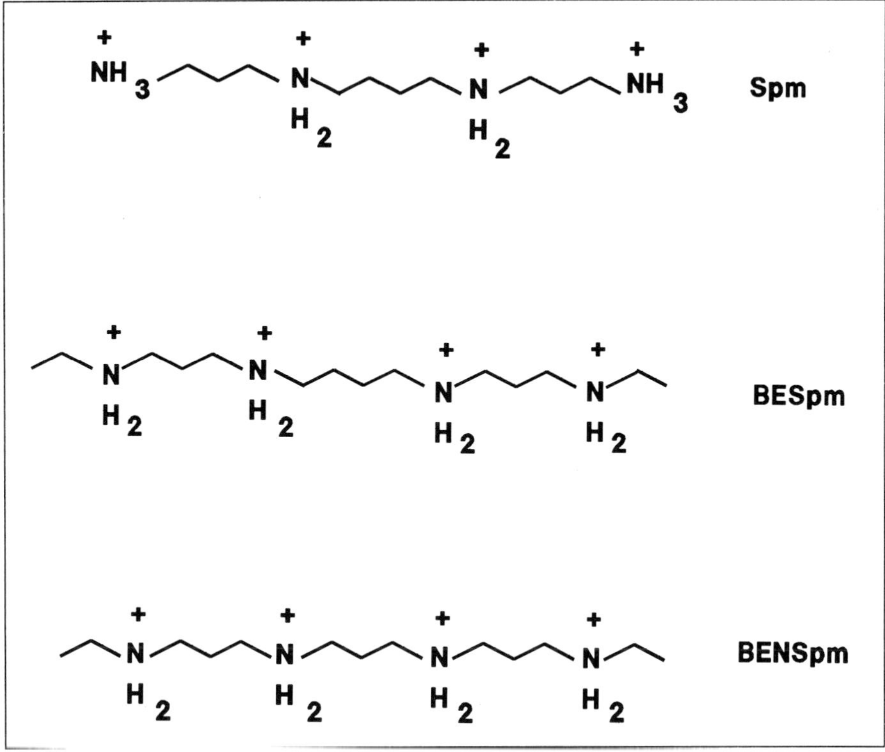

*Fig. 4.3. Structures of spermine and its analogs N^1,N^{12}-*bis*(ethyl)spermine (BESpm) and N^1,N^{11}* bis*(ethyl)norspermine (BENSpm).*

The characterization of the human SSAT gene has revealed an interesting structural feature of the 5' untranscribed region. Two long-stretches of alternating purine-pyrimidine (PuPy) sequences with extremely high A-T contents (>80%) are found in regions between -3,230 and -2,690 bp relative to the transcriptional start site. Recent studies have implicated the possibility of specific polyamine-DNA interactions. Computer-modeling analyses suggest that spermine produces a closing of the major groove around the spermine molecule and widens the minor groove only in the alternating Pu-Py sequences.[24] Of potentially greater interest is the recent finding that spermine induces structural changes specific to alternating A-T sequences.[20] The nuclear localization of spermine[80] suggests that spermine, in some cases, may have a direct role of regulating gene expression in a sequence-specific manner. Therefore, the alternating A-T sequences in the SSAT gene could be a potential cis-elements which mediate polyamine-regulated gene expression. One sequence motif in the SSAT gene consists of four identical repeats 5' ATATATGTGTATATATAT 3' repeats. This motif has also been found in several other genes in the genome of many species. This motif is found in the 5' untranscribed region of several human genes including the estrogen receptor gene.[66] Preliminary results from transient transfection studies suggest that the 18-mer repeat may be functionally involved in the BESpm-mediated transcriptional activation, which highlights a potential importance of this sequence motif in regulation of gene expression in general.

The expression of many genes is known to be affected by the methylation status of what are referred to as CpG islands.[3,13,67] The human SSAT gene has two regions of high CpG content. A CpG island in the 5' region of the human SSAT gene spans the entire first exon. This region is completely unmethylated on the single copy of SSAT in male derived cells.[49] There also is a 3' CpG island, downstream from the coding region which is unmethylated in all superinducing cell lines and generally methylated in the nonsuperinducing lines.[49] Although, these methylation differences do not appear to tightly control the transcription of SSAT in the various cell types, further study will be necessary to determine if CpG methylation plays an active role in the differential response of SSAT to the cytotoxic analogs. Regardless of the mechanisms involved, SSAT methylation status may prove to be an important prognostic indicator of response. It is also important to note that in cells of female origin containing two copies of the SSAT gene, one copy of SSAT appears to be inactivated as indicated by near complete methylation of the 5' island (Mank, Xiao, Berkey and Casero—unpublished observation). This pattern of inactivation of one copy of X-specific genes is typical for nonpseudoautosomal genes on the inactive X-chromosome in human females.[41] However, the potential for activation of a normally inactive SSAT gene in a diseased state and the potential consequences of increased gene dosage must be considered.

The SSAT gene is uniquely regulated among the other genes of the polyamine metabolic pathway. Unlike ODC and AdoMetDC, which appear to be primarily regulated translationally and posttranslationally in response to changes in intracellular polyamine levels,[39] the steady-state level of SSAT mRNA responds in a positive manner to the polyamines and their analogs.[7,9] This has particularly interesting implications since it has been suggested that polyamines have a direct or indirect roles in regulating the expression of genes at transcription levels.[14,16] The SSAT gene provides a useful system as a model gene for investigating the general role of polyamines in regulation of gene expression.

V. REGULATION/DEREGULATION OF SSAT EXPRESSION AND PHYSIOLOGIC CONSEQUENCES

A. The Superinduction of SSAT by Some Polyamine Analogs Is Associated with Cytotoxicity

Although SSAT is highly inducible in response to many stimuli[10] including the polyamines themselves,[77] the most powerful inducers of SSAT examined are structural analogs of the polyamines (Fig. 4.3).[6,45,64,69] These analogs can produce levels of SSAT amounting to almost 1% of the total soluble protein[7] (superinduction), and mimic the effects of natural polyamines in down-regulating ODC and AdoMetDC activities without substituting functionally for the natural polyamines.[6,8,70] Additionally, the analogs act as competitive inhibitors of polyamine uptake. Superinduction, therefore, leads to a complete loss of normal polyamines from the treated cells and, more interestingly, it is associated with a phenotype-specific cytotoxicity.[7,69]

A correlation between SSAT superinduction and the unusual cytotoxic effect of the polyamine analog, N^1, N^{12} *bis*(ethyl)spermine (BESpm), was first observed in the human lung carcinoma lines.[8] Differential sensitivity of human lung cancer lines to the analogs appears to be a functional difference related to SSAT induction between the responding and nonresponding cells. By comparing a human small cell lung carcinoma (SCLC) line H82 (analog insensitive) with a human large cell lung carcinoma line H157 (analog sensitive), it was found that the large cell exhibited a massive dose- and time-dependent increase in the activity of SSAT, whereas in the small cell there was no significant change in SSAT with increasing time or concentration.[6] A similar association between increased expression of SSAT and the cell growth inhibition was also observed in other human solid tumor lines including melanoma, pancreatic and breast[17,19,69] and Chinese hamster ovary (CHO) cells.[62] Expanded studies in multiple human lung cancers[9] and human melanoma cell lines[78] have revealed a general correlation between increased SSAT activity and cytotoxicity in response to BESpm. Thus, a pattern has emerged with the bis(ethyl)polyamine analogs that suggests cell types which respond to analog treatment with superinduced SSAT activity (>100-fold induction within 24 hr) go on to die. Those cell types which are more moderately induced or are not induced at all, are generally only growth inhibited.

Studies in human breast cancer lines demonstrate that BESpm modestly induces SSAT as compared to non-SCLC and melanoma lines and produces significant growth inhibition without net cell loss.[19] Importantly, the estrogen dependent and drug resistant status of the individual lines did not correspond to the response to the analog. It appears that the kinetics of induction of SSAT activity, both rate and rapidity of onset, may be important determinants of analog effects on cell growth, although the direct role of SSAT induction in the subsequent cytotoxicity has not been demonstrated.

The induction of SSAT by polyamine analogs varies among different cell lines[71] as well as the different cell types originating from the same tissue.[9,78] This heterogeneity in response is not a result of differences in analog accumulation. Therefore the basis for differential induction must occur by another mechanism. Detailed studies in the human lung carcinoma model and in the human melanoma system demonstrate that the SSAT superinduction accompanied increase in steady-state SSAT mRNA is the result of a combination of the increases in gene transcription rate and in message half-life.[78,89] Since the analog-induced accumulation of SSAT-specific mRNA in the lung and melanoma models does not entirely account for the entire observed increase in SSAT activity resulting from analog exposure, mRNA processing and translational/posttranslational controls must also be involved in the superinduction of SSAT. It is likely that the initial increase in transcription is necessary for the down stream events of superinduction to occur. However, in contrast to the non-SCLC lung lines and melanoma lines, a good correlation exists between the increase in steady-state mRNA levels and the increased SSAT activity by BESpm in human breast cancer cell lines.[19] These results suggest that in some specific instances gene transcription may be the principle control point of induction. It is possible to postulate that the posttranscriptional control is necessary for the SSAT superinduction phenotype and that the absence of those mechanisms is associated with the failure of the analogs to superinduce SSAT. It is clear from current data that the initial response to analog treatment occurs at the transcription level. Even cell lines which do not respond in a cytotoxic manner, but do induce the enzyme demonstrate increased SSAT mRNA levels. An interesting exception to this is the unresponsive SCLC NCI-H82 line. The sensitive ribonuclease protection assay suggests that H82 does not express any SSAT mRNA in response to BESpm treatment. Recently completed nuclear run-on assays demonstrate no SSAT transcription in control or BESpm-treated H82 cells suggesting that the SSAT gene is inactive in these cells. Although more work is required to verify these findings, they suggest the possibility that SSAT activity is not required for survival in NCI-H82 cells. Further, the inactivity of this gene may protect these cells from the cytotoxic activity of BESpm.

It should be noted that currently there is no proof that SSAT is directly involved in the cell type-specific cytotoxicity produced by the bis(ethyl)polyamine analogs. There are polyamine analogs which are cytotoxic to treated cells that do not induce SSAT significantly.[50] However, there are no reports of SSAT superinduction by the bis(ethyl)polyamines without cytotoxicity. Therefore, although SSAT activity may not be necessary for cytotoxicity, it may serve as a potentially useful prognostic indicator of cellular response.

B. Chromosomal Conformation May Control Induction Response

The susceptibility of chromatin to DNase I cleavage has been associated with actively expressed genes. DNase I hypersensitive sites are often related to a nucleosome free region of DNA that may reflect an underlying sequence-specific DNA-protein interaction. Studies of DNase I hypersensitivity of the human SSAT gene have revealed multiple cell-specific hypersensitive sites which exist in the responsive cell line H157 and appear to be absent in the nonresponsive cell line H82 cells.[89] These results strongly suggest a difference in chromatin structure between the sensitive and insensitive cell types and the response to the polyamine analog may primarily be controlled at the chromatin level.

Three H157-cell-specific hypersensitive sites (DHS) have been identified in the 5' untranscribed region (DHS A) and the third intron (DHS B and C) of the human SSAT gene (Fig. 4.4). Preliminary results from transient transfection assays suggest that the region around DHS A in the 5' flanking sequence of the SSAT gene are important for BESpm-mediated transcriptional activation. DHS B and C in the third intron may be related with the DNA methylation differences between the two cell types. It was found that several methylation-sensitive restriction sites around these DHS were partially methylated in H82 cells, but completely unmethylated in H157 cells.[49] It seems that the presence or absence of methylation in this region reflects alternative chromatin conformation. Methylation status of eukaryotic sequences is frequently correlated with gene expression,[3] particularly in the 5' regulatory regions. However, our preliminary results suggest that methylation changes in areas 3' to the coding region may have significant effects on the expression of SSAT. It is possible that the regions defined by DHS B and C must remain unmethylated as prerequisites for the SSAT transcription in H157 cells although this remains to be determined by functional analyses. It is likely that the cell type-specific promoter/enhancer utilization may result in the differential expression of the SSAT gene among the various cell types.

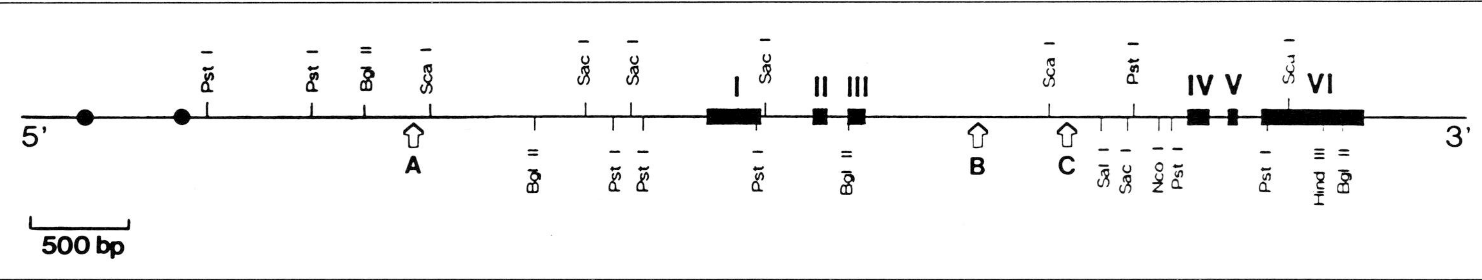

Fig. 4.4. Structure of the human SSAT gene. The six exons are numbered and represented by filled boxes. The open arrows indicate the H157-cell-specific DNAse I hypersensitive sites (A,B, and C). The filled circles represent two long-stretches of alternating purine-pyrimidine sequences with very high A-T contents.

C. Stress-Induced SSAT Expression and Its Physiological Relevance

A common feature of the response of eukaryotic and prokaryotic organisms to environmental stress is the rapid production of a defined set of proteins, among these are heat shock or other stress-induced proteins.[4] Studies at the molecular level have revealed an overlap in the patterns of gene expression in response to a variety of stresses. Recent findings suggest that SSAT is one of the stress-induced proteins.[29,30,58]

The stimulation of SSAT by heat shock and chemical stress such as diethyldithiocarbamate (DDC) appears to be under a control of polyamine-dependent mechanisms.[29,30,38] Depletion of intracellular putrescine and spermidine pools with α-difluoromethylornithine (DFMO) inhibited the heat-induction of SSAT activity. The effect of DFMO on heat-shock induction of SSAT can be reversed by the addition of putrescine and spermidine, indicating that intracellular polyamine levels are essential for the stress-induced SSAT activity. Since polyamines induce SSAT, it is possible that stress disrupts the intracellular polyamine localization[31,38] releasing polyamines from their normal binding compartments and allowing them to become initiators for the induction of SSAT. The effect of hyperthermia on the elevation of enzyme activities seems to be specific in the polyamine pathway for SSAT because the activity of ODC was suppressed by this effect.[58] Therefore, the stimulation of polyamine catabolism by increasing SSAT activity may be a part of the general cellular response to stress.

The mechanisms underlying the stress-induced SSAT activity appears to differ from classical heat-shock genes. A common mechanism of the previously described heat-shock genes is that they are primarily controlled at the level of transcription.[47] Although the increase in SSAT activity by stressors was inhibited by actinomycin D, the steady-state levels of SSAT mRNA remained essentially unchanged under stress conditions.[30,58] It should be noted that the timing of the treatment with actinomycin D seems to be important for its effect on the inhibition of SSAT activity. It was found that the simultaneous addition of actinomycin D with ethanol or hyperthermia did not inhibit an increase in SSAT activity, whereas the enhancement of the enzyme activity was suppressed when actinomycin D was added to the culture 30 min before stress treatments.[58] It appears that actinomycin D suppresses the increase in SSAT activity by a mechanism not affecting the transcription of SSAT, suggesting that a posttranscriptional regulation of SSAT expression is involved in response to stresses. However, these results do not discount the possibility that the transcription of another gene is required for the up-regulation of SSAT.

Protein synthesis inhibition studies with cycloheximide suggest that stress-induced protein synthesis is required for the simulation of SSAT activity. Furthermore, the decrease in the half-life of SSAT activity by cycloheximide during stress indicates that other stress-induced proteins may stabilize SSAT. The stabilization of SSAT would result in an increase in enzyme activity. The induction of SSAT is known to be associated with a stabilization of enzyme protein in some cases.[65] Therefore, it is likely that the posttranslational regulation is involved in the stress-related induction of SSAT activity. However, the induction of SSAT in response to lithium chloride in Erlich ascites cells was attributed to the increased translation rate.[54] These observations suggest that SSAT can be regulated by translation and/or posttranslational modifications that may be cell- and stimuli-specific.

As previously stated, catabolism of polyamines is a two-step reaction involving acetylation by a highly inducible SSAT and subsequent oxidation by a constitutively expressed, flavin-dependent polyamine oxidase (PAO). Certain stresses, such as heat shock, stimulate polyamine catabolism by the activation of SSAT. Studies in Chinese hamster ovary (CHO) cells demonstrate that endogenous spermidine oxidation proceeds in a heat-dose dependent manner via induction of SSAT activity.[37] Like rodent cells, many human cells of both normal and neoplastic origin express PAO activity and catabolize polyamines by a mechanism which includes PAO,[5] suggesting that the hyperthermic stress may produce an oxidative stress in mammalian cells via induction of polyamine oxidation. The oxidation of polyamines generates reactive oxygen species such as hydrogen peroxide H_2O_2 which has been suggested as one of the principal toxic candidates and appears to be capable of affecting cell viability.[56] Therefore, it is possible to speculate that the induction of SSAT and subsequent stimulation of polyamine catabolism may play a role in the stress-induced cytotoxicity.

One interesting and related phenomenon is that certain mildly adverse conditions which induce SSAT activity can also result in the injured cell dying by programmed cell death (apoptosis). These conditions include hyperthermia, hypothermia, ischemia, exposure to radiation, toxins, and various chemicals.[75] Apoptosis is an important physiological process in mammalian cells, essential for many aspects of normal development and required for maintaining homeostasis. The similarities in the morphological and biochemical patterns defined as apoptosis within different cell types and species, during normal development and as a response to external stimuli, suggest a common pathway of cellular mortality. Presumably, some common mediators are involved intracellularly that directly induce apoptosis or cause the cell to undergo apoptosis in response to the induced damages. One of the known mediators is H_2O_2, an important reactive oxygen species (ROS). It was found that cell death resulting from low levels of oxidative stress by treatment with exogenous H_2O_2 resembled apoptosis.[40] Several lines of evidences suggest that hydrogen peroxide generated by oxidation of polyamines may be an important mediator of apoptosis in the development of mammalian embryos[18,32] and c-*myc* induced apoptosis.[59] As mentioned above, the polyamine catabolism can generate H_2O_2 via oxidative reactions. A direct consequence of SSAT activity is the production of N[1]-acetylated polyamines which are much better substrates of PAO than parent polyamines. The increased SSAT activity, therefore, enhances the oxidation of polyamines. It raises the possibility that the induction of SSAT activity by mildly adverse conditions can be a "trigger" to activate the cell death pathway. However, the induction of SSAT activity alone may not be sufficient to result in apoptosis. Other intracellular changes are likely to be involved in the biochemical mechanism for regulation of apoptosis, including increases in intracellular calcium, activation of protein kinase A and C and changes in other gene expression. It is, however, important to consider that DNA fragmentation leading to irreversible changes in cells is thought to be an important step in the progression of apoptosis. Polyamines, particularly spermine, stabilize chromatin structure and may inhibit DNA fragmentation.[86] It is conceivable that induction of SSAT activity and the subsequent drop in intracellular polyamines could have a role in the regulation of apoptosis. The possibility that SSAT induction may be part of a common cellular pathway in programmed cell death is an intriguing one which will require further study.

D. Involvement of SSAT Gene Expression in Neoplastic Growth

The intracellular polyamine content and activity of the enzymes of the metabolic pathway are regulated precisely according to growth status. The activity of the biosynthetic enzymes can be rapidly compensated for by the activity of SSAT and polyamine catabolism. A growing body of evidence correlates the development of tumors to the aberrant polyamine metabolism.[61,85] A high spermidine/spermine ratio and an elevated concentration of N[1]-acetylspermidine have been found in many kinds of cancers,[43,82] suggesting that the polyamine metabolism, specifically polyamine acetylation, may be altered in cancer cells.

A rapid and remarkable induction of SSAT mRNA after cell activation from quiescence[23] suggests that increased expression of SSAT, like ODC, may be critical for the growth of mammalian cells. Studies in growing Yoshida AH-130 ascites hepatoma cells demonstrated that SSAT activity progressively increased during tumor growth.[22] The increased SSAT activity was accompanied by an accumulation of the steady-state SSAT mRNA without changes in transcription rate suggesting mechanisms other than transcriptional control were involved in the SSAT induction. Moreover, it was found that SSAT activity and N[1]-acetylspermidine levels gradually increase both in the ascitic fluid and hepatoma cells throughout tumor growth. Additionally, N[1]-acetylspermidine levels greatly increased after the administration of a specific PAO inhibitor. These data indicate that the increased SSAT activity during neoplastic growth results in acceleration of the interconversion of polyamines and subsequent polyamine efflux, which is one likely mechanism important for the control of intracellular polyamine pools. More importantly, a similar regulation of SSAT expression was observed in tumor-bearing rats. The hepatic involution during the quasi-stationary phase of growth is characterized by enhanced apoptosis, which may account, in part, for the enhanced rate of protein turnover in the liver of Yoshida AH-130 hepatoma-bearing rats.[83] The coincidence of increased levels of SSAT activity at the beginning of the quasi-stationary phase in Yoshida AH-130 hepatoma cells supports the hypothesis that induction of SSAT activity could have a role in the regulation of apoptosis.

The elevation of SSAT activity and the resulting enhanced polyamine catabolism have been implicated to be important steps in carcinogenesis.[55] It was found that the activity of SSAT increases and N^1-acetylspermidine accumulates in colonic tumors of rats treated with the carcinogen 1,2-dimethylhydrazine.[35] Importantly, growth of these tumors is suppressed by treatment with PAO inhibitors[36] suggesting that complete polyamine catabolism is important for carcinogenesis in this model. In vivo studies of chemical carcinogenesis have demonstrated very high levels of SSAT and ODC activities in two kinds of bladder tumors, N-butyl-N-(4-hydroxybutyl)nitrosamine-induced transitional cell carcinoma (TCC) and melamine-induced papillomatosis, but not in normal mucosa. It has been suggested that elevated putrescine levels are critical for tumor development.[61] The above data suggest that the formation of some types of tumors may require increases in putrescine beyond that produced by increases in ODC. Therefore, the requirement could be met by a concurrent increase in SSAT activity. These findings suggest that the alteration of polyamine catabolism may have a unique role from a benign to malignant phenotype conversion in vivo.

VI. SUMMARY

SSAT is the key regulatory and rate limiting enzyme in polyamine catabolism. The discovery of cell types which superinduce this enzyme in response to treatment with polyamine analogs has facilitated the cloning of the SSAT gene and the further detailed study of its regulation. Preliminary results indicate that the expression of SSAT is controlled at every level from transcription to stabilization of the protein, and the regulation appears to be cell- and/or stimuli-specific.

The results of current studies indicate that in some cases the overexpression of SSAT in response to various polyamine analogs is closely associated with their cytotoxic response. Because of these and other recent data the potential importance of SSAT in programmed cell death is now being examined. That the by-products of the acetylase/oxidase pathway can be mediators of the cell death program is an intriguing possibility that will require further study. Similarly, the induction of SSAT in response to various stressors suggests a role for SSAT in the general cellular response to stress.

Finally, the altered regulation of SSAT in tumor cells suggests that this enzyme may be an important player in transformation and progression in carcinogenesis. So that in addition to being a target for the development of antineoplastic agents SSAT may be a logical target for the production of chemopreventive agents.

Continued study of the regulation and roles of SSAT should provide considerable insight into the role of this enzyme in cellular homeostasis and provide a clearer picture of the specific molecular roles in which polyamines are involved.

ACKNOWLEDGMENTS

Parts of the work described were funded by NIH grants CA51085, CA58184, and CA63552. The authors thank Dr. Diane Krause for her suggestions.

REFERENCES

1. Basu HS, Schwietert HCA, Feuerstein BG and Marton LJ. Effects of variation in the structure of spermine on the association with DNA and the induction of DNA conformational changes. Biochem J 1990;269:329-334.
2. Basu HS, Sturkenboom CJM, Delcros J-G, et al. Effect of polyamine depletion on chromatin structure in U-287 MG human brain tumor cells. Biochem J 1992;282:723-727.
3. Bird AP. CpG-rich islands and the function of DNA methylation. Nature 1986;321:209-213.
4. Burdon RH. Heat shock proteins in relation to medicine. Mol Aspects Med 1993;14:83-165.
5. Carper SW, Tome ME, Fuller DJM, et al. Polyamine catabolism in rodent and human cells in culture. Biochem J 1991;280:289-294.
6. Casero RA, Celano P, Ervin SJ, et al. Differential induction of spermidine/spermine N^1-acetyltransferase in human lung cancer cells by the bis(ethyl)polyamine analogs. Cancer Res 1989;49:3829-3833.
7. Casero RA, Celano P, Ervin SJ, et al. High specific induction of spermidine/spermine N^1-acetyltransferase in a human large cell lung carcinoma. Biochem J 1990;270:615-620.
8. Casero RA, Ervin SJ, Celano P, et al. Differential response to treatment with bis(ethyl)polyamine analogs between human small cell lung carcinoma and undifferentiated large cell lung carcinoma in culture. Cancer Res 1989;49:639-643.
9. Casero RA, Mank AR, Xiao L, et al. Steady-state messenger RNA and activity correlates with sensitivity to N^1, N^{12}-Bis(ethyl)spermine in human cell lines representing the major forms of lung cancer. Cancer Res 1992;52:5359-5363.
10. Casero RA and Pegg AE. Spermidine/spermine N^1-acetyltransferase-the turning point in polyamine metabolism. FASEB J 1993;7:653-661.
11. Casero RA Jr and Baylin SB. Concepts for deriving specific inhibitors of polyamine biosynthesis - human lung cancer cells as a model system. In:

In: Palfreman, M.P., McCann, P.P., Lovenberg, W., Temple, J.G. and Sjoerdsma, A., eds. Enzymes as Targets for Drug Design, San Diego: Academic Press, 1989, p. 185-200.

12. Casero RA J., Celano P, Ervin SJ, et al. Isolation of characterization of a cDNA clone that codes for human spermidine/spermine N^1-acetyltransferase. J Biol Chem 1991;266:810-814.
13. Cedar H. DNA methylation and gene activity. Cell 1988;53:3-4.
14. Celano P, Baylin SB and Casero RA. Polyamines differentially modulate the transcription of growth-associated genes in human colon carcinoma cells. J Biol Chem 1989;264:8922-8927.
15. Celano P, Baylin SB, Giardiello FM, et al. Effect of polyamine depletion on c-*myc* expression in human colon carcinoma cells. J Biol Chem 1988;263:5491-5494.
16. Celano P, Berchtold CM, Giardiello FM and Casero RA. Modulation of growth gene expression by selective alteration of polyamines in human colon carcinoma cells. Biochem Biophys Res Comm 1989;165:384-390.
17. Chang BK, Bergeron RJ, Porter CW, et al. Regulatory and antiproliferative effects of N-alkylated polyamine analogs in human and hamster pancreatic adenocarcinoma cell lines. Cancer Chem and Pharm 1992; 30:183-188.
18. Coffino P and Poznanski A. Killer Polyamines? J Cell Biochem 1991;45:54-58.
19. Davidson NE, Mank AR, Prestigiacomo LJ, et al. Growth inhibition of hormone-responsive and -resistant human breast cancer cells in culture by N^1,N^{12}-*bis*(ethyl)spermine. Cancer Res 1992;53:2071-2075.
20. Delcros J-C, Sturkenboom MCJM, Basu HS, et al. Differential effects of spermine and its analogs on the structures of polynucleosides complexed with ethidium bromide. Biochem J 1993;291:269-274.
21. Denstman SC, Ervin SJ and Casero RA. Comparison of the effects of treatment with the polyamine analog N^1,N^8-bis(ethyl)spermidine (BESpd) or difluoromethylornithine (DFMO) on the topoisomerase II-mediated formation of 4'-(9-acridinylamino)methanesulfon-M-anisidide (m-AMSA)-induced cleavable complex in the human lung carcinoma line NCI H157. Biochem Biophys Res Comm 1987;149:194-202.
22. Desiderio MA and Bardella L. Expression of spermidine/spermine N^1-acetyltransferase in growing yoshida AH-130 hepatoma cells. Hepatology 1994;19:728-734.
23. Desiderio MA, Mattei S, Biondi G and Colombo MP. Cytosolic and nuclear spermidine acetyltransferase in growing NIH 3T3 fibroblasts stimulated with serum or polyamines: relationship to polyamine-biosynthetic decarboxylases and histone acetyltransferase. Biochem J 1993;293:475-479.
24. Feuerstein BG, Pattabiraman N and Marton LJ. Molecular mechanics of the interactions of spermine with DNA: DNA bending as a result of ligand binding. Nucleic Acids Research 1990;18:1271-1282.
25. Ficker E, Taglialatela M, Wible BA, et al. Spermine and spermidine as

gating molecules for inward rectifier K^+ channels. Science 1994; 266:1068-1072.
26. Fogel-Petrovic M, Kramer DL, Ganis B, et al. Cloning and sequence analysis of the gene and cDNA encoding mouse spermidine/spermine N^1-acetyltransferase-a gene uniquely regulated by polyamines and their analogs. Biochem Biophys Acta 1993;1216:255-264.
27. Fogel-Petrovic M, Shappell NW, Bergeron RJ and Porter CW. Polyamine and polyamine analog regulation of spermidine/spermine N^1-acetyltransferase in MALME-3M human melanoma cells. J Biol Chem 1993; 268:19118-19125.
28. Fukuchi J, Kashiwag K, Tokio K and Igarashi K. Properties and structure of spermidine acetyltransferase in *Escherichia coli*. J Biol Chem 1994;269:22581-22585.
29. Fuller DJ, Carper SW, Clay L, et al. Polyamine regulation of heat-shock-induced spermidine N^1-acetyltransferase activity. Biochem J 1990; 267:601-605.
30. Gerner EW, Kurtis TA, Fuller DJM and Casero RA. Stress induction of the spermidine/spermine N^1-acetyltransferase by post-transcriptional mechanism in mammalian cells. Biochem J 1993;294:491-495.
31. Gerner EW, Stickney DG, Herman TS and Fuller DJM. Polyamines and polyamine biosynthesis in cells exposed to hyperthermia. Radiation Res. 1983;93:340-352.
32. Gramzinski RA, Parchment RE and Pierce GB. Evidence linking programmed cell death in the blastocyst to polyamine oxidation. Differentiation 1990;43:59-65.
33. Gross DS and Garrard WT. Nuclease hypersensitive sites in chromatin. Ann Rev Biochem 1988;57:159-197.
34. Gross M and Rubino MS. Regulation of eukaryotic initiation factor-2B activity by polyamines and amino acid starvation in rabbit reticulocyte lysate. J Biol Chem 1989;264:21879-21884.
35. Halline AG, Dudeja PK and Brasitus TA. 1,2-Dimethylhydrazine-induced alterations in N^1-acetylspermidine levels and spermidine N^1-acetyltransferase activity in rat colonic mucosa. Cancer Res 1989;49:633-638.
36. Halline AG, Dudeja PK, Jacoby RF, et al. Effect of polyamine oxidase inhibition on the colonic malignant transformation process induced by 1,2-dimethylhydrazine. Carcinogenesis 1990;11:2127-2132.
37. Harari PM, Fuller DJM and Gerner EW. Heat shock stimulates polyamine oxidation by two distinct mechanisms in mammalian cell cultures. Int J Radiation Onc Biol Phys 1989;16:451-457.
38. Harari PM, Tome ME, Fuller DJM, et al. Effects of diethyldithiocarbamate and endogenous polyamine content on cellular responses to hydrogen peroxide cytotoxicity. Biochem J 1989; 260:487-490.
39. Heby O and Persson L. Molecular genetics of polyamine synthesis in eukaryotic cells. TIBS 1990;15:153-158.
40. Hockenbery DM, Oltvai ZN, Yin X-M, et al. Bcl-2 functions in an antioxidant pathway to prevent apoptosis. Cell 1993;75:241-251.

41. Hornstra IK and Yang TP. Multiple in vivo footprints are specific to the active allele of the X-linked human hypoxanthine phosphoribosyltransferase gene 5' region: Implications for X chromosome inactivation. Mol Cell Biol 1992;12:5345-5354.
42. Jain S, Zon C and Sundaralingam M. Base only binding of spermine in the deep groove of the A-DNA octamer d(GTGTACAC). Biochemistry 1989;28:2360-2364.
43. Kingsnorth AN and Wallace HM. Elevation of monoacetylated polyamines in human breast cancer. Eur J Cancer Clin Oncol 1985;21:1057-1062.
44. Kramer DJ, Khomutov RM, Bukim YV, et al. Cellular characterization of a new irreversible inhibitor of S-adenosylmethionine decarboxylase and its use in determining the relative abilities of individual polyamines to sustain growth and viability of L1210 cells. Biochem J 1989;259:325-331.
45. Libby PR, Bergeron RJ and Porter CW. Structure-function correlations of polyamine analog-induced increases in spermidine/spermine N^1-acetyltransferase activity. Biochem Pharmacology 1989;38:1435-1442.
46. Libby PR, Ganis B, Bergeron RJ and Porter CW. Characterization of human spermidine/spermine N^1-acetyltransferase purified from cultured melanoma cells. Arch Biochem Biophys 1991;284:238-244.
47. Lindquist S. The heat-shock response. Ann Rev Biochem 1986; 55:1151-1191.
48. Lopatin AN, Makhina EN and Nichols CG. Potassium channel block by cytoplasmic polyamines as the mechanism of intrinsic rectification. Nature 1994;372:366-369.
49. Mank AR, Xiao L and Casero RA. DNA methylation differences in the human spermidine/spermine N^1-acetyltransferase gene correlates with lung tumor phenotype-specific response to treatment with N^1, N^{12}-(*bis*)ethylspermine. Proc Am Assoc Cancer Res 1994;35:548.
50. Marton LG and Pegg AE. Polyamines as targets for therapeutic intervention. Ann Rev Pharmacol Toxicol 1995;35:55-91.
51. Marton LJ and Morris DR. Inhibition of polyamine metabolism: Biological significance and basis for new therapies. In:, eds. McCann, P.P., Pegg, A.E. and Sjoerdsma, A. Orlando: Academic Press, 1987, p. 79-105.
52. Mathews HR. Polyamines, chromatin structure and transcription. BioEssays 1994;15:561-566.
53. Matsui I and Pegg AE. Effect of inhibitors of protein synthesis on rat liver spermidine N^1-acetyltransferase. Biochem Biophys Acta 1981; 675:373-378.
54. Matsui-Yuasa I, Obayashi M, Hasuma T and Otani S. Enhancement of spermidine/spermine N^1-acetyltransferase activity by treatment with lithium chloride in Erlich ascites tumor cells. Chem-Biol Interactions 1992; 81:233-242.
55. Matsui-Yuasa I, Otani S, Yano Y, et al. Spermidine/spermine N^1-acetyltransferase, a new biochemical marker for epithelial proliferation in rat bladder. Jpn J Cancer Res 1992;83:1037-1040.
56. Morgan DMW. Polyamine oxidases and cellular interactions. Adv Polyamine Res 1981;3:65-73.

57. Morgan JE, Blankenship JW and Matthews HR. Polyamines and acetylpolyamines increase the stability and alter the conformation of nucleosome core particles. Biochemistry 1987;26:3643-3649.
58. Obayashi M, Matsui-Yuasa I, Kitano A, et al. Post-translational regulation of spermidine/spermine N1-acetyltransferase with stress. Biochem Biophys Acta 1992;1131:41-46.
59. Packham G and Cleveland JL. Ornithine decarboxylase is a mediator of c-myc-induced apoptosis. Mol Cell Biol 1994;14:5741-5747.
60. Pegg AE. Recent advances in the biochemistry of polyamines in eukaryotes. Biochem J 1986;234:249-262.
61. Pegg AE. Polyamine metabolism and its importance in neoplastic growth and as a target for chemotherapy. Cancer Res 1988;48:759-774.
62. Pegg AE, Pakala R and Bergeron RJ. Induction of spermidine/spermine N^1-acetyltransferase activity in Chinese-hamster ovary cells by N^1,N^{11}-*bis*(ethyl)norspermidine and related compounds. Biochem J 1990; 267:331-338.
63. Pegg AE, Stanely BA, Wiest L and Casero RA. Nucleotide sequence of hamster spermidine/spermine N^1-acetyltransferase cDNA. Biochem Biophys Acta 1992;1171:106-108.
64. Pegg AE, Wechter R, Pakala R and Bergeron RJ. Effect of N^1,N^{12}--*bis*(ethyl)spermine and related compounds on growth and polyamine acetylation, content, and excretion in human colon tumor cells. J Biol Chem 1989;264:11744-11749.
65. Persson L and Pegg AE. Studies of the induction of spermidine/spermine N^1-acetyltransferase using a specific antiserum. J Biol Chem 1984; 259:12364-12367.
66. Piva R, Gambari R, Zorzato F, et al. Analysis of upstream sequences of the human estrogen receptor gene. Biochem Biophys Res Commun 1992;183:996-1002.
67. Plump AS, Smith JD, Hayek T, et al. Severe hypercholesterolemia and atherosclerosis in apolipoprotein E-deficient mice created by homologous recombination in ES cells. Cell 1992;71:343-353.
68. Porter CW, Bernacki RJ, Miller J and Bergeron RJ. Antitumor activity of N^1,N^{11}-*bis*(ethyl)norspermine against human melanoma xenografts and possible biochemical correlates of drug action. Cancer Res 1993; 53:581-586.
69. Porter CW, Ganis B, Libby PR and Bergeron RJ. Correlations between polyamine analog-induced increases in spermidine/spermine N^1-acetyltransferase activity, polyamine pool depletion, and growth inhibition in human melanoma cell lines. Cancer Res 1991;51:3715-3720.
70. Porter CW, McManis J, Casero RA and Bergeron RJ. The relative abilities of bis(ethyl) derivatives of putrescine, spermidine, and spermine to regulate polyamine biosynthesis and inhibit cell growth. Cancer Res 1987;47:2821-2825.
71. Porter CW, Regenass U and Bergeron RJ. Polyamine inhibitors and analogs as potential anticancer agents. In: Falk Symposium on Polyamines, Lancaster: Kluwer Press, 1992; 301-322.

72. Pryciak PM, Miller HP and Varmus HE. Simian virus 40 minichromosomes as targets for retroviral integration in vivo. Proc Natl Acad Sci USA 1992;89:9237-9241.
73. Rogers S, Wells R and Rechsteiner M. Amino acid sequences common to rapidly degraded proteins: the PEST hypothesis. Science 1986;234:364-368.
74. Schulze-Lohoff E, Fees H, Zanner S, et al. Inhibition of immediate-early-gene induction in renal mesangial cells by depletion of intracellular polyamines. Biochem J 1994;298:647-653.
75. Schwartzman RA and Cidlowski JA. Apoptosis: The biochemistry and molecular biology of programmed cell death. Endocr Rev 1993;14:133-151.
76. Seiler N. Potential roles of polyamine interconversion in the mammalian organism. Ad Exp Med Biol 1988; 250:127-145.
77. Shappell NW, Fogel-Petrovic MF and Porter CW. Regulation of spermidine/spermine N^1-acetyltransferase by intracellular polyamine pools. Evidence of a functional role in polyamine homeostasis. FEBS 1993;321:179-183.
78. Shappell NW, Miller JT, Bergeron RJ and Porter CW. Differential effects of the spermine analog, N^1, N^{12}-bis(ethyl)spermine, on polyamine metabolism and cell growth in human melanoma cell lines and melanocytes. Anticancer Res 1992;12:1083-1090.
79. Shinki T and Suda T. Purification and characterization of spermidine N^1-acetyltransferase from chick duodenum. Eur J Biochem 1989; 183:285-290.
80. Snyder RD. Polyamine depletion is associated with altered chromatin structure in Hela cells. Biochem J 1989;260:697-704.
81. Tabor CW and Tabor H. Polyamines. Ann Rev Biochem 1984; 53:749-790.
82. Takenoshita S, Matsuzaki S, Nakano G, et al. Selective elevation of the N^1-acetylspermidine level in human colorectal adenocarcinomas. Cancer Res 1984;44:845-847.
83. Tessitore L, Valente G, Bonelli G, et al. Regulation of cell turnover in the livers of tumor-bearing rats: occurrence of apoptosis. Int J Cancer 1989;44:697-700.
84. Thomas TJ and Messner RP. Structural specificity of polyamines in left-handed Z-DNA formation. Immunological and spectroscopic studies. J Mol Biol 1988;201:463-467.
85. Verma A and Boutwell RK. Inhibition of carcinogenesis by inhibitors of putrescine biosynthesis. In: Inhibition of polyamine metabolism. Biological significance and basis for new therapies, eds McCann, P.P., Pegg, A.E. and Sjoerdsma, A. Orlando: Academic Press, 1987, p. 249-258.
86. White E. Death-defying acts: a meeting review on apoptosis. Genes and Dev 1993;7:2277-2284.
87. Xiao L, Celano P, Mank AR, et al. Structure of the human spermidine/spermine N^1-acetyltransferase gene (Exon/intron gene organization and localization to Xp22.1). Biochem Biophys Res Comm 1992; 187: 1493-1502.

88. Xiao L, Celano P, Mank AR, Pegg AE and Casero RA, Jr. Characterization of a full-length cDNA which codes for the human spermidine/spermine N^1-acetyltransferase. Biochem Biophys Res Comm 1991; 179:407-415.
89. Xiao L, Mank AR and Casero RA. The role of cell-type specific chromatin structure in transcriptional activation of the human spermidine/spermine N^1-acetyltransferase (SSAT) gene. Proc Am Assoc Cancer Res 1994;35:557.

CHAPTER 5

Biological and Therapeutic Implications of the Effects of Polyamines on Chromatin Condensation

Hirak S. Basu, Laurence J. Marton

I. SUMMARY

Studies of polyamine-DNA interactions have led to the development of polyamine analogs that deplete all three cellular polyamines and inhibit cell growth. Analog treated cells showed abnormal DNA-nuclear matrix interactions and impaired chromatin condensation. These changes in chromatin structure may induce supercoiling stress during DNA synthesis, inhibit cell division and may lead to a loss of nucleosomes.

Approaches to cancer treatment include the use of agents that attempt to prevent cell division by modifying cellular DNA. Some of these agents, such as *cis*-diamminedichloroplatinum (CDDP) and chloroethylnitrosourea (CENU), preferentially modify linker DNA and/or relaxed chromatin as compared with nucleosomal DNA and/or condensed chromatin. Inhibitors of DNA topoisomerases, the enzymes that reside at the DNA-nuclear-matrix attachment sites, are also extensively used in cancer chemotherapy. The therapeutic efficacy of DNA-binding agents and topoisomerase inhibitors may be enhanced by combining them with treatment with polyamine analogs that modify chromatin structure and DNA-matrix association.

Polyamines: Regulation and Molecular Interaction, edited by Robert Casero. © 1995 R.G. Landes Company.

II. INTRODUCTION

The scientific history of the polyamines begins almost three centuries ago when Antony van Leuwenhoek detected spermine crystals in seminal fluids. Since then the polyamine spermidine and its precursor diamine putrescine have been detected in almost all living systems.[1,2] In addition, spermine is found in all mammalian cells. It is now understood that the polyamines are among those biological molecules which are required for growth and proliferation of all living cells with the possible exception of Halo- and Methanobacteriales.[3] Our understanding of the polyamine biosynthetic pathway and its intricate regulation in both pro- and eukaryotic cells continues to progress rapidly.[4] A highly regulated biosynthetic pathway exists in almost all living cells. The intracellular levels of the biosynthetic enzymes, including ornithine decarboxylase (ODC) which catalyzes the production of putrescine, and S-Adenosylmethionine decarboxylase (AdoMetDC) which is involved in the formation of both spermidine and spermine, increase followed by an increase in cellular polyamine levels when cells are stimulated to grow and proliferate.[5] However, little is known of the molecular mechanism(s) by which the polyamines, the products of this well-regulated biosynthesis, participate in cell division.

III. POLYAMINE-DNA INTERACTIONS

The naturally occurring polyamines spermidine and spermine contain primary and secondary amino groups with pK values ranging from 8.9 to 10.2.[6] They behave like polycations under normal physiological conditions. It is logical, therefore, to assume that physiological anions such as nucleic acids should be primary binding sites for the polyamines in vivo. The modes of interaction of the polyamines with DNA in vitro have been extensively studied utilizing computer aided molecular modeling and by physical-chemical methods.[7] It is now well established that spermidine and spermine interact with and induce structural changes in isolated DNA and in specific polynucleotides in cell free systems. Spermidine and spermine cause DNA to condense and aggregate and induce B-Z and B-A transitions in certain DNA sequences in vitro.[8-10] Computer modeling and physical-chemical studies indicate that spermine can also specifically interact with and induce bends in defined DNA sequences before DNA aggregation actually takes place.[11-16] It has been hypothesized that similar structural changes occur in vivo as well and one of the major biological functions of the polyamines may relate to the induction of such structural changes in DNA.[17] Attempts to obtain accurate information about the in vivo localization of polyamines, however, have been frustrated because of the tendency of the polyamines to relocalize very quickly with the slightest perturbation of cellular integrity. In spite of this difficulty, indirect evidence is now available that shows the importance of the polyamine-DNA interactions in cell division.[7,17]

It has been shown that analogs of polyamines that interact with DNA differently than do the natural polyamines, inhibit tumor cell growth in culture.[18,19] Conversely, polyamine analogs that mimic natural polyamine-DNA interactions may not inhibit cell growth even though they deplete cellular polyamines.[20] Additional evidence that suggests the importance of polyamine-DNA interactions in cell growth is provided by the work of Ghoda et al[21] Chinese hamster ovary (CHO) cells that have been selected for a lack of ODC activity were transfected with either mouse or trypanosomal ODC. While mouse ODC is suppressed by polyamines or their analogs, trypanosomal ODC is not. Thus two identical cell lines which differ only in their sensitivity to the polyamine/analog-induced ODC inhibition were obtained. Two polyamine analogs—one with an affinity for DNA stronger than and the other with an affinity for DNA weaker than that of spermine—were used. The polyamine analog which interacts with DNA more strongly than does spermine, inhibited cell growth in the presence of significant cellular polyamine levels probably by displacing the cellular polyamines from their binding sites. The ability of the analog to inhibit polyamine biosynthetic enzymes and thereby to deplete cellular polyamines may augment its growth inhibitory effects. On the other hand, the analog that has an affinity for DNA weaker than that of spermine, inhibited cell growth only when it was able to deplete cellular polyamines. Subsequently, Albanese et al[22] have also shown that the growth inhibitory effects of polyamine analogs are not due to ODC inhibition or polyamine depletion. Such data make it imperative to investigate the exact molecular mechanism(s) by which polyamine-DNA interactions and the resultant structural changes in DNA may be involved in the regulation of cell division.

IV. POLYAMINE-DNA INTERACTIONS AND CELL DIVISION

During each eukaryotic cell division, DNA is replicated at its sites of attachment on the nuclear matrix[23,24] and the DNA strands condense into daughter chromatids. Then, during mitosis, interphase chromatin condenses and nuclear structure is disrupted to form chromosomal scaffolds and sister chromatids which finally segregate into daughter cells.[25-27] Among the possible biological functions of the cationic polyamines related to cell division may be the stabilization of replication complexes between polyanions such as DNA and the nuclear matrix and the condensation and packaging of newly synthesized DNA into nucleosomes and chromatin. We will first discuss the effects of polyamines and their analogs on DNA-matrix association and on chromatin structure as two unrelated events. Then we will consider some recent indirect evidence that may imply a causal relationship between these two events.

A. DNA-Matrix Association

Although it has been known for decades that a protein-rich nuclear fraction remains after the extraction of nuclear membrane and chromosomes,[28] the importance of this structure, now known as the nuclear matrix, was not realized until a combination of autoradiography and electron microscopic studies demonstrated that DNA synthesis takes place at the DNA attachment sites on the nuclear matrix.[29-32] Subsequently, it has also been shown that actively transcribing genes are associated with the nuclear matrix.[23,33,34] During the last decade, numerous reports on the role of periodic attachment of specific DNA sequences to the nuclear matrix and the role of such attachment in DNA replication and gene expression have been published.[23,35-38]

The manner in which DNA attaches to the matrix can be divided into two general categories: (i) stable association of matrix attachment regions (MAR) of DNA to matrix proteins; and (ii) transient attachment of regions of DNA that are not yet clearly defined probably during transcription and replication processes.[24,35-40] Some data regarding the specific DNA sequences that attach to the matrix are now available.[36,38] Specific DNA-matrix attachments are generally believed to prevent transcription of inactive genes which would be caused by the supercoiling or uncoiling of adjacent DNA due to transcription of active genes in their vicinity.[25,34] It has recently been reported that the MAR sequence in a chicken lysozyme gene is an uncurved region that resides at some distance from a strongly curved sequence.[41] It has also been shown that DNA cruciform structures reside at or near replication origins[42] or at promoter sites of specific genes.[43] These sites are believed to be close to the MARs. Almost nothing is known about the nature of transient attachment of DNA with the nuclear matrix and the specificity of DNA sequences, if any, involved in such attachment. Also little is known about the cellular components that mediate and stabilize matrix attachment in general and transient attachment in particular. Such mediation is a key factor in the regulation of biologic processes such as transcription and replication which control cell division and proliferation. Because polyamines are known to induce bending of specific sequences[14-16] and to stabilize DNA triple-helices and cruciform structures in vitro,[45-47] they may play key roles in DNA-matrix association and thereby be involved in the regulation of those cellular processes related to gene expression and cell division.

Roti-Roti and coworkers[48,49] have developed a simple method of studying relative changes in matrix-attachment and the supercoiling status of DNA in nuclei that employs propidium-iodide titration of dehistonized nucleoids coupled with fluorescence image analysis. Propidium iodide intercalates into DNA bases and uncoils negative supercoils. The uncoiled DNA is seen as a fluorescent halo surrounding the periphery of the nucleoid. The diameter of the halo increases with the propidium concentration, as more and more negative supercoils unfold. Eventually the halo diameter reaches a maximum, after which further increases in the propidium concentration cause positive supercoiling and a decrease of the halo diameter. This method was used to study the effect of cytotoxic polyamine analogs on DNA-matrix association in HeLa cells.[50] The halo diameter at its maximum was smaller in analog treated cells than in control cells and it decreased with increased duration of drug treatment. These data were interpreted to suggest an increase in the number of matrix-DNA attachment sites in the analog treated cells. The abnormal DNA-matrix attachment noted in analog treated cells may not only provide a plausible mechanism for the growth inhibitory effects of certain polyamine analogs but may also have therapeutic consequences (see below).

B. Chromatin Structure and Condensation

In eukaryotic nuclei, DNA exists as a nucleoprotein complex known as chromatin. Since the early 1970s, there have been many studies of chromatin structure and its possible role(s) in such processes as DNA replication and the regulation of gene expression.[51-53] Considerable progress has been made in defining the molecular organization of chromatin. Based on nuclease digestion and electron micrographic results[54-59] the "bead and string" model of chromatin was proposed.[57-59] In this model, smaller subunits called "nucleosomes" are connected by spacer DNA called "linkers." During subsequent years, investigations into the constituents, size, shape, and DNA content of nucleosomes[60,61] resulted in the crystallization of nucleosome core particles and the elucidation of their three-dimensional structure.[62]

Nucleosomes are found in all eukaryotic organisms and are composed of DNA and five basic proteins: histone H1 (or H5), H2A, H2B, H3, and H4. Dimers of histone H2A, H2B, H3, and H4 associate to form a cylindrical octamer with a diameter of 10nm. 146 + 2 base pairs of DNA are wrapped around the octamer in 1.75 negative superhelical turns to form the nucleosome core particle. Histone H1 (or H5) probably associates with DNA at one or both ends of the core to form the nucleosome.[51,63] The nucleosomes are connected by strands of linker DNA 0-80 bp long[64] to form the basic chromatin structure of approximately 10 nm diameter. This structure further condenses into 30 nm chromatin fibers in vivo or in vitro in the presence of divalent and polyvalent cations. Most efficient among such cations are the polyamines spermidine and spermine.[63,65]

C. Structure of DNA Incorporated into Nucleosomes

While the secondary structure of nucleosomal DNA has some features of B-DNA, the nucleosomal DNA, however, differs considerably from the classic B-DNA structure in solution.[66] By hydroxyl radical footprinting, Wolffe and his coworkers[67] showed that DNA twist shifts from 10.5 + 0.1 bp/turn in solution to only 10.18 + 0.05 bp/turn in nucleosomes. They proposed that DNA bending and/or kinking in nucleosomes contributes to the transition of DNA from its solution phase structure to its nucleosomal structure. Four sharply bent structures at the 30, 60, 80, and 110 bp regions of DNA have been found in nucleosome crystals.[62] The anomalous imino proton peaks found in ^{1}H-nuclear magnetic resonance (NMR) studies of nucleosome core particles in solution have been attributed to regions of DNA bending.[68] The pattern of photochemically induced DNA strand breakage is consistent with the notion of at least two bent or kinked sites in the nucleosomes 1.5 helical turns apart.[69] Such bending or kinking at specific, but as yet not defined, sites may result from interaction of the core histones and various intracellular cations with nucleosomal DNA.[51,65]

During the past decade, methods such as thermal denaturation, circular dichroism (CD) and NMR spectroscopy and gel electrophoresis have been applied to study the DNA structure in nucleosomes.[51,70-72] Although some studies show that peptide fragments of core histone H4 induces DNA bending,[73] nucleosomes are more likely to form on existing bends in DNA structure.[74,75] Apart from bending, other DNA secondary structural features are also recognized by the nucleosomes; for example, nucleosomes are not formed on left-handed Z-DNA.[76,77] Therefore, it is reasonable to assume that intracellular spermine and spermidine might induce bending or B-Z transitions at specific DNA sequences and thereby participate in the condensation and positioning of the nucleosomes. Proper positioning of the nucleosomes at specific sites may regulate the transcription of specific genes and the replication of genomic DNA.[51,53,78-83]

D. Effects of Polyamines on Chromatin Condensation

Several reports describe the effects of spermidine and spermine on the condensation of chromatin in vitro. Hewish and Burgoyne[56] first showed, using electron microscopy, that the nucleosomal structure of rat liver chromatin is better preserved when polyamines are added to the isolation buffer. Electric dichroism studies[84] have clearly shown that polyvalent cations, including spermidine and spermine, induce a specific compact DNA structure in purified chicken erythrocyte chromatin before the onset of aggregation. Of the cations tested, spermine was the most efficient at inducing such a structure. This observation has been supported by X-ray diffraction, electron microscopy, and analytical ultracentrifugation data on isolated chromatin,[65] and by polarized fluorescence recovery after photobleaching of intact nuclei.[85]

Some degree of specificity has been observed in polyamine-induced chromatin condensation. Ultraviolet (UV) and CD melting studies of DNA in the nucleosomal core particle have shown that the polyamine-induced stabilization of nucleosomal DNA differs from the stabilization induced by Mg^{2+} or hexamine cobalt,[86] and electric dichroism studies have shown that the interaction of polyamines with chromatin cannot be explained by counterion condensation theory alone.[84] This suggests that the interaction is not driven only by the electrostatic interaction between two oppositely charged species. Studies on the interaction of photoactive analogs of spermine with nucleosomal DNA suggest that polyamines may have specific binding sites on DNA in nucleosomes.[87,88] However, the attachment of the photoactive group to spermine alters the binding characteristics of this analog. Flow linear dichroism and UV light-scattering studies have shown that, although spermidine and spermine condense chromatin in vitro, spermidine analogs with shorter carbon chain lengths and putrescine do not do so.[89] Thus, in addition to net charge, the length of the carbon chains of polyamines and the distribution of the positive charges on the chain affect polyamine-induced chromatin condensation in vitro. We have reported that spermine analogs with shorter chain lengths do not induce bending and condensation of DNA very effectively in cell free systems and that they inhibit the proliferation of brain tumor cells in tissue culture.[18] Using antibodies against polyamines to map the intracellular binding sites of polyamines and fluirescence microscopy, Larsson and his coworkers[90] showed that polyamines are intimately associated with highly condensed chromatin in fixed cells. The integration of retrovirus to the SV-40 genome is sensitive to the state of condensation of the SV-40 minichromosome.[91] Varmus and his coworkers[92] have shown that the in vitro integration pattern of mouse leukemia virus into SV-40 minichromosome resembles the in vivo pattern only when spermidine is added to the in vitro reaction mixture. This suggests that spermidine induced chromatin condensation and nucleosomal positioning in vitro are similar to what takes place in vivo.

Snyder[93] used sucrose density gradient centrifugation and nick translation procedures to show that nuclei isolated from HeLa cells after putrescine and spermidine depletion caused by treatment with alpha-difluoromethylornithine (DFMO), an inhibitor of polyamine biosynthesis, are more susceptible to digestion by DNase I, DNase II, and micrococcal nuclease (MNase). These results suggest possible changes in chromatin structure in DFMO treated cells. Similar enhancement of MNase and DNase I sensitivity of nuclei isolated from polyamine depleted U-87 MG human brain tumor cells treated with a cytotoxic polyamine analog bis-ethylhomospermine (BE-4-4-4) as compared to untreated controls was observed.[94] This effect was noted within 72-96 hrs of treatment, before the onset of cytotoxic effects, suggesting that the changes induced in chromatin structure by treatment with the polyamine analog is a possible cause and certainly not an effect of cell killing. It was also reported that the nuclease sensitivities of the nuclei were inversely correlated with the concentrations of intracellular polyamines.

Densitometric analysis of SDS-polyacrylamide gel electrophorograms of total histones in control untreated and BE-4-4-4 treated HeLa cell nuclei showed that the total histone content of treated cells is less than that in approximately the same number of nuclei from control untreated cells.[50] Histones are more readily displaced by propidium iodide from the chromatin of analog treated cells than from the chromatin of the control untreated cells. These data suggest a weaker DNA-histone complex in the nuclei of analog treated cells which may lead to a destabilization of nucleosomal and chromatin structure resulting in the observed enhancement of MNase sensitivity of nuclei isolated from these cells.[94] A model that shows the possible effects of analogs on nuclear DNA and chromatin structure discussed above is shown schematically in Figure 5.1.

If the major growth inhibitory effects of polyamine depletion result secondary to the effects of polyamines on chromatin condensation, cell growth should be arrested at the end of the G2 phase or at the beginning of the M phase of the cell cycle, coincident with the step of chromatin condensation. However, depletion of cellular polyamines most frequently arrest cells in the G1 S phase.[94,95] One possible explanation for this anomaly may be due to an interrelationship between DNA synthesis at the matrix attachment sites during the S phase of the cell cycle and the formation and phasing of nucleosomes. Similar to what was observed with HeLa cells,[50] the number of DNA-nuclear matrix attachment sites increased in 1,19-*bis*(ethylamino)-5,10,15-triazanonadecane (BE-4-4-4-4) treated U-251 MG (NCI) human brain tumor cells when compared with control untreated cells (Basu, HS et al, unpublished data). These data also suggest that, unlike the "normal" attachment sites, the "abnormal" attachment sites in the treated cells may not contain active topoisomerases (see below). If this observation extends to other analogs and to other cell lines, it may be hypothesized that the absence of active topoisomerases at the DNA-nuclear matrix attachment sites may result in a high supercoiling stress in DNA during replication. Jackson[96] has shown that it is difficult to reconstitute nucleosomes in vitro using DNA that is under a positive supercoiling stress, and that those nucleosomes are thermally unstable and have an "open" structure which is more accessible for digestion by MNase or DNase I. He and his coworkers have also reported[97,98] a loss of nucleosomes from the positively supercoiled end of minichromosomes during the in vitro transcription. Thus abnormal DNA-matrix attachment in polyamine depleted cells may lead to both an inhibition of DNA synthesis, causing a G1 arrest, as well as the observed instability of the nucleosomal complex.[50,93,94] Detailed studies of the structure and stability of nucleosomal DNA using UV and CD melting techniques should more convincingly demonstrate the in vivo existence of the "open" nucleosomal structure, as observed by Jackson's group in vitro.[97-99] These studies should provide indirect evidence of "abnormal" supercoiling stress in DNA in polyamine depleted cells.

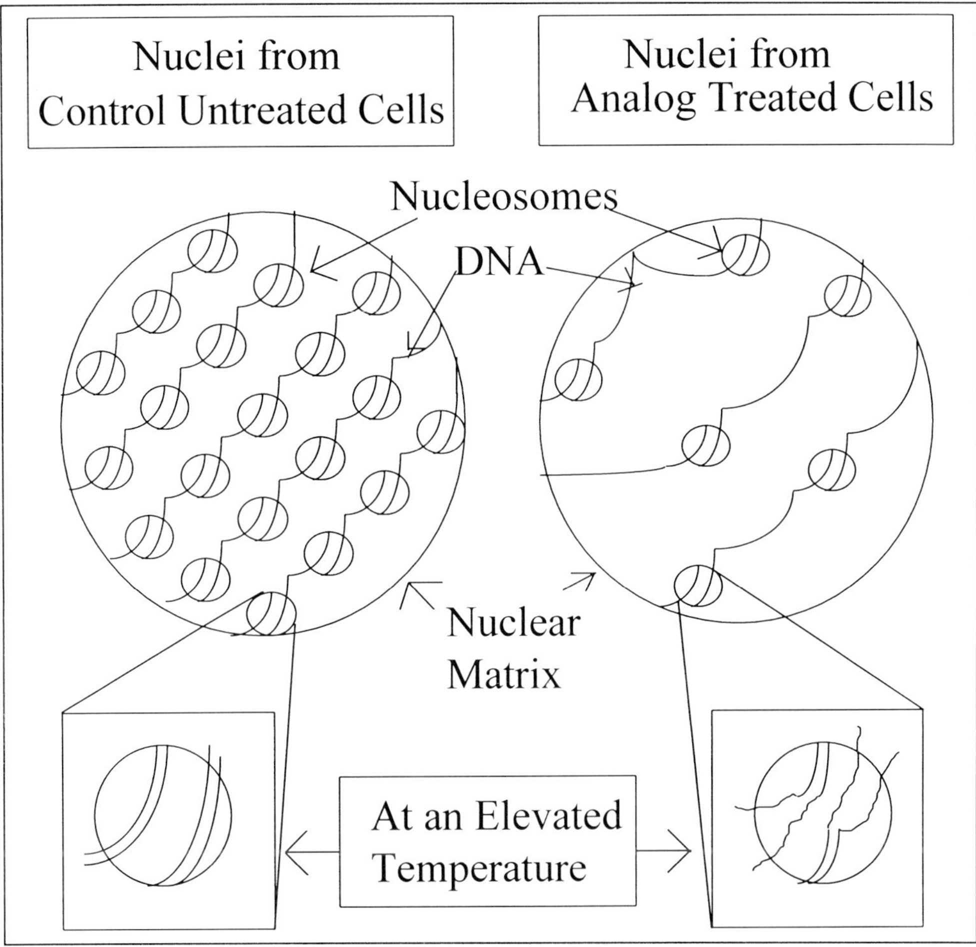

Fig. 5.1. Schematic diagram of possible chromatin structure in polyamine analog treated cell nuclei.

V. POSSIBLE THERAPEUTIC SIGNIFICANCE OF POLYAMINE ANALOG INDUCED CHANGES IN CHROMATIN STRUCTURE

While the effects of polyamine analogs and biosynthesis inhibitors on chromatin structure are still being evaluated, the information obtained so far may be utilized to develop new therapeutic protocols. During the past 2 decades, there have been efforts to develop antineoplastic polyamine analogs that inhibit polyamine biosynthesis and thereby deplete intracellular polyamines.[5,100] Only during the past 5 years have analogs been developed that can deplete all three cellular polyamines.[5,7,17-19,100-102] Surprisingly, some analogs are ineffective as growth inhibitors despite their ability to deplete all cellular polyamines.[20]

These analogs themselves function as polyamines. We propose that certain polyamine analogs inhibit cell growth by interacting with appropriate intracellular site(s) in a modified fashion. Specifically, we believe that these analogs interact with cellular DNA in a way that produces critical alterations of DNA-matrix attachment and/or chromatin structure. This hypothesis is strengthened by the fact that polyamine analogs which have marked growth-inhibitory effects against several tumor cell lines both in tissue culture as well as in nude mouse xenografts, also affect chromatin structure.[7,17,50,93,94] Several cytotoxic polyamine analogs as well as DFMO are now being evaluated clinically as anti-neoplastic agents.

Cancer chemotherapy often involves treatment with more than one drug. Extensive reviews on the strategies and reasoning for combination treatments are now available.[103,104] One or more mechanism(s), such as spatial cooperation, mutually exclusive modes of action, protection of normal cells, and/or synergistic enhancement of tumor response may be responsible for the enhanced efficacy of any specific combination. Impaired chromatin condensation in analog treated and polyamine depleted cells might be utilized therapeutically by combining treatment of analogs together with DNA binding anticancer agents that prefer binding to relaxed over condensed chromatin.

A. DNA Alkylating Anti-Cancer Agents

There are several commonly used anti-cancer drugs that act by alkylating DNA bases with subsequent formation of DNA intra- and inter-strand cross-links. Cis-diamminedichloroplatinum (CDDP) and its analogs and analogs of chloroethylnitrosourea (CENU) are two popular classes of drugs that fall into this category.

1. *Cis*-diamminedichloroplatinum (CDDP)

CDDP, which forms at the platinum electrodes of electrolysis units, was originally introduced as an anti-microbial agent.[105] Rosenberg and his coworkers[106] first demonstrated its anti-cancer properties. It is now widely used as an anti-cancer drug. CDDP has two halide atoms separated by a distance of 3.3 angstroms.[107] This distance is perfectly suited for a nucleophilic attack at two consecutive DNA bases which are 3.4 angstroms apart in the classical B-DNA structure.[108] Such nucleophilic attack forms an intra-strand cross-link between two adjacent bases. While CDDP forms both intra- and inter-strand DNA cross-links,[109] the intra-strand DNA cross-linking activity of CDDP is generally correlated with its cytotoxicity.[110]

Changes in DNA secondary and superhelical structures that modify the distance between two DNA bases would be expected to alter the DNA cross-linking activity of CDDP. CDDP has been shown to discriminate between different DNA structures in vitro.[111] B-DNA conformation transits from 10.6 bp per turn of the helix to 10.18 + 0.05 bp per turn when compacted into nucleosomes (see above). This transition changes the inter-base distance in DNA. Supercoiling of DNA around the nucleosome core will also change mutual base orientations. These changes would be expected to alter the DNA cross-linking activity of CDDP. Foka and Paoletti[112] treated isolated chromatin with CDDP. They measured the amount of platinum incorporation as representative of CDDP adduct formation in MNase sensitive and resistant fractions of the chromatin. At a drug concentration where CDDP primarily interacts with DNA bases, as opposed to other potential sites of interaction, platinum preferentially incorporates into the MNase sensitive linker DNA. This observation has been confirmed by MNase digestion kinetic studies of chicken erythrocyte nuclei treated with CDDP.[113] The treated nuclei are relatively resistant to MNase digestion as compared to the untreated control nuclei. Resistance to MNase digestion suggests specific modifications of linker DNA, the preferred substrate for MNase.

2. Chloroethylnitrosoureas (CENU)

CENUs are another class of commonly used anti-cancer agents that interact with and alkylate DNA bases.[114] Some of the initial monoalkylated products eventually form either inter- or intra-strand DNA cross-links both in vitro and in vivo.[115] Unlike CDDP, the formation of DNA inter-strand cross-links is correlated with the cytotoxicity of these agents.[116] As for intra-strand cross-linking, inter-strand cross-link formation is also sensitive to DNA secondary and superhelical structures.[117] Therefore, one would anticipate that, as for CDDP, CENUs would also interact with nucleosomal DNA differently than with linker DNA.

Tew et al[118] first studied the effects of 1,(2-chlorethyl)-3-cyclohexyl nitrosourea (CCNU) on the MNase and DNase I sensitivity of HeLa cell nuclei. They observed that, unlike CDDP, CCNU preferentially interacts with nucleosomal DNA. This has also been observed for 1-(4-amino-2-methyl-5-pyrimidinyl)methyl-3-(2-chloroethyl)-3-nitrosourea (ACNU) and CCNU treatments of L1210 and bone marrow cells, respectively.[119] CENUs, however, preferentially bind to transcriptionally active and relaxed regions of chromatin rather than to transcriptionally inactive and relatively condensed regions.[120] HeLa cells which have been treated with hydrocortisone to transform inactive and condensed euchromatin into active and relatively decondensed heterochromatin, showed greater CCNU sensitivity and DNA inter-strand cross-linking than untreated control cells. Two other nitrosoureas, methyl-nitrosourea (MNU) and ethyl-nitrosourea (ENU), form DNA monoadducts similar to those formed by BCNU but without subsequent cross-linking.[121] Using isolated chromatin that was condensed in vitro with NaCl or spermine and then treated with either MNU or ENU,[122,123] it was shown that these agents modify the condensed chromatin less than uncondensed chromatin. The depletion of cellular polyamines may relax chromatin in vivo as well and thereby, increase alkylation and consequently the cytotoxicity of CENUs.

B. Nuclear-Matrix Bound Enzymes

Several key enzymes such as DNA polymerase, RNA polymerase, DNA repair and recombination enzymes and topoisomerases reside on the nuclear matrix and are localized at or near the DNA-matrix attachment sites.[124] Knowledge of the effects of polyamine analogs on matrix-DNA attachment may be utilized in developing therapeutic protocols for the combination of analogs with inhibitors of topoisomerases. Several topoisomerase inhibitors such as Camptothecin and its analogs CPT-11 and Topotecan and the podophyllotoxin analogs Etoposide (VP-16) and Teneposide (VM-26) are now in clinical trials.[125-127] Abnormal matrix-DNA attachment may also affect enzymes involved in the repair of DNA damage.

1. Topoisomerases

The enzymes topoisomerase (topo) I and II are found on the nuclear matrix[128] and assist in DNA transcription, replication, repair, recombination and sister chromatid segregation by releasing supercoiling stress and topological constraints during these processes.[129] Although yeast mutants without topo I activity grow normally,[130] it is now generally believed that either topo I or topo II activity is essential for eukaryotic DNA replication.[131] Spermidine modulates both the eukaryotic and prokaryotic activity of topo I[132] and enhances the binding of mammalian topo II to DNA in vitro.[133] Polyamine-induced structural changes in DNA might enhance topoisomerase binding, activity, or both. Both topo I and II enzymes are also involved in the attachment of DNA to the nuclear matrix and proper chromatin condensation.[134,135] Therefore, an increased number of attachment sites of DNA to the nuclear matrix may be accompanied by an increase in topo I and II content, activity, or both, which may enhance cellular sensitivity to the action of topoisomerase inhibitors.[136,137] Most topoisomerase inhibitors inhibit the enzyme at the DNA strand rejoining step and thereby, cause DNA strand breaks.[138,139] DNA strand breaks are also observed in nuclei of cells exposed to ionizing radiation.[140] DNA double-strand breaks are correlated with the cytotoxicity of both topo inhibitors and ionizing radiation.[141,142]

2. Ionizing Radiation Induced DNA Damage and Repair

In the past several years, studies have shown that radiation induced DNA double-strand breaks (DSB) and cell survival are sensitive to changes in chromatin structure.[143] Pulsed field gel electrophoresis has been employed to measure DSB in nuclei at various stages of chromatin unfolding.[144] Isolated nuclei treated with varying salt concentrations to remove nuclear histones before being subjected to ionizing radiation, showed that radiation induced DSB increased in proportion to the degree of dehistonization of the nuclei. Quantitation of the DNA damage caused by ionizing radiation showed that deproteinized DNA is about 60- to 70-fold more sensitive to radiation induced DSB than DNA from intact nuclei.[145] Polyamine analogs that impair chromatin condensation should, therefore, sensitize cells to ionizing radiation.

The role of chromatin structure in the repair of radiation induced DNA DSB is less understood. It has been established that in transcriptionally active, relaxed chromatin DNA is repaired more effectively than in inactive, condensed chromatin.[146,147] However, ionizing radiation also changes the transcriptional status of chromatin and specifically activates several repair proteins.[148] Therefore, it is difficult to predict the role of chromatin structure in the repair of DNA damage caused by ionizing radiation.

C. Combination Treatment of Polyamine Analogs and Inhibitors with Other Cytotoxic DNA-binding Agents and with Ionizing Radiation

The study of the effects of polyamine analogs and biosynthesis inhibitors on the cytotoxicity of DNA binding anti-cancer agents that are sensitive to changes in chromatin structure is in its infancy. DFMO pretreatment of 9L cells increases the cytotoxicity of BCNU by 1.3-fold and decreases the cytotoxicity of CDDP to a similar extent.[149-151] These changes in cytotoxicity correlate with changes in the number of interstrand crosslinks.[152] Attempts have been made to interpret these results on the basis of the effects of polyamines on naked DNA structure in vitro.[151,153,154] However, a good correlation between the in vitro effects of polyamines on DNA structure and the effects of polyamine depletion on CDDP or CENU cytotoxicity has not yet been established. The effects of polyamines on the nucleoprotein complex of DNA in chromatin may be more relevant to the cytotoxicity of DNA-binding agents than the effects on the naked DNA. DFMO has very little effect on intracellular spermine levels and a very small effects on the state of chromatin condensation in most cell lines.[93,94,100] Among the naturally-occurring polyamines, spermine has the most profound effects on DNA structure in chromatin.[9] Therefore, agents that can deplete spermine along with other intracellular polyamines need to be tested in combination with DNA-binding anti-cancer drugs to determine their ability to enhance the therapeutic efficacy of these drugs. Very recently, it has been observed that BE-3-3-3 pretreatment enhanced the cytotoxicity of CDDP by 2.3-fold in cultured human tumor cells (H.S. Basu et al, unpublished data). It has also been shown that BE-4-4-4-4 pretreatment enhanced the cytotoxicity of BCNU by approximately 4-fold (A. Sarkar et al, personal communication) in BCNU resistant human brain tumor cell lines. Surprisingly, these enhancements of BCNU and CDDP cytotoxicity were not coupled with the usual enhancement of DNA cross-linking. Further studies on the mechanism(s) for the enhancement of the cytotoxicity of these drugs are ongoing.

One polyamine analog bis(ethyl)spermidine (BES) has been used in combination with the topo II inhibitor m-AMSA.[155] BES pretreatment of cells enhances m-AMSA induced DNA strand breaks. The extent of enhancement is similar to that seen for DFMO pretreatment.[156] While DFMO has been shown to enhance the cytotoxicity of m-AMSA and Etoposide,[157] the extent to which BES affects the cytotoxicity of m-AMSA has not yet been reported. Preliminary studies on the effects of the polyamine analog BE-4-4-4-4 on the cytotoxicity of the topo II inhibitor Etoposide in U-251 MG (NCI) human brain tumor cells showed that BE-4-4-4-4 neither increased Etoposide induced DNA strand breakage nor enhanced the cytotoxicity of Etoposide.[158] However, initial BE-4-4-4-4 pretreatment conditions were chosen to minimize its cytotoxicity and not to optimize drug uptake.

During the past several years experiments have been carried out in several laboratories in which cultured tumor cells pretreated with polyamine analogs were exposed to ionizing radiation.[159,160] Results showed that analog pretreatment enhanced radiation induced DNA damage and cell kill. The reasons for such enhancement remain controversial. Williams et al[160] argue that the enhancement of radiation induced DSB rather than the repair of DNA damage in analog treated cells is responsible for the observed enhancement of cell kill. However, Deen and his coworkers (personal communication) have recently observed that treatment with a series of analogs does inhibit potential lethal damage repair (PLDR) in some human brain tumor cells in culture.

Irrespective of the exact mechanism(s) underlying the enhancement of cytotoxicity caused by DNA binding agents and ionizing radiation, replacement of intracellular polyamines with selected polyamine analogs definitely opens up a new avenue for designing new and effective protocols for cancer therapy.

VI. UNANSWERED QUESTIONS AND UNSOLVED MYSTERIES

Studying the effects of polyamine analogs and biosynthesis inhibitors on chromatin structure has increased our understanding of specific functions of the polyamines. However, many unanswered questions remain. It is quite possible that the answer to any one of these questions will open new horizons for polyamine research.

Certain biological functions of the polyamines may relate to their abilities to interact at specific sites with DNA or with other cellular macromolecules. However, polyamine function may also be secondary to a more generalized, nonspecific role as highly regulatable members of the cellular cationic pool. In order to pin-point the specific intracellular site(s) of action of the polyamines, it is imperative that we gain further insight into the subcellular localization of the polyamines. Attempts to identify the subcellular localization site(s) of the polyamines in vivo has thus far failed to yield definitive results with the exception of fairly specific data for *Neurospora crassa*.[4] Davis and his coworkers[161-163] have shown that in *N. crassa* excess intracellular polyamines are either bound to cellular macromolecules or are sequestered in vacuoles in vivo. They argue that because of the tight binding of polyamines to cellular anions, the total intracellular levels of polyamines may not reflect the concentrations of polyamines actually required for their biological functions which may be carried out by a small but freely diffusable cellular polyamine pool. This may help explain why the tightly regulated polyamine biosynthetic machinery produces millimolar levels of polyamines in cycling cells while minuscule amounts of polyamines in the growth medium are sufficient to induce cell division in growth arrested polyamine deficient organisms.[164,165] The development of monoclonal antibodies against spermine have been reported.[166] Microinjection of antibodies into viable cells and their monitoring by fluorescence microscopy could help to determine the subcellular localization of spermine in vivo. Attempts thus far have met with limited success. Extension of these studies, when feasible, may also help address the question of redistribution of intracellular polyamines during the course of the cell cycle.

Although the cytotoxic polyamine analogs are very efficient in inhibiting mammalian cell growth, many of them are ineffective in inhibiting yeast or trypanosomal growth in culture (Basu et al and Fairlamb et al, unpublished results). Both yeast and trypanosomal cell chromatin lack histone H1 which is present in all mammalian cells. If polyamine analogs bind at the H1 binding site(s) on DNA and exert their effects on chromatin structure and cell proliferation by interfering in "normal" H1-DNA interaction, these analogs may well be inactive in arresting yeast or trypanosomal growth. On the other hand, the trypanosomal mitochondrion is different from its mammalian counterpart and there are reports of polyamine analogs and biosynthesis inhibitors affecting the structure, number, and integrity of mitochondria in mammalian cells.[167-169] Although the works of Ghoda et al[21] and Albanese et al[22] suggest that the growth inhibitory effects of the polyamine analogs are probably not due to their effects on mitochondria, further studies of the effects of analogs both on mitochondria and on [histone] H1-DNA interactions in mammalian cells are needed to definitively answer this question.

It has been shown that polyamines can efficiently stabilize the triple-helical DNA structure.[46,47] It has been suggested that DNA triplex structure exists at chromosomal recombination sites[170] and may be involved in the interaction of anti-sense oligonucleotides with cellular DNA.[171] Spermine was routinely used by Dervan et al[172] while studying oligonucleotide triplexes. Polyamine analogs that are more efficient than spermine in stabilizing triple-helices might be utilized in studies with DNA triplexes and perhaps even in anti-sense research and gene therapy.

It is known that the N-methyl-D-aspartate (NMDA) receptor has a specific polyamine binding site and that spermine and spermidine modulate intracellular ion transport across neuronal cell membranes containing NMDA receptors.[173] It has also been shown that polyamines inhibit mitochondrial Ca^{2+} flux.[174] However, very little is known about the role of polyamines in the transmembrane cation transport of cells which do not have NMDA receptors. Koenig et al[175] first showed that polyamine biosynthesis and the intracellular presence of polyamines are essential for androgen induced Ca^{2+} flux in renal cells. Subsequently, Feuerstein et al[176] showed that platelet-derived-growth-factor induced Ca^{2+} flux in human brain tumor cells in culture is inhibited by DFMO pretreatment. Other studies have shown that neuronal Ca^{2+} and K^+ transport are regulated by polyamines[177] and that in electrically stimulated myocytes, intracellular Ca^{2+} uptake is blocked by the addition of polyamines to the culture medium.[178] Very recently, polyamines have been shown to have negative regulatory effects on intracellular K^+ transport.[179] Garrard and his coworkers[180,181] have shown that yeast mutants defective in Ca^{2+} metabolism have abnormal nuclear-matrix organization and DNA-matrix interactions. It is clear that the relationship of polyamines to cellular cation flux needs to be ascertained in detail, to determine whether or not the observed effects of polyamine analogs on DNA-matrix association and changes in nucleosomal and chromatin structure are related to impaired calcium uptake in polyamine depleted cells. Definition of the relationship of the polyamines to cellular ion transport may also provide leads to the use of analogs or inhibitors in the treatment of cardiovascular or other diseases related to ion transport.

While information related to the radiosensitizing effects of polyamine depletion in tumor cells is still being accumulated,[159,160] it has been shown that polyamine depletion can protect normal cells from radiation injury and reduce the vasogenic edema associated with radiation therapy.[182,183] It is curious that polyamine depletion can apparently enhance the radiation sensitivity of tumor tissue while at the same time protecting normal cells from radiation damage. Interestingly, a difference in matrix-DNA interaction exists between normal and tumor tissue.[184] Such differences, along with possible differences in chromatin structure between normal and tumor tissue may provide clues for the difference observed in their radioresponsiveness. The involvement of apoptotic pathways for radiation induced cell kill and the role of polyamines in these pathways may also provide answers.

Lastly, in this article we discussed polyamine-DNA interactions and their biological and therapeutic implications. It must be kept in mind that the cationic polyamines can interact very efficiently with another major cellular polyanion-RNA. NMR studies of the interactions of polyamines with yeast $tRNA^{phe}$ showed specific sites of interaction for polyamines with the double-stranded section of the RNA.[185] Analysis of the A-DNA crystal structure in the presence of spermine also showed tight binding of spermine at the major groove.[10] Since crystals of double stranded RNA and ribopolynucleotides often show an A-DNA type structure, tight binding of polyamines with double-stranded viral RNA is also likely. With increasing interest in retrovirus containing double-stranded RNA, such as HTLV and HIV, detailed studies of polyamine-RNA interactions may lead to additional scientific insights and therapeutic possibilities.

Acknowledgment

A part of the work described in this article was supported by National Institutes of Health grant (CA-49409) and the Wendy Will Case Cancer Fund.

References

1. Tabor CW and Tabor H. Polyamines. Ann Rev Biochem 1984;53:749-790.
2. Pegg AE and McCann PP. Polyamine metabolism and function. Am J Physiol 1982;243: C212-C221.
3. Hamana K and Matsuzaki S. Polyamines as a chemotaxonomic marker in bacterial systematics. Crit Rev Microbiol 1992;18:261-283.
4. Davis RH, Morris DR, Coffino P. Sequestered end products and enzyme regulation: the case of ornithine decarboxylase. Microbiological Reviews 1992;56:280-290.
5. Pegg AE. Polyamine metabolism and its importance in neoplastic growth and as a target for chemotherapy. Cancer Research 1988;48:759-774.
6. Hague DM and Moreton AD. Protonation sequence of aliphatic polyamines by ^{13}C-NMR spectroscopy. J Chem Soc Perkin Trans II 1994;265-270.
7. Feuerstein BG, Williams LD, Basu HS and Marton LJ. Implications and concepts of polyamine-nucleic acid interactions. J Cell Biochem 1991;46:37-47.
8. Gosule LC and Schellman JA. DNA condensation with polyamines. J Mol Biol 1978;121:311-326.
9. Behe M and Felsenfeld G. Effects of methylation on a synthetic polynucleotide—The Z-transition in poly(dG-me^5dC)poly(dG-me^5dC). Proc Natl Acad Sci (USA) 1981;78:1619-1623.
10. Jain S, Zon G and Sundaralingam M. Hexagonal crystal structure of A-DNA octamer d(GTGTACAC) and its comparison with the tetragonal Structure correlated variations in helical parameters. Biochem 1989; 28:2360-2364.

11. Feuerstein BG, Pattabiraman N and Marton LJ. Spermine-DNA interaction: A theoretical study. Proc Natl Acad Sci (USA) 1986; 83:5948-5952.
12. Feuerstein BG, Pattabhiraman N and Marton LJ. Molecular dynamics of spermine-DNA interaction sequence specificity and DNA bending for a simple ligand. Nuc Acid Res 1989;17:6883-6892.
13. Feuerstein BG, Pattabiraman N and Marton LJ. Molecular mechanics of the interactions of spermine with DNA: DNA bending as a result of ligand binding. Nuc Acid Res 1990;18:1271-82.
14. Basu HS, Shafer RH and Marton LJ. A stopped-flow H-D exchange kinetic study of spermine-polynucleotide interaction. Nuc Acid Res 1987;15:5873-5886.
15. Marquet R and Houssier C. Different binding modes of spermine to A-T and G-C base pairs modulate the bending and stiffening of the DNA double helix. J Biomol Str Dyn 1988;6:235-246.
16. Plum GE and Bloomfield VA. Effects of spermidine and hexaamine cobalt on thymine imino proton exchange. Biochemistry 1990; 29:5934-594.
17. Basu HS, Feuerstein BG and Marton LJ. Polyamine-DNA interactions and their biological significance. In "Proceedings of Polyamines in the Gastrointestinal Tract—Falk Symposium # 62. Oct. 6-8, 1991. Lancaster, England. Kluwer Publishing, 1992.
18. Basu HS, Feuerstein BG, Deen DF, et al. Correlation between the effects of polyamine analogs on DNA conformation and cell growth. Cancer Res 1989;49:5591-5597.
19. Basu HS, Pellarin M, Feuerstein BG, et al. Interaction of a polyamine analog 1,19-*bis*-(ethylamino)-5,10,15-triazanonadecane (BE-4-4-4-4) with DNA and effect on growth survival and polyamine levels in seven human brain tumor cell lines. Cancer Res 1993;53:3948-3955.
20. Porter CW, Cavanaugh PF(Jr), Stolowich N, et al. Biological properties of N^4—and N^4, N^8—spermidine derivatives in cultured L1210 leukemia cells. Cancer Res 1985;45:2052-2057.
21. Ghoda LY, Basu HS, Porter CW, et al. Role of ornithine decarboxylase suppression and polyamine depletion in the antiproliferative activity of polyamine analogs. Mol Pharm 1992;42:302-306.
22. Albanese L, Bergeron RJ and Pegg AE. Investigations of the mechanism by which mammalian cell growth is inhibited by N^1N^{12}-*bis*(ethyl)spermine. Biochem J 1993;291:131-7.
23. Nelkin B, Pardoll D, Robinson S, Small D and Vogelstein B. Nuclear structure and DNA organization In Tumor Cell Heterogeneity. Acad Press, 1982: 441-457.
24. Razin SV. DNA interactions with the nuclear matrix and spatial organization of replication and transcription. Bioessays 1987;6:19-23.
25. Roberge M, Dahmus ME and Bradbury EM. Chromosomal loop/nuclear matrix organization of transcriptionally active and inactive RNA polymerases in HeLa nuclei. J Mol Biol 1988;201:545-555.

26. Nagl W. Nuclear structure during cell cycle. In Rost, TL, Gifford, EM, Jr., eds. Mechanism and Control of Cell Division. Dowden Hutchinson and Ross Stroudsberg PA 1977.
27. Bekers AGM, Gijzen HJ, Taalman RDF and Wanka F. Ultrastructure of the nuclear matrix from *Phyusarum polycephalum* during the mitotic cycle. J Ultrastructural Research 1981;75:352-363.
28. Zbarsky IB, Dmitrev NP and Yermolayeva LP. On the structure of tumor cell nuclei. Exp Cell Res 1962;27:573-576.
29. Pardoll DM, Vogelstein B and Coffey DS. A fixed site of DNA replication in eucaryotic cells. Cell 1980;19:527-536.
30. McCready SJ, Godwin J, Mason DW, et al. DNA is replicated at the nuclear cage. J Cell Sci 1980;46:365-386.
31. Vaughn JP, Dijkwel PA, Mullenders LHF and Hamlin JL. Replication forks are associated with the nuclear matrix. Nuc Acid Res 1990;18:1965-69.
32. Vogelstein B, Pardoll DM and Coffey D. Supercoiled loops and eucaryotic DNA replication. Cell 1980;22:79-85.
33. Ciejak SM, Tsai MJ, O'Malley BW. Actively transcribed genes are associated with nuclear matrix. Nature 1983;306:607-609.
34. Ip YT, Jackson V, Meier J and Chalkley R. The separation of transcriptionally engaged genes. J Biol Chem 1988;263:14044-52.
35. Tsuitsui K, Tsuitsui K and Muller MT. The nuclear scaffold exhibits DNA-binding sites selective for supercoiled DNA. J Biol Chem 1988;263:7235-41.
36. Dijkwel PA and Hamlin JL. Matrix attachment regions are positioned near replication initiation sites genes and an interamplicaon junction in the amplified dihydrofolate reductase domain of chinese hamster ovary cells. Mol Cell Biol 1988;8:5398-5409.
37. He D, Martin T and Penman S. Localization of heterogeneous nuclear ribonucleoprotein in the interphase nuclear matrix core filaments and on perichromosomal filaments at mitosis. Proc Natl Acad Sci (USA) 1991;88:7469-7473.
38. Blasquez VC, Sperry AO, Cockerill PN and Garrard WT. Protein:DNA interactions at chromosomal loop attachment sites. Genome 1989;31:503-9.
39. de Jong L, van Driel R, Stuurman N, et al. Principles of nuclear organization. Cell Biol Int Report 1990;14:1051-74.
40. Hirose S and Ohta T. DNA supercoiling and eukaryotic transcription - cause and effect. Cell Str and Funct 1990;15:133-135.
41. von Kriess JP, Phi-Van L, Diekmann S and Stratling WH. A non-curved chicken lysozyme 5' matrix attachment site is 3' followed by a strongly curved DNA sequence. Nuc Acid Res 1990;18:3881-5.
42. Ward GK, Shihab-el-Deen A, Zannis-Hadjopoulos M and Price GB. DNA cruciforms and the nuclear supporting structure. Exp Cell Res 1991; 195:92-98
43. Pestov DC, Dayn A, Siyanova E-Yu, et al. H-DNA and Z-DNA in the mouse c-*Ki-ras* promoter. Nuc Acid Res 1991;19:6527-32.

44. Han H and Dervan PB. Different conformational families of pyrimidine-purinepyrimidine helices depending on backbone composition. Nuc Acids Res 1994;22:2837-44.
45. Morse HE and Dervan PB. Sequence-specific cleavage of double-helical DNA by triple-helix formation. Science 1987;238:645-50.
46. Thomas T and Thomas TJ. Selectivity of polyamines in triplex DNA stabilization. Biochemistry 1993;32:14068-74.
47. Shafer RH, Pilch D and Scaria PV. The structure and stability studies on short DNA triple helices. In Proceed VIIth Conv Biomol Str Dyn, 1992: 15-29.
48. Roti-Roti JL and Wright WD. Visualization of DNA loops in nucleoids from HeLa cells: Assays for DNA damage and repair. Cytometry 1987;8:461-467.
49. Wright WD, Higashikubo R and Roti-Roti JL. Fluorescent methods for studying subnuclear particles. In Methods in Cell Biology. NY: Academic Press, 1990:353-362.
50. Basu HS, Wright WD, Deen DF, Roti-Roti J and Marton LJ: Treatment with polyamine analog alters DNA matrix association in HeLa cell nuclei: A nucleoid halo assay. Biochemistry 1993;32:4073-4076.
51. van Holde KE. In Rich, A, ed. Chromatin. Springer Series in Mol Biol NY: Springer-Verlag, 1989.
52. Simpson RT. Nucleosome positioning in vivo and in vitro. BioEssays 1986;4:172-176.
53. Simpson RT. Nucleosome positioning: Occurrence mechanisms and functional consequences. Prog Nuc Acid Res Mol Biol 1991;40:143-184.
54. Olins AL and Olins DE. Spheroid chromatin units. Science 1974; 183:330-332.
55. Woodcock CLF. Ultrastructure of inactive chromatin. J Cell Biol 1973;59:368a.
56. Hewish DR and Burgoyne LA. Chromatin sub-structure The digestion of chromatin DNA at regularly spaced sites by a nuclear deoxyribonuclease. Biochem Biophys Res Comm 1973;52: 504-510.
57. Noll M. Subunit structure of chromatin. Nature 1974;251:249-251.
58. van Holde KE, Saharsrabuddhe CG, Shaw BR, et al. Electron microscopy of chromatin subunit particles. Biochem Biophys Res Comm 1974; 60:1365-1370.
59. Kornberg R. Chromatin structure: A repeating unit of histones and DNA. Science 1974;184:868-871.
60. Kornberg R. Structure of Chromatin. Ann Rev Biochem 1977;46:931-954.
61. Felsenfeld G. Chromatin. Nature 1978;271:115-122.
62. Richmond TJ, Finch JT, Rushton B, Rhodes D and Klug A. Structure of nucleosome core particle at 7A resolution. Nature 1984;311:532-537.
63. Pederson DS, Thomas F, Simpson RT. Core particle fiber and transcriptionally active chromatin structure. Ann Rev Cell Biol 1986;2:117-47.
64. Widom J. Toward a unified model of chromatin folding. Ann Rev Biophys Chem 1989;18:365-395.

65. Widom J. Physico-chemical studies of the folding of the 100 A nucleosome filament into the 300 A filament. J Mol Biol 1986;190:411-424.
66. Morse RH and Simpson RT. DNA in nucleosome. Cell 1988;54:285-287.
67. Hayes JJ, Tullius TD and Wolffe AE. The structure of DNA in nucleosomes. Proc Natl Acad Sci (USA) 1990;87:7405-7409.
68. McMurray CT, van Holde KE, Jones RL and Wilson WD. Proton NMR investigation of the nucleosome core particle: Evidence for regions of altered hydrogen bonding. Biochem 1985;24:7037-7044.
69. Hogan ME, Rooney TF, Austin RH. Evidence for kinks in DNA folding in the nucleosome. Nature 1987;328:554-557.
70. Tatchell K and van Holde KE. Reconstitution of chromatin core particle. Biochemistry 1977;16:5295-5303.
71. Cowman MK and Fasman GD. Circular dichroism analysis of mononucleosome DNA conformation. Proc Natl Acad Sci (USA) 1978;75:4759-4763.
72. Cowman MK and Fasman GD. Dependence of mononucleosome deoxyribonucleic acid conformation on the deoxyribonucleic acid length and H1/H5 content. Circular dichroism and thermal denaturation studies. Biochemistry 1980;19:532-541.
73. Ebralidse KK, Grachev SA and Mirzabekov AD. A highly basic histone H4 domain bound to the sharply bent region of nucleosomal DNA. Nature 1988;333:365-367.
74. Shrader TE and Crothers DM. Artificial nucleosome positioning sequence. Proc Natl Acad Sci (USA) 1989;86:7418-7422.
75. Pennings S, Muyldermans S, Meersseman G and Wyns L. Formation stability and core histone positioning of nucleosomes reassembled on bent and other nucleosome-dervied DNA. J Mol Biol 1989;207:183-192.
76. Garner M and Felsenfeld G. Effect of Z-DNA on nucleosome placement. J Mol Biol 1987;196:581-590.
77. Ausio J, Zhou G, van Holde KE,. A re-examination of the reported B-Z transition in nucleosomes reconstituted with poly(dG-me5dC). Biochemistry 1987;26:5595- 5599.
78. Paranjape S, Kamakaka RT and Kadonaga JT. Role of chromatin structure in the regulation of transcription by RNA polymerase II. Ann Rev Biochem 1994;63:265-97.
79. Pina B, Bruggemeier U and Beato M. Nucleosome positioning modulates accessibility of regulatory proteins to the mouse mammary tumor virus promoter. Cell 1990;60:719-31.
80. Simpson RT, Roth SY, Morse RH, et al. Nucleosome positioning and transcription. Cold Spring Harbor Symp Quant Biol 1994;58:237-245.
81. Lu Q, Wallrath LL and Elgin SCR. Nucleosome positioning and gene regulation. J Cell Biochem 1994;55:83-92.
82. Wolffe AP. Nucleosome positioning and modification: chromatin structures that potentiate transcription. Trends in Biol Sci 1994;19:240-44.
83. Wolffe AP, Almouzni G, Pruss DU and Hayes JJ. Transcription factor access to DNA in the Nucleosome. Cold Spring Harbor Symp Quant Biol 1994;58:225-235.

84. Sen D and Crothers DM. Condensation of chromatin. Role of multivalent cations. Biochemistry 1986;25:1495-1503.
85. Selvin PR, Scalettar BA, Langmore JP, et al. A polarized photobleaching study of chromatin reorientation in intact nuclei. J Mol Biol 1990;214:911-922.
86. Morgan JE, Blankenship JW and Matthews HR. Polyamines and acetylpolyamines increase the stability and alter the conformation of nucleosome core particles. Biochemistry 1987;26:3643-3649.
87. Clark E, Swank RA, Morgan JE, et al. Two new photoaffinity polyamines appear to alter the helical twist of DNA in nucleosome core particles. Biochemistry 1991;30:4009-4020.
88. Matthews HR Polyamines chromatin structure and transcription. Bioessays 1993;15:561-6.
89. Smirnov IV, Dimitrov SI and Makarov VL. Polyamine-DNA interactions Condensation of chromatin and naked DNA. J Biomol Str Dyn 1988;5:1149-1161.
90. Hougaard DM, Del Castillo AM, Larsson L-I. Endogenous polyamines associate with DNA during its condensation in mammalian tissue. A fluorescence cytochemical and immunocytochemical study of polyamines in rat liver. Eur J Cell Biol 1988;45:311-314.
91. Pryciak PM and Varmus HE. Nucleosomes, DNA-binding proteins and DNA sequence modulate retroviral integration target site selection. Cell 1992;69:769-80.
92. Pryciak PM and Muller HP and Varmus HE. Simian virus 40 minichromosomes as targets for retroviral integration in vivo. Proc Natl Acad Sci (USA) 1992;89:9237-41.
93. Snyder RD. Polyamine depletion is associated with altered chromatin structure in HeLa cells. Biochem J 1989;260:697-704.
94. Basu HS, Sturkenboom MCJM, Feuerstein BG and Marton LJ. Effect of polyamine depletion on the chromatin structure of U-87 MG human brain tumor cells. biochem J 1992;282:723-727.
95. Koza RA and Herbst EJ. Deficiencies in DNA replication and cell cycle progression in polyamine depleted HeLa cells. Biochem J 1992;281:87-93.
96. Jackson V. Influence of positive stress on nucleosome assembly. Biochemistry 1993;32: 5901-12.
97. Pfaffle P, Gerlach V, Bunzel L and Jackson V. In vitro evidence that transcription-induced stress causes nucleosome dissolution and regeneration. J Biol Chem 1990;265:16830-40.
98. Jackson V. In vivo studies on the dynamics of histone-DNA interaction: evidence for nucleosome dissolution during replication and transcription and a low level of dissolution independent of both. Biochemistry 1990;29:719-31.
99. Pfaffle P and Jackson V. Studies on rates of nucleosome formation with DNA under stress. J Biol Chem 1990;265:16821-9.
100. McCann PP, Pegg AE and Sjoerdsma A. Inhibition of Polyamine Metabolism. FL: Academic Press Orlando,1987.
101. Marton LJ, Pegg AE and Morris DR. Direction for polyamine research. J Cell Biochem 1991;45:7-8.

102. Porter CW and Bergeron RJ. Enzyme regulation as an approach to interference with polyamine biosynthesis - An alternative to enzyme inhibition. Adv Enz Reg 1988;27:57-79.
103. Johnson CS. Modulation of chemotherapy antineoplastic agents: enhancement of antitumor activities by interleukin-1. Current Op Oncol 1992;4:1108-15.
104. Thatcher N, Lorrigan P, Burt P and Stout R. Intensive combined modality therapy in small cell lung cancer. Seminars in Oncology 1994;21:9-22.
105. Howell HB (ed). Platinum and other metal coordination compounds in cancer chemotherapy. Plenum Press NY 1991.
106. Kociba RJ, Sleight SD and Rosenberg B. Inhibition of Dunning ascitic leukemia and Walker 256 carcinosarcoma with *cis*-diamminedichloroplatinum (NSC-119875) Cancer Chemotherapy Reports Part I 1970;54:325-8.
107. Rosenberg B. Platinum complex-DNA interactions and anticancer activity. Biochimie 1978;60:859-867.
108. Watson JD and Crick FHC. A structure for deoxyribose nucleic acid. Nature 1953;171:737-738.
109. Zwelling LA, Anderson T and Kohn KW. DNA-protein and DNA interstrand cross-linking by cis- and trans-platinum(II) diamminedichloride in L1210 mouse leukemia cells and relation to cytotoxicity. Cancer Res 1979;39:365-9.
110. Zwelling LA and Kohn KW. Mechanism of action of *cis*-dichlorodiammineplatinum(II) Cancer Treat Reports 1979;63:1439-44.
111. Pinto AL and Lippard SJ. Binding of the antitumor drug *cis*-diamminedichloroplatinum(II) (cisplatin) to DNA. Acta Biochim Biophys Hung 1985;780:167-180.
112. Foka M and Paoletti J. Interaction of *cis*-diamminedichloro-platinum(II) to chromatin: specificity for the drug distribution. Biochem Pharmacol 1986;35:3283-3291.
113. Hayes J and Scovell WM. *Cis*-diamminedichloro-platinum(II) modified chromatin and nucleosomal core particle. Biochim Biophys Acta 1991;1089:377-385.
114. Cheng CJ, Fujimura S, Grunberger D and Weinstein IB. Interaction of 1-(2-chloroethyl)-3-cyclohexyl-1-nitrosourea (NSC 79037) with nucleic acids and proteins in vivo and in vitro. Cancer Res 1972;32:22-7.
115. Kohn KW. Interstrand cross-linking of DNA by 1,3-*bis*(2-chloroethyl)-1-nitrosourea and other 1-(2-haloethyl)-1-nitrosoureas. Cancer Res 1977;37:1450-4.
116. Erickson LC, Bradley MO, Ducore JM, et al. DNA crosslinking and cytotoxicity in normal and transformed human cells treated with antitumor nitrosoureas. Proc Natl Acad Sci (USA) 1980;77:467-471.
117. Ludlum DB. DNA alkylation by the haloethylnitrosoureas: nature of modifications produced and their enzymatic repair or removal. Mutation Res 1990;233:117-26.
118. Tew KD, Sudhakar S, Schein PS and Smulson ME. Binding of

chlorozotocin and 1-(2-chloroethyl)-3-cyclohexyl-1-nitrosourea to chromatin and nucleosomal fractions of HeLa cells. Cancer Res 1978; 38:3371-8.

119. Green D, Tew KD, Hisamatsu T and Schein PS. Correlation of nitrosourea murine bone marrow toxicity with deoxyribonucleic acid alkylation and chromatin binding sites. Biochem Pharm 1982;31:1671-9.
120. Tew KD, Schein PS, Lindner DJ, et al. Influence of hydrocortisone on the binding of nitrosourea to nuclear chromatin subfractions. Cancer Res 1980;40:3697-3703.
121. Thielmann HW, Schroder CH, O'Neil JP, et al. Relationship between DNA alkylation and specific-locus mutation induction by N-methyl- and N-ethyl-N-nitrosourea in cultured chinese hamster ovary cells. Chemico-Biol Int 1979;26:233-43.
122. Nehl P and Rajewsky MF,. Ethylation of nucleophilic sites in DNA by N-ethyl-N-nitrosourea depends on chromatin structure and ionic strength. Mutation Res 1985;150:13-21.
123. Marushige K and Marushige Y. Alkylation of isolated chromatin with N-methyl-N-nitrosourea and N-ethyl-N-nitrosourea. Chemico-Biol Int 1983;46:165-77.
124. Vemuri MC, Raju NN and Malhotra SK. Recent advances in nuclear matrix function. Cytobios 1993;76:117-28.
125. Slichenmyer WJ, Rowinsky EK, Donehower RC and Kaufmann SH. The current status of camptothecin analogs as antitumor agents. J Natl Cancer Inst 1993;85:271-91.
126. Smit EF, Ousterhuis BE, Berendsen HH, et al. Phase I study of oral teniposide (VM-26). Seminars in Oncology 1992;19(2 Suppl 6):35-9.
127. Budman DR, Igwemezie LN, Kaul S, et al. Phase I evaluation of a water-soluble etoposide prodrug etoposide phosphate given as a 5-minute infusion on days 1, 3 and 5 in patients with solid tumors. J Clin Oncology 1994;12:1902-9.
128. Berrios M, Osheroff N and Fisher PA. In situ localization of DNA topoisomerase II a major polypeptide component of the *Drosophila* nuclear matrix fraction. Proc Natl Acad Sci (USA);82:4142-6.
129. Osheroff N. Biochemical basis for the interactions of type I and type II topoisomerases with DNA. PharmTher 1989;41:223-41.
130. Uemura T and Yanagida M. Isolation of type I and II DNA topoisomerase mutants from fission yeast: single and double mutants show different phenotypes in cell growth and chromatin organization. EMBO J 1984;3:1737-44.
131. Charcosset JY, Soues S and Laval F. Poisons of DNA topoisomerases I and II. Bulletin du Cancer 1993;80:923-54.
132. Srivenugopal KS and Morris DR. Differential modulation by spermidine of reactions catalyzed by type 1 prokaryotic and eukaryotic topoisomerases. Biochemistry 1985;24:4766-71.
133. Pommier Y, Kerrigan D and Kohn K. Topological complexes between DNA and topoisomerase II and effects of polyamines. Biochemistry

1989;28:995-1002.
134. Heck MM, Hittelman WN and Earnshaw WC. Differential expression of DNA topoisomerases I and II during the eukaryotic cell cycle. Proc Natl Acad Sci (USA) 1988;85:1086-90.
135. Gasser SM, Walter R, Dang Q and Cardenas ME. Topoisomerase II: its functions and phosphorylation. Antonie Van Leeuwenhoek 1992;62:15-24.
136. Chresta CM, Hicks R, Hartley JA and Souhami RL. Potentiation of etoposide-induced cytotoxicity and DNA damage in CCRF-CEM cells by pretreatment with non-cytotoxic concentrations of arabinosyl cytosine. Cancer Chemo Pharm 1992;31:139-45.
137. Lorico A, Boiocchi M, Rappa G, et al. Increase in topoisomerase-II-mediated DNA breaks and cytotoxicity of VP16 in human U937 lymphoma cells pretreated with low doses of methotrexate. Int J Cancer 1990;45:156-62.
138. Liu LF. DNA topoisomerase poisons as antitumor drugs. Ann Rev Biochem 1989;58:351-75.
139. Liu LF and D'Arpa P. Topoisomerase-targeting antitumor drugs: mechanisms of cytotoxicity and resistance. Important Adv Oncology 1992;79-89.
140. Ward JF. DNA damage and repair. Basic Life Sciences 1991;58:403-21.
141. Tanizawa A, Fujimori A, Fujimori Y and Pommier Y. Comparison of topoisomerase I inhibition DNA damage and cytotoxicity of camptothecin derivatives presently in clinical trials. J Natl Cancer Inst 1994;86:836-42.
142. D'Incalci M. DNA-topoisomerase inhibitors. Current Op Oncol 1993;5:1023-8.
143. Lett JT. Damage to DNA and chromatin structure from ionizing radiation and the radiation sensitivities of mammalian cells. Progr Nucl Acid Res Mol Biol 1990;39:305-52.
144. Elia MC and Bradley MO. Influence of chromatin structure on the induction of DNA double strand breaks by ionizing radiation. Cancer Res 1992;52:1580-6.
145. Warters RL and Lyons BW. Variation in radiation-induced formation of DNA double-strand breaks as a function of chromatin structure. Rad Res 1992;130:309-18.
146. Terleth C, van de Putte P and Brouwer J. New insights in DNA repair: preferential repair of transcriptionally active DNA. Mutagenesis 1991;6:103-11.
147. Hanawalt PC. Selective DNA repair in active genes. Acta Biologica Hungarica 1990;41:77-91.
148. Boothman DA, Majmudar G and Johnson T. Immediate X-ray-inducible responses from mammalian cells. Rad Res 1994;138:S44-6.
149. Alhonen-Hongisto L, Deen DF and Marton LJ. Time dependence of the potentiation of 1,3-*bis*(2-chloroethyl)-1-nitrosourea cytotoxicity caused by alpha-difluoromethylornithine induced polyamine depletion in 9L rat brain tumor cells. Cancer Res 1984;44:1819-22.
150. Hung DT, Deen DF, Seidenfeld J and Marton LJ. Sensitization of 9L rat brain gliosarcoma cells to 1,3-bis(2-chlooethyl)-1-nitrosourea by alpha-

difluoromethylornithine an ornithine decarboxylse inhibitor. Cancer Res 1981;41:2783-2785.

151. Oredsson SM, Deen DF and Marton LJ. Decreased cytotoxicity of cis-diamminedichloroplatinum(II) by alpha-difluormethylornithine depletion of polyamines in 9L rat brain tumor cells in vitro. Cancer Res 1982;42:1296-9.

152. Hunter KJ, Deen DF, Pellarin M and Marton LJ. Effect of alpha-difluoromethylornithine on 1,3-bis(2-chloroethyl)-1-nitrosourea and cis-diamminedichloroplatinum(II) cytotoxicity DNA interstrand crosslinking and growth in human brain tumor cell lines in vitro. Cancer Res 1990;50:2769-72.

153. Oredsson SM, Pegg AE, Alhonen-Hongisto L, et al. Possible factors in the potentiation of 1-(2-chloroethyl)-3-trans-4-methylcyclohexyl-1-nitrosourea cytotoxicity by alpha-difluoromethylornithine in 9L rat brain tumor cells. EuJ Cancer Clin Oncol 1984;20:535-42.

154. Hung DT, Oredsson SM, Pegg AE, et al. Potentiation of 1,3-bis(2-chloroethyl)-1-nitrosourea cytotoxicity in 9L rat brain tumor cells by methylglyoxal-bis(guanylhydrazone) an inhibitor of S-adenosyl-L-methionine decarboxylase. Eur J Cancer Clin Oncol 1984;20:417-20.

155. Denstman SC, Ervin SJ and Casero RA Jr. Comparison of the effects of treatment with the polyamine analog N^1,N^8 bis(ethyl)spermidine (BESpd) or difluoromethylornithine (DFMO) on the topoisomerase II-mediated formation of 4'-(9- acridinylamino) methanesulfon-m-anisidide (m-AMSA) induced cleavable complex in the human lung carcinoma line NCI H157. Biochem Biophysi Res Comm 1987;149:194-202.

156. Zwelling LA, Kerrigan D and Marton LJ. Effect of difluoromethylornithine an inhibitor of polyamine biosynthesis on the topoisomerase II-mediated DNA scission produced by 4'-(9-acridinylamino)methanesulfon-m-anisidide in L1210 murine leukemia cells. Cancer Res 1985;45:1122-6.

157. Dorr RT, Liddil JD and Gerner EW. Modulation of etoposide cytotoxicity and DNA strand scission in L1210 and 8226 cells by polyamines. Cancer Res 1986;46:3891-5.

158. Smirnov IV, Feuerstein BG, Pellarin M, et al. Pretreatment with the polyamine analog 1,19-*bis*(ethylamino)-5,10,15,-triaza-nonadecane (BE-4-4-4-4) inhibits etoposide cytototxicity in U-251 MG (NCI) human brain tumor cells. Cell Mol Biol (in Press).

159. Chen CZ, Hu LJ, Bergeron RJ, et al. Radiopotentiation of human brain tumor cells by the spermine analog N^1,N^{14}-*bis*(ethyl)homospermine. Inter J Rad Oncology Biology Physics 1994;29:1041-7.

160. Williams JR, Casero RA and Dillehay LE. The effect of polyamine depletion on the cytotoxic response to PUVA gamma rays and UVC in V79 cells in vitro. Biochem Biophys Res Comm 1994;201:1-7.

161. Pitkin J and Davis RH. The genetics of polyamine synthesis in *Neurospora crassa*. Arch Biochem Biophys 1990;278:386-391.

162. Davis RH and Ristow JL. Polyamine toxicity in *Neurospora crassa*: protective role of the vacuole. Arch Biochem Biophys 1991;285:306-11.

163. Davis RH. Management of polyamine pools and the regulation of ornithine decarboxylase. J Cell Biochem 1990;44:199-205.
164. Tabor CW and Tabor H. Polyamines in microorganisms. Microbiol Reviews 1985;49:81-99.
165. Balasundaram D, Tabor CW and Tabor H. Spermidine or spermine is essential for the aerobic growth of *Saccharomyces cerevisiae*. Proc Natl Acad Sci (USA) 1991;88:5872-6.
166. Garrthwaite I, Stead AD and Rider CC. Assay of the polyamine spermine by a monoclonal antibody-based ELISA. J Immunol Methods 1993; 162:175-8.
167. Giffin BF, McCann PP, Bitonti AJ and Bacchi CJ. Polyamine depletion following exposure to DL-alpha-difluoromethylornithine both in vivo and in vitro initiates morphological alterations and mitochondrial activation in a monomorphic strain of *Trypanosoma brucei brucei*. J Protozoology 1986;33:238-43.
168. Snyder RD, Beach DC and Loudy DE. Anti-mitochondrial effects of bisethyl polyamines in mammalian cells. Anticancer Res 1994;14:347-56.
169. He Y, Suzuki T, Kashiwagi K, et al. Correlation between the inhibition of cell growth by *bis*(ethyl)polyamine analogs and the decrease in the function of mitochondria. Eur J Biochem 1994;221:391-8.
170. Jain SK, Inman RB and Cox MM. Putative three-stranded DNA pairing intermediate in recA protein-mediated DNA strand exchange: no role for guanine N-7. J Biol Chem 1992;267:4215-22.
171. Maher LJ, Wold B and Dervan PB. Oligonucleotide-directed DNA triple-helix formation: an approach to artificial repressors? Antisense Res Dev 1991;1:277-81.
172. Singleton SF, Dervan PB. Equilibrium association constants for oligonucleotide-directed triple helix formation at single DNA sites: linkage to cation valence and concentration. Biochemistry 1993;32:13171-9.
173. Williams K, Romano C, Dichter MA and Molinoff PB. Modulation of the NMDA receptor by polyamines. Life Sciences 1991;48:469-98.
174. Lapidus RG and Sokolove PM. Spermine inhibition of the permeability transition of isolated rat liver mitochondria: an investigation of mechanism. Arch Biochem Biophys 1993;306:246-53.
175. Koenig H, Goldstone A and Lu CY. Polyamines regulate calcium fluxes in a rapid plasma membrane response. Nature 1983;305:530-4.
176. Feuerstein BG, Szollosi J, Basu HS and Marton LJ. alpha-Difluoromethylornithine alters calcium signaling in platelet-derived growth factor-stimulated A172 brain tumor cells in culture. Cancer Res 1992;52:6782-9.
177. Drouin H and Hermann A. Intracellular action of spermine on neuronal Ca^{2+} and K^+ currents. Eur J Neuroscience 1994;6:412-9.
178. Ventura C, Ferroni C, Flamigni F, et al. Polyamine effects on [Ca^{2+}] homeostasis and contractility in isolated rat ventricular cardiomyocytes. Am J Physiol 1994;267:H587-92.
179. Ficker E, Taglialatela M, Wible B, et al. Spermine and spermidine as gating molecules for inward rectifier K^+ channels. Science 1994; 266:1068-1072.

180. Sperry AO, Fishel BR and Garrard WT. Mutations that affect nuclear organization in yeast. Meth Cell Biol 1991;35:525-41.
181. Fishel BR, Sperry AO and Garrard WT. Yeast calmodulin and a conserved nuclear protein participate in the in vivo binding of a matrix association region. Proc Natl Acad Sci (USA) 1993;90:5623-7.
182. Gobbel GT, Marton LJ, Lamborn K, et al. Modification of radiation-induced brain injury by alpha-difluoromethylornithine. Rad Res 1991;128:306-15.
183. Fike JR, Gobbel GT, Marton LJ and Seilhan TM. Radiation brain injury is reduced by the polyamine inhibitor alpha-difluoromethylornithine. Rad Res 1994;138:99-106.
184. Pienta KJ, Partin AW and Coffey DS. Cancer as a disease of DNA organization and dynamic cell structure. Cancer Res 1989;49:2525-32.
185. Frydman L, Rossomando PC, Frydman V, et al. Interactions between natural polyamines and tRNA: an ^{15}N NMR analysis. Proc Natl Acad Sci (USA) 1992;89:9186-90.
186. Jain S Zon G and Sundaralingam M. Hexagonal crystal structure of the A-DNA octamer d(GTGTACAC) and its comparison with the tetragonal structure: correlated variations in helical parameters. Biochemistry 1991;30:3567-76.

CHAPTER 6

Modulation of NMDA Receptors by Polyamines

Keith Williams

I. OVERVIEW

In 1885 a basic substance called neuridine was identified in human brain and subsequently shown to be identical to spermine.[1] Over a century later in 1995 the functions of spermine in the brain remain largely unknown. However, research into the neurobiology of polyamines took on a new lease of life in the late 1980s with the discovery that spermine and spermidine can modulate *N*-methyl-D-aspartate (NMDA) receptors in mammalian brain.[2,3] It is still not known whether endogenous polyamines act on NMDA receptors in vivo to affect neuronal function but there is now a large body of evidence supporting the existence of discrete polyamine recognition sites on the NMDA receptor. Voltage-dependent Ca^{2+} channels[4] and K^+ channels[5,6] represent other recently described targets for polyamines in the nervous system. Ongoing work in a number of laboratories is concerned with the site and mechanism of action of polyamines on NMDA receptors studied at a cellular and molecular level. That work, together with a consideration of the role of NMDA receptors in physiological and pathological processes, is the focus of this chapter.

II. INTRODUCTION

A. Glutamate Receptors

Much of the information signaling, storage and retrieval in the brain involves chemical synaptic transmission. Fast synaptic transmission is mediated by simple amino acids such as glutamate, aspartate, glycine and γ-aminobutyric acid (GABA). Glutamate is the major fast

Polyamines: Regulation and Molecular Interaction, edited by Robert Casero. © 1995 R.G. Landes Company.

excitatory neurotransmitter in the mammalian central nervous system (CNS). Effects of glutamate involve activation of cell surface receptors located on neurons and, in a few cases, on glial cells.[7,8] As with other neurotransmitter systems, glutamate receptors function as primary signal transducers. An understanding of the structure, localization, function and regulation of glutamate receptors is essential for understanding normal synaptic transmission at glutamatergic synapses and for understanding how signaling processes may be deranged in pathological conditions.

Glutamate receptors fall into two classes: ligand-gated ion channels and the so-called "metabotropic" receptors.[7,9,10] Ligand-gated ion channels (also called "ionotropic" receptors) are multimeric complexes that contain an integral cation-selective ion channel whereas metabotropic receptors are composed of a single polypeptide that couples via G-proteins to phospholipases or adenylyl cyclase[9-11] (Fig. 6.1). Glutamate receptors that are ligand-gated ion channels are classified on the basis of agonists that selectively activate them. Three classes of receptors have been described. These are the NMDA, α-amino-3-hydroxy-5-methyl-4-isoxazolepropionic acid (AMPA) and kainate receptors.[7,12] Multiple cDNAs encoding subunits of NMDA, AMPA and kainate receptors have been cloned in recent years[9] (Fig. 6.1). AMPA and kainate receptors mediate fast depolarization at most glutamatergic synapses (Fig. 6.2). NMDA receptors are also involved in normal synaptic transmission, but activation of NMDA receptors is more often associated with various forms of synaptic plasticity rather than with fast point-to-point signaling in the brain.[7,13] AMPA or kainate receptors and NMDA receptors may be colocalized at many glutamatergic synapses[14,15] (Fig. 6.2).

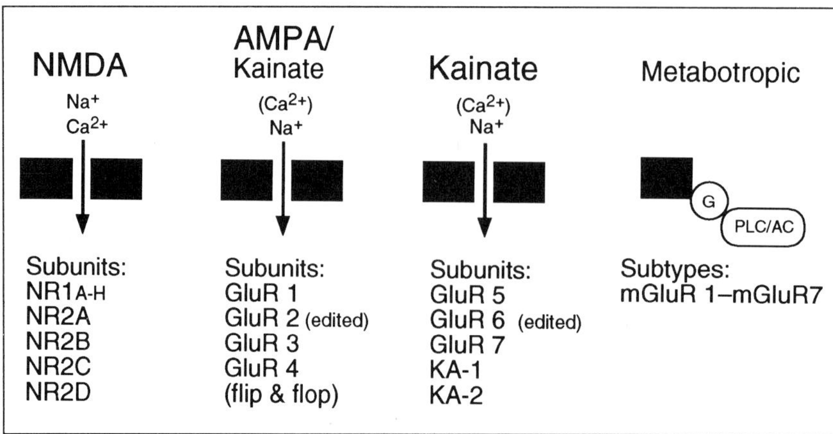

Fig. 6.1. Subtypes and cloned subunits of mammalian glutamate receptors.

B. NMDA Receptors: Physiology and Pathophysiology

NMDA receptors have received an enormous amount of attention in recent years. This is largely because of the discovery that the receptors play a pivotal role in the induction of synaptic plasticity, including processes that may underlie learning and memory, and because excessive activation NMDA receptors leads to neuronal cell death.[7,8,16-23] The role of NMDA receptors in physiological and pathological processes is still being unraveled. NMDA receptors are expressed on virtually all neurons in the brain[24,25] and it is likely that numerous functions for these receptors remain to be discovered.

A hallmark feature of NMDA receptors is that they are blocked by Mg^{2+} at resting membrane potentials and this block is relieved by depolarization.[26,27] Thus, activation of NMDA receptors requires not only binding of synaptically released glutamate but simultaneous depolarization of the postsynaptic membrane. This is achieved by activation of AMPA/kainate receptors at the same synapse or at nearby synapses receiving inputs from different neurons. Thus, NMDA receptors may function as "coincidence detectors" being activated only when there

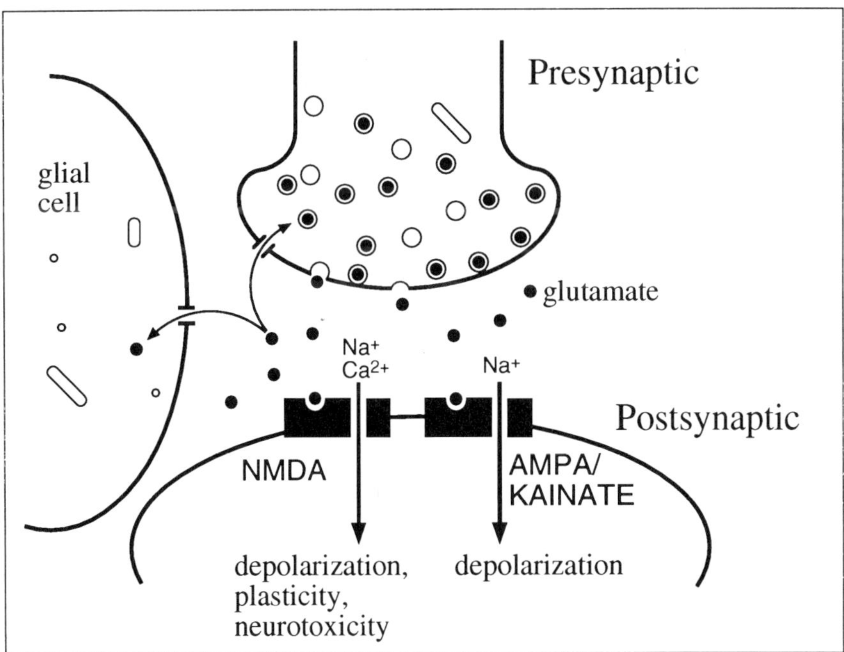

Fig. 6.2. Excitatory synaptic transmission. A cartoon depiction of a CNS synapse that uses glutamate as a neurotransmitter. When the presynaptic neuron fires, glutamate is released into the synaptic cleft where it activates glutamate receptors on the surface of the postsynaptic neuron. Glutamate is rapidly cleared from the cleft by re-uptake.

is simultaneous firing of two or more neurons. A well-characterized phenomenon that involves NMDA receptors is the induction of long-term potentiation (LTP). LTP refers to a prolonged (hours-days) increase in the size of a postsynaptic response to a presynaptic stimulus of given strength. Various forms of LTP have been described, some of which involve NMDA receptors.[16,28] Activation of NMDA receptors triggers one type of associative LTP in the hippocampus, an area of the brain involved in memory formation. The flip-side of LTP is long-term depression (LTD), a process in which the strength of synaptic signaling is weakened for a prolonged period. Recent studies have shown that NMDA receptors are involved in the induction of LTD at the same synapses where the receptors control the induction of LTP.[29-31] The mechanisms underlying the induction and maintenance of LTP and LTD are not fully understood. However, it seems that the frequency and pattern of synaptic stimulation, and thus the pattern of activation of glutamate receptors, controls whether a synapse undergoes LTP or LTD.[32] LTP and LTD are fascinating experimental phenomena and may provide a model for how synapses are strengthened or weakened in vivo. An understanding of LTP and LTD is important because these or similar processes occurring in the brain have been proposed as a cellular substrate for some forms of learning and memory.[16,28,32] Thus, if LTP is part of the cellular basis for memory then LTD may be the cellular basis for forgetting (or at least for not remembering).

The ion channel of the NMDA receptor is highly permeable to Ca^{2+} (Fig. 6.3). Many of the responses mediated by NMDA receptors involve increases in intracellular Ca^{2+}, which may activate protein kinases and phosphatases and enzymes such as nitric oxide synthase that are involved in trans-synaptic signaling.[33-35] A gradation of intracellular Ca^{2+} levels, dependent on the frequency of activation of NMDA

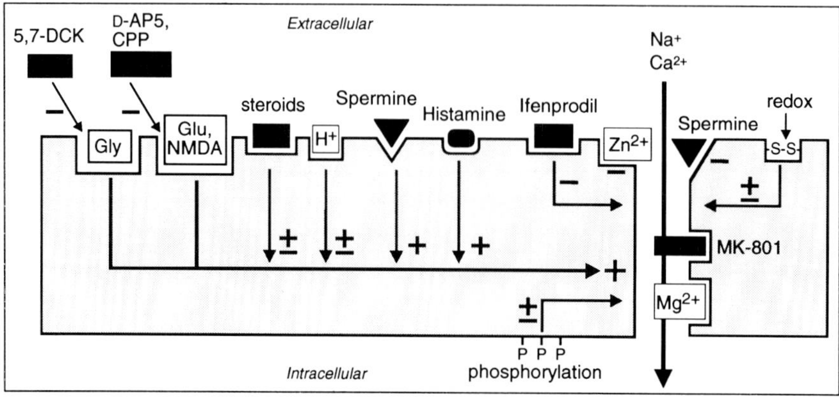

Fig. 6.3. Schematic model of the NMDA receptor.

receptors, may differentially activate kinase and phosphatase enzymes including CaM kinase II and calcineurin that are involved in the induction and maintenance of LTP and LTD.[35]

In addition to functional plasticity such as LTP and LTD, NMDA receptors also control morphological plasticity of synapses, particularly in the developing nervous system. These forms of plasticity include the consolidation of ocular dominance patterns in the visual cortex and whisker barrels in the somatosensory cortex[36,37] and elimination of supernumery synapses in the cerebellum.[38] Many of these processes are activity-dependent, and the strengthening and retention of synapses may depend on coordinated neuronal activity, with inactive synapses being eliminated.[22,23,36,37]

An area that has attracted considerable attention from basic and applied research concerns the role of NMDA receptors in neurotoxicity and neurodegeneration. It is now well established that exposure of cultured neurons to high concentrations of glutamate for only a few minutes can lead to neuronal cell death.[18,21,39] The cascade of events leading to neuronal death is triggered by excessive activation of glutamate receptors and influx of Ca^{2+}. This process is thought to be similar to the neurotoxicity that occurs after ischemia or hypoglycemia in the brain where a massive release and impaired re-uptake of glutamate leads to excess stimulation of glutamate receptors and subsequent cell death.[18,20,21] Although activation of AMPA, kainate and metabotropic receptors may contribute to neuronal cell death, the major villain appears to be the NMDA receptor. NMDA receptor antagonists can attenuate or block neurotoxicity and may therefore have considerable clinical utility to prevent brain damage that occurs after ischemia following stroke.

NMDA receptors may also be involved in the development of susceptibility to epileptic seizures and in the occurrence of seizure activity.[40] In animal models, NMDA receptor antagonists have anticonvulsant activity and may find clinical use as anticonvulsants.[41] It is conceivable that abnormal expression, function or regulation of glutamate receptors may be involved in the etiology of some neurological disorders. A role for NMDA receptors or disordered glutamatergic transmission in the etiology of chronic neurodegenerative diseases and in schizophrenia has been postulated.[20,42-45]

A major obstacle to the development of clinically useful NMDA receptor antagonists is the occurrence of side-effects associated with these antagonists. These include psychotomimetic effects and impaired motor function. A range of side effects is hardly surprising given the widespread distribution of NMDA receptors in the brain and spinal cord. Broad-spectrum antagonists that block most or all of the NMDA receptors in the CNS would be predicted to have a range of effects on normal CNS function. However, the identification and ongoing characterization of multiple subunits and subtypes of NMDA receptors (see below) may lead to the development of antagonists to selectively target subpopulations of NMDA receptors that are involved in specific physiologic functions.

A variety of endogenous substances including polyamines, histamine and neurosteroids have been found to potentiate activity of NMDA receptors in vitro. If these compounds modulate NMDA receptors in the intact brain then their binding sites may represent targets for therapeutic approaches to reduce activity of NMDA receptors. Conceptually, this is an attractive strategy because reducing NMDA receptor activity by allosteric modulation may provide more effective long-term therapies than by using channel blockers or competitive antagonists that have an all-or-none effect on receptor function. Conversely, if NMDA receptor activity is reduced in some pathologies, then the host of modulatory sites on the receptor may provide therapeutic targets for enhancing receptor activity to improve cognitive function.

C. Modulation of NMDA Receptors and How to Study It

NMDA receptors contain a number of distinct recognition sites for endogenous and exogenous ligands.[7,13,46] These include binding sites for glutamate (or NMDA), glycine, Mg^{2+}, Zn^{2+}, polyamines, histamine, neurosteroids, and open-channel blockers such as phencyclidine (PCP) and MK-801. The receptor is also modulated by protons, redox reagents and arachidonic acid (Fig. 6.3). Channel activity and blockade by Mg^{2+} are also influenced by phosphorylation catalyzed by protein kinase C and protein-tyrosine kinases.[47-50] There is an absolute requirement for glycine for the channel to be opened by NMDA or glutamate.[51] Thus, glycine can be considered a "co-agonist" at the NMDA receptor (Fig. 6.3). In this regard, NMDA receptors appear to be unique among neurotransmitter receptors in requiring two different amino acid agonists to activate the receptor.

A number of radioligands have been used to study the properties of NMDA receptors on broken membranes prepared from brain tissue or from cultured neurons. These include ligands such as [^{3}H]CGP39653 and [^{3}H]CPP that label the glutamate recognition site,[52,53] [^{3}H]5,7-DCK, which labels the glycine site,[54] and open-channel blockers such as [^{3}H]MK-801, ^{125}I-MK-801, and [^{3}H]TCP.[55-59] Binding of MK-801 is greatly enhanced by glutamate and glycine, which increase the rates of both association and dissociation of binding presumably by increasing the accessibility of the ion channel binding site (Fig. 6.3). Because radiolabeled MK-801 binds to the open-channel state of NMDA receptors it has been used extensively to study the effects of various modulators, including polyamines, that affect receptor/channel properties. This approach is technically quite straightforward and numerous different compounds can be analyzed at various concentrations under different assay conditions. A drawback of binding assays is that they provide no direct information as to the effects of modulators on the function of NMDA receptors, and the results of ligand binding experiments can be difficult to interpret because binding of MK-801 exhibits complex kinetics of association and dissociation. Furthermore, in assays with receptors on brain membranes it is likely that there are various subtypes of NMDA receptors that may be differentially sensitive to effects of modulators.

A number of approaches have been used to study the functional properties of native NMDA receptors. The most widely used techniques involve electrophysiological recording or fluorescent imaging of intracellular Ca^{2+} to study NMDA receptors on cultured neurons and in brain slices.[7,8,60] Cultured neurons are usually prepared from the neocortex, hippocampus, striatum, cerebellum or spinal cord of embryonic or neonatal animals, although the use of acutely dissociated neurons from mature brain has also been described. A wealth of information regarding the functional and biophysical properties of NMDA receptors has come from studies using cultured neurons. However, neurons from different brain regions express different subtypes of NMDA receptors and the expression of these receptors is influenced by time in culture and by the culture conditions. For example, when neurons from fetal rat neocortex are maintained in culture the expression of NMDA receptor subtypes is altered over days 7-21 in vitro. These changes, which are similar to changes that occur during development in vivo, are dependent on the concentration of serum in the medium.[61,62] Thus, the NMDA receptors that one studies on cultured neurons represent particular forms of the receptor having particular functional and pharmacological properties. Nowhere is this more apparent than in studies of the effects of polyamines. Apparently conflicting results concerning the effects of spermine on NMDA receptors are probably due to the existence of NMDA receptor subtypes on cultured neurons (see below).

Oocytes from the African clawed toad *Xenopus laevis* have also been used to study the functional properties of NMDA receptors. Oocytes will synthesize many types of receptors and ion channels and insert them in the plasma membrane after injection of mRNA extracted from tissues or cRNA prepared from cDNA clones.[63] The receptors can then be studied using two-electrode voltage-clamp recording (Fig. 6.4). NMDA receptors are expressed in oocytes after injection of rat brain mRNA,[64] and the biophysical and pharmacological properties of such receptors have been well-documented.[49,65-68] However, it is now known that there are multiple subunits of NMDA receptors and that the subunit mRNAs are regionally and developmentally regulated. Thus, mRNA from rat brain may produce various subtypes of NMDA receptors in oocytes depending on the brain region and age of the animal used to prepare mRNA.[61,69] The oocyte system does, however, remain extremely useful for studying recombinant NMDA receptors expressed from cloned subunits including mutant and chimeric subunits constructed to study structure-function relationships of the receptors (see below).

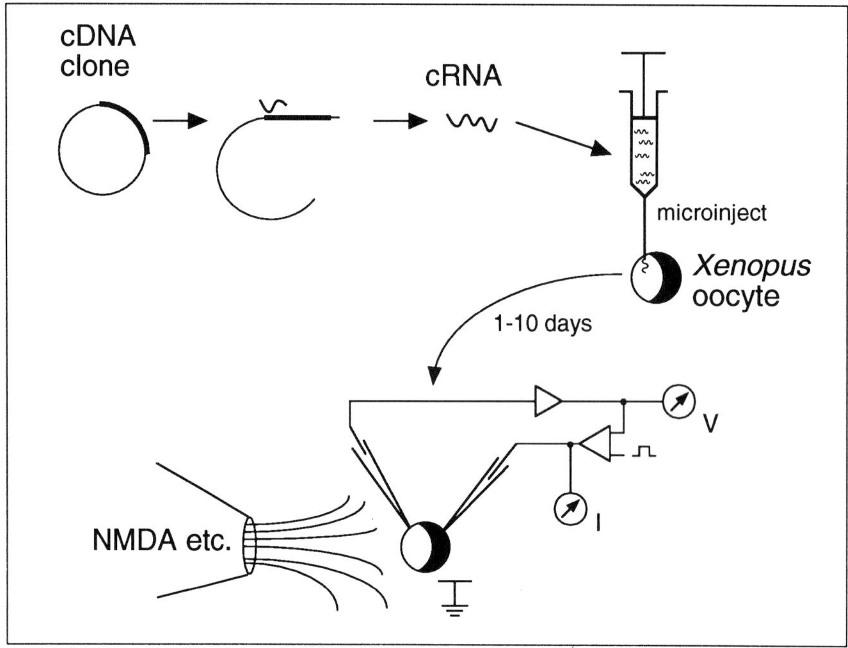

Fig. 6.4. Schematic of the methodology used for injection and two-electrode voltage-clamp recording of Xenopus *oocytes.*

III. EFFECTS OF POLYAMINES ON NATIVE NMDA RECEPTORS

A. Ligand-Binding Studies

Modulation of NMDA receptors by polyamines was first reported by Ransom and Stec[2] who showed that spermine and spermidine increased the binding of [^{3}H]MK-801 to NMDA receptors on rat brain membranes. It was proposed that the stimulatory effect of polyamines was mediated at a novel site associated with the NMDA receptor.[2,3] The effects of spermine and spermidine on the binding of radiolabeled channel blockers have been confirmed in a number of laboratories (e.g. refs. 3, 70-75). Modulatory effects seen in binding assays have been reviewed elsewhere[46,76] and only the cardinal points are described here.

The concentration-effect curves for spermine and spermidine are biphasic, with stimulation of [^{3}H]MK-801 binding at concentrations of 1-100 μM and inhibition at higher concentrations[3,72,74] (Fig. 6.5). This suggests that polyamines have at least two effects on NMDA receptors, presumably mediated through two or more distinct polyamine binding sites. The effects of polyamines have a marked structural specificity and are dependent on the chain length and the number of amino and imino groups in linear polyamine analogs.[3,46,74,77]

The apparent affinity for $[^3H]$MK-801 is increased 2- to 3-fold in the presence of 10-100 μM spermine or spermidine.[2,3,73,74] A more pronounced effect of spermine is to increase the rates of association and dissociation of $[^3H]$MK-801 binding.[72,75,78] This is similar to the effect of glutamate and is thought to reflect an increased accessibility of the ion channel binding site for MK-801, a biochemical correlate of channel opening.[78,80] Thus, part of the mechanism of action of spermine may be to increase accessibility of the ion-channel binding site. Many of the properties of the stimulatory polyamine site are retained after detergent solubilization of NMDA receptors, suggesting that this site is tightly associated with the other components of the NMDA receptor complex and is probably localized on the receptor protein.[81-83]

In addition to effects on the binding of open-channel blockers, spermine increases the affinity of the receptor for $[^3H]$glycine and for glutamate-site antagonists such as CPP and D-AP5.[84-89] Notably, the pharmacological profiles for effects of polyamines on the glutamate and glycine sites are different from effects on the binding of MK-801 and may therefore involve different polyamine binding sites.[76] A similar conclusion has come from functional studies of native and recombinant NMDA receptors (see below).

A number of polyamine analogs have been suggested to act as competitive antagonists or as inverse agonists at the stimulatory polyamine site.[46,78,90] However, many of these compounds have multiple actions

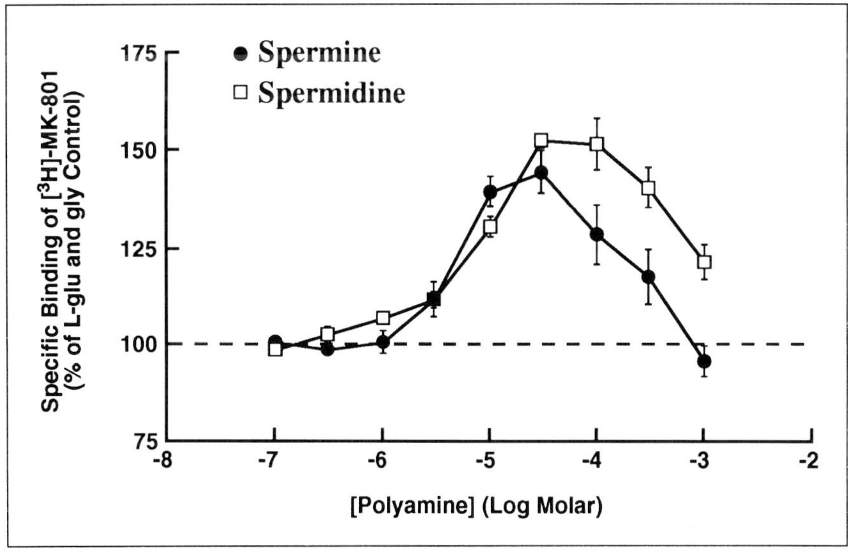

Fig. 6.5. Effects of polyamines on $[^3H]$MK-801 binding. The effects of spermine and spermidine on binding of $[^3H]$MK-801 to NMDA receptors on rat brain membranes were measured in the presence of 100 μM L-glutamate (L-glu) and glycine (gly). Reproduced, with permission, from K Williams et al, Mol Pharmacol 36:575-581, 1989.

on the NMDA receptor, and interpretation of their effects at the stimulatory polyamine site is not straightforward.[76] In functional assays many polyamine analogs act as voltage-dependent blockers with a mechanism apparently unrelated to antagonism at the stimulatory polyamine site but similar to the voltage-dependent block seen with spermine.[13,46,76] These compounds, which include arcaine, diethylenetriamine and 1,10-diaminodecane are probably of limited value for studying the stimulatory polyamine site but may be useful for studying the mechanism of the voltage-dependent block by polyamines.[91,92]

B. Electrophysiological Studies

Spermine has been found to enhance NMDA-induced whole-cell currents in cultured cortical, hippocampal, and spinal cord neurons. Stimulatory effects of spermine are seen at concentrations of 10-1000 μM, similar to concentrations that increase the binding of [^{3}H]MK-801. The degree of potentiation by spermine varies widely between individual neurons, with some neurons showing up to 200% enhancement of whole-cell currents and some showing no enhancement.[93-95] This suggests that NMDA receptors expressed on different neurons are differentially sensitive to stimulatory effects of spermine. This may be due to the existence of different molecular forms of the NMDA receptor.

A number of mechanisms contribute to the stimulatory effects of spermine as measured electrophysiologically. One effect involves an increase in the affinity of NMDA receptors for glycine ("glycine-dependent" stimulation) and probably corresponds to the increase in affinity for glycine seen in binding assays.[68,86,93] An increase in the affinity of the NMDA receptor for glycine cannot wholly account for the ability of spermine to increase NMDA currents because stimulatory effects of spermine are seen in the presence of saturating concentrations of glycine ("glycine-independent" stimulation).[68,78,93,95,96] One mechanism that underlies glycine-independent stimulation by spermine is an increase in the frequency of channel opening of NMDA receptors.[95,96] Spermine also reduces desensitization of NMDA receptors.[94] The ability of spermine to enhance NMDA currents has a rapid onset and a rapid offset and appears to involve an extracellular polyamine binding site.[94] It seems likely that the stimulatory effects of spermine seen in whole-cell and single-channel recordings of NMDA receptors are mechanistically related to those underlying the increase in [^{3}H]MK-801 binding and are probably mediated through the same polyamine recognition site.

In almost all systems where the effects of spermine have been investigated, an inhibition of NMDA currents (in addition to potentiation) has been reported. Inhibition is voltage-dependent and, at physiological resting membrane potentials, inhibition is seen at concentrations of spermine similar to or slightly higher than those required for stimulation. Thus, stimulatory and inhibitory dose-response curves for spermine overlap. In cells where stimulation by spermine is prominent, the inhibition is manifest as a decrease in the degree of stimulation seen with high concentrations of spermine. Alternatively, in neurons where there is minimal stimulation by spermine, a direct inhibition of NMDA currents is observed.[93]

Inhibition by spermine is strongly voltage-dependent, being seen at membrane potentials more negative than about -30 mV. Single-channel patch clamp recording has been used to study the mechanism of this effect. Spermine (10-1000 μM) decreased the single channel conductance and decreased the average channel open-time when currents were measured at hyperpolarized potentials (-70 to -100 mV) but not at depolarized potentials (+40 to +50 mV).[95-97] Voltage-dependent block by spermine may be due to a fast open-channel block similar to that of Mg^{2+}, involving binding of spermine to a site within the membrane-spanning regions of the ion channel. However, the observation that spermine but not Mg^{2+} decreases single-channel conductance suggests a different site or mechanism of action of spermine compared with Mg^{2+}. It has been proposed that the blocking site for spermine does not lie deep within the ion channel pore and may be more peripheral than the Mg^{2+} binding site. Alternatively, the voltage-dependence of the spermine block may be due to a charge-screening effect rather than binding to a site within the membrane spanning regions of the channel.[95,97] That is, spermine may interact with negatively charged residues on the NMDA receptor/channel protein, possibly close to the channel mouth, to impede ion flow. A decreased rate of cation permeation caused by charge-screening would account for the decrease in single channel conductance. It seems likely that spermine has both channel-blocking and charge-screening effects at the NMDA receptor. Synthetic polyamine analogs may show predominantly channel-block or charge-screening effects depending on their structure and should therefore be useful for studying the sites and mechanisms of block by polyamines.[91,97,98]

A particularly intriguing class of voltage-dependent blockers are the polyamine-conjugated spider and wasp toxins which include Joro toxins isolated from *Nephila clavata*, argiotoxins from *Argiope lobata* and *Argiope trifasciata*, philanthotoxins from *Philanthus triangulum*, and α-agatoxins from *Agelenopsis aperta*.[99-101] The toxins have structural similarities, consisting of a polyamine moiety coupled to a phenyl or indole group, in some cases via an amino acid linkage (Fig. 6.6). The toxins are potent antagonists of invertebrate glutamate receptors including receptors that mediate neuromuscular transmission in insects. Toxins that block glutamate receptors will therefore paralyze insects, and it is this property that presumably confers an evolutionary advantage to spiders that produce toxins and prey on insects. Several of the toxins, including argiotoxin$_{636}$ and the α-agatoxin Agel-505 (Fig. 6.6), have been shown to inhibit NMDA receptors in a strongly voltage-dependent manner, probably acting as open-channel blockers.[102,103] These toxins are considerably more potent than polyamines, inhibiting NMDA responses with nanomolar affinities, and have 50- to 100-fold selectivity for NMDA receptors over AMPA and kainate receptors.[102-105] Thus, polyamine-containing spider toxins are useful tools for studies of the ion channel of NMDA receptors.

IV. MOLECULAR BIOLOGY OF NMDA RECEPTORS

A. NR1 Subunits

Nakanishi and coworkers first reported the cloning of an NMDA receptor subunit designated NMDAR1 (NR1) from a rat brain cDNA library by expression cloning using *Xenopus* oocytes.[106] Many of the

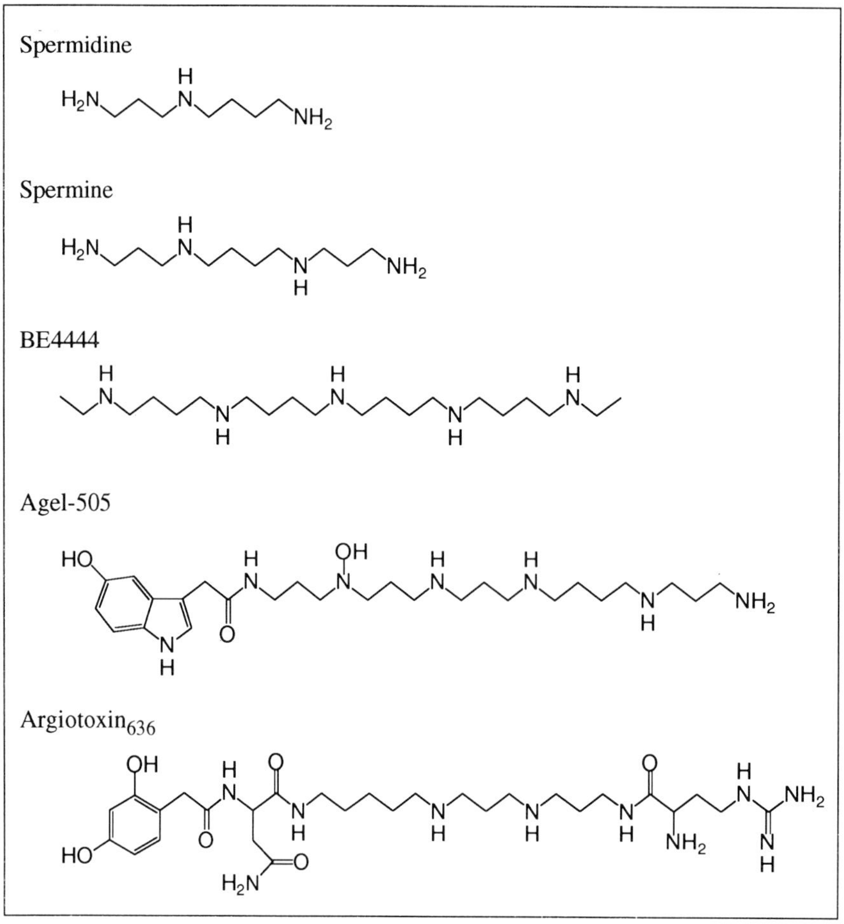

Spermidine

Spermine

BE4444

Agel-505

Argiotoxin$_{636}$

Fig. 6.6. Structures of polyamines, BE4444 and two polyamine-conjugated spider toxins.

properties of native NMDA receptors were seen with homomeric NR1 receptors expressed in oocytes. The receptors were sensitive to glycine and were shown to gate Ca^{2+} and be blocked in a voltage-dependent manner by Mg^{2+}.[106] Two other laboratories independently cloned NR1 subunits from rat[107] and mouse[108] brain. The mouse homologue of NR1 was termed ζ1 (Fig. 6.7). The NR1 clone codes for a polypeptide with a protein molecular weight (104 kDa) and structural topography similar to subunits of AMPA and kainate receptors but with only limited amino acid identity (20-25%) to those subunits. The proposed topography of NR1, based in part on analogy to other ligand-gated channels like the nicotinic acetylcholine receptor,[109] includes four transmembrane regions (TMI-TMIV) and a large extracellular N-terminal domain (Figs. 6.7 and 6.8). The presence of consensus phosphorylation sites for protein kinase C on the C-terminus of NR1 suggests that the C-terminus is located intracellularly.[110] The subunit may thus contain five rather than four transmembrane regions (Fig. 6.8). The binding sites for agonists and for some modulators may be located on the large extracellular N-terminal domain. Amino acid residues that are involved in forming the glycine binding site are located in this region.[111]

NR1 mRNA is widely distributed throughout the brain[106] as is the NR1 protein detected by immunocytochemistry.[24,25] The NR1 subunit represents a single gene product[9,112] but the properties of NR1 are controlled by alternative splicing of mRNA. Eight functionally active

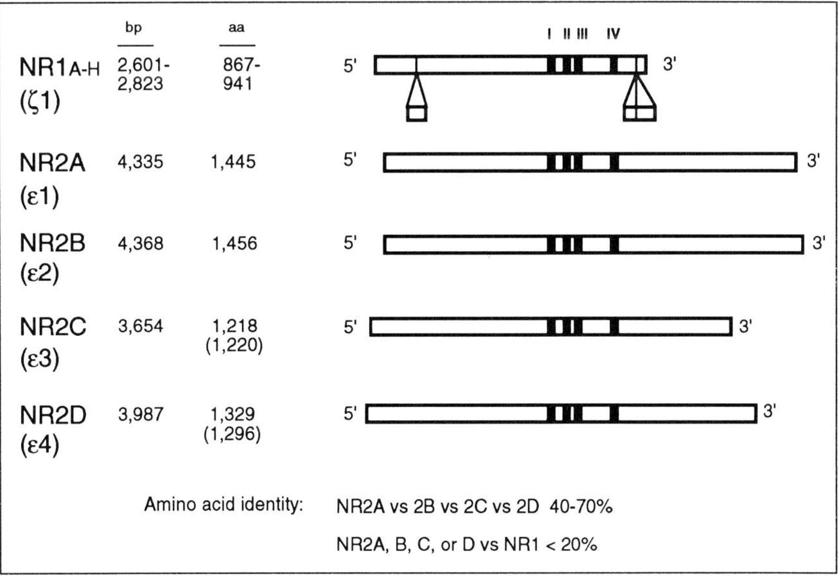

Fig. 6.7. Cloned NMDA receptor subunits. Schematic of the cDNAs encoding rat (NR1, NR2A-D) and mouse (ζ1, ε1-4) NMDA receptor subunits. bp, base-pairs; aa, amino acids.

splice variants have been described based on the presence or absence of an N-terminal (5') insert and one or two adjacent C-terminal (3') inserts[9,112,113] (Fig. 6.7). Several sets of nomenclature have been used for the splice variants.[9] The nomenclature used here, NR1A-NR1H, corresponds to that of Sugihara et al.[113] All of the NR1 splice variants produce functional homomeric NMDA receptors when expressed in *Xenopus* oocytes, but differ in their sensitivity to phosphorylation by protein kinase C and modulatory effects of polyamines, Zn^{2+} and histamine.

There are complex regional and developmental changes in the splicing pattern of NR1 mRNA in the brain.[114] Only splicing of the N-terminal insert is dealt with here because this affects sensitivity to polyamines (see below). Variants of NR1 such as NR1A that lack the N-terminal insert are the predominant forms of NR1 (80-90%) in the cerebral cortex and hippocampus.[114,115] However, variants containing the insert may be expressed in subsets of neurons that also express variants without the insert.[114] A more complex pattern of expression is seen in the cerebellum where NR1 variants that lack the insert predominate in neonatal animals whereas variants that contain the insert predominate in adult animals.[114] Variants that contain the insert are also expressed at high levels in the olfactory bulb and thalamus in adult animals.[114] Differences in regional and developmental splicing of NR1 mRNA undoubtedly contribute to functional diversity of NMDA receptors. It is conceivable that alternative splicing may also dictate assembly of particular NR1/NR2 heteromers and that receptors containing different splice variants may have different turnover rates or synaptic localizations.

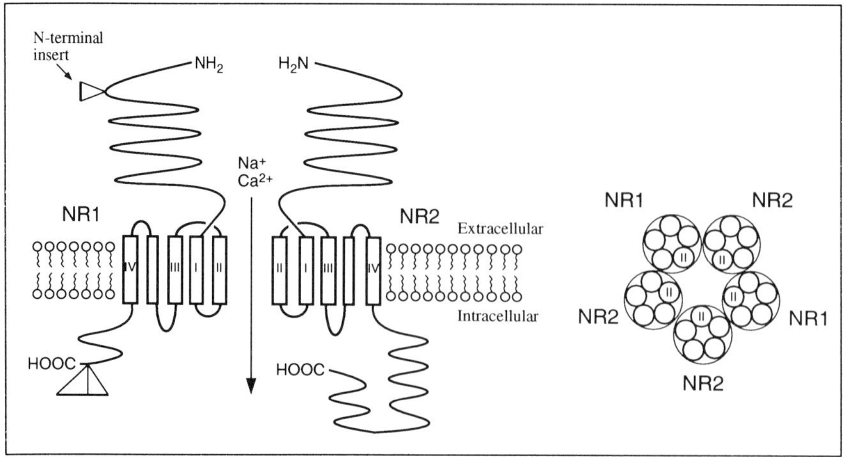

Fig. 6.8. Hypothetical topography and subunit arrangement of NMDA receptors. Left, NR1 and NR2 subunits may each contain five transmembrane domains with large extracellular N-termini. Right, a hypothetical NMDA receptor is composed of five NR subunits with the TMII region of each subunit lining the ion channel pore.

B. NR2 Subunits

Four rat brain cDNA clones designated NR2A, NR2B, NR2C, and NR2D (Fig. 6.7) were isolated by homology to NR1 and GluR1-4.[116,117] Equivalent cDNAs termed ε1-ε4 were isolated from mouse brain.[118-120] The NR2 subunits are large polypeptides (≈160 kDa) and are 40-70% homologous to each other. The amino acid sequences of the NR2 clones are only 18% identical to NR1.

NR2 subunits show distinct regional and developmental patterns of expression. In adult rat and mouse brain, NR2A and NR2B are the major NR2 subunits expressed in the telencephalon including the cerebral cortex and hippocampus.[10,115,116,121,122] In neonatal animals NR2B is expressed at high levels in these areas whereas expression of NR2A begins at about postnatal day 7 and increases progressively to adult levels that are roughly equivalent to levels of NR2B.[115,122-124] In the striatum of adult rats NR2B is expressed at much higher levels than NR2A.[125] The NR2c subunit is found almost exclusively in the cerebellum of adult animals. In the cerebellum of neonatal animals NR2C is absent and NR2B is the predominant NR2 subunit. During development expression of NR2B in the cerebellum ceases, and NR2B is replaced by NR2C.[115,122,123,126] The NR2D subunit is transiently expressed at high levels in cortex and hippocampus during development and is present mainly in midbrain regions such as the thalamus in adult animals.[121-123,125]

Injection of *Xenopus* oocytes with NR2 subunits in the absence of NR1 does not produce functional NMDA receptors. However, co-expression of NR2 and NR1 subunits generates NMDA receptors that produce 10- to 100-fold larger whole-cell currents than do homomeric NR1 receptors.[116-119] Co-expression of NR1 and NR2 subunits in mammalian cells such as HEK 293 cells also produces functional NMDA receptors.[116,122,127] This suggests that the two kinds of subunits form heteromeric multisubunit complexes. Although the subunit composition of native NMDA receptors has not yet been defined, they are likely to consist of combinations of NR1 and NR2 subunits. It is possible that some native NMDA receptors contain more than one type of NR2 subunit[124,128,129] or NR1 splice variant. A hypothetical subunit arrangement is shown in Fig. 6.8. It is important to note that this model is speculative. There is, as yet, no direct information about the topography of individual subunits or the arrangement of subunits within a receptor complex. Estimates of the sizes of native and recombinant NMDA receptors range from 630 kDa to 850 kDa, consistent with at least a pentameric receptor complex containing NR1 and NR2 subunits.[24,129] In Fig. 6.8 the receptor is shown containing two NR1 subunits because native NMDA receptors probably have two binding sites for glutamate and two sites for glycine on each receptor complex.[130,131] A detailed understanding of the structure of NMDA receptors will undoubtedly come only after X-ray crystallograpic studies of the receptor proteins.

The proposed topography of NR1 and NR2 subunits includes four or five membrane-spanning domains,[11] four of which (TMI-TMIV) are analogous to the proposed transmembrane domains in subunits of other ligand-gated ion channels. The TMII regions of NR1 and NR2 have a conserved asparagine residue in a position analogous to a glutamine or arginine residue in TMII of GluR subunits that controls Ca^{2+} permeability of AMPA receptor channels.[132-134] Mutation of this asparagine in NR1 or NR2 to glutamine or arginine reduces or abolishes Ca^{2+} permeability and voltage-dependent block by Mg^{2+}.[135-137] This suggests that the TMII region of NR1 and NR2 subunits may be involved in forming the ion-channel pore of NMDA receptors (Fig. 6.8) as has been proposed for other ligand-gated ion channels, and that the conserved asparagine residue is critical for the control of permeability to divalent cations.[11]

Heteromeric NMDA receptors containing NR1 and NR2 subunits have different pharmacological and functional properties depending on the type of NR2 subunit in the receptor. Receptors containing NR2C are less sensitive to blockade by Mg^{2+} and MK-801 than are receptors containing NR2A or NR2B.[116-118,138] NR1/NR2B receptors have a 10-fold higher affinity for glycine (see Fig. 6.11) and a 400-fold higher affinity for ifenprodil than do NR1/NR2A receptors.[118,139,140] Differences in sensitivity to antagonists acting at the glutamate and glycine sites have also been observed.[117,118,141,142] In some cases the properties of recombinant NMDA receptors match those of subtypes of native NMDA receptors and, in turn, the distribution of those native receptors matches the distribution of particular NR2 subunits in the brain. For example, recombinant NR1A/NR2B receptors have a high affinity for ifenprodil whereas NR1A/NR2A receptors have a low affinity for this antagonist;[140] during postnatal development there is a delayed expression of native NMDA receptors having a low affinity for ifenprodil that corresponds to the delayed expression of the NR2A subunit.[61,115] In mature rat brain the different regional expression of NR2A, NR2B, NR2C, and NR2D corresponds to the regional expression of four subtypes of NMDA receptor having different affinities for agonists and competitive antagonists.[141,142]

The subunit composition of NMDA receptors may have an important influence on the role of the receptor in synaptic transmission. The concentration of glutamate during synaptic activation may initially be as high as 1 mM but decline to background levels within a few milliseconds.[143] NMDA receptor responses typically have a slow offset (tens to hundreds of ms). Thus, the duration of synaptic responses mediated by NMDA receptors may be controlled by the kinetics of channel activation/inactivation and the rate of dissociation of glutamate from the receptor rather than by clearance of glutamate from the synaptic cleft.[143-145] Studies with recombinant NMDA receptors have shown that the rate of offset of whole-cell currents is different in receptors containing different NR2 subunits.[122] Currents gated by NR1/NR2B and NR1/NR2C receptors have a 3- to 4-fold slower offset decay than do currents gated by NR1/NR2A receptors. Even more striking is the decay rate at NR1/NR2D receptors which is 10- to 40-fold slower than the decay at receptors containing NR2A, NR2B, or NR2C.[122] This is consistent with the idea that the response decay is controlled by unbinding of glutamate because NR1/NR2D receptors have a higher affinity for glutamate than do receptors containing NR2A, NR2B or NR2C. Thus, the subunit composition of native NMDA receptors may influence their kinetic properties and, consequently, the time course of synaptic responses.

In single-channel studies it was reported that the unitary conductance of NR1/NR2A receptor channels is similar to that of NR1/NR2B, and the properties of both receptor types are similar to those of native NMDA receptors on hippocampal neurons.[127,146,147] NR1/NR2C channels have a lower unitary conductance than do NR1/NR2A or NR1/NR2B.[127,146] Thus, NR2 subunits may control the conductance/gating properties of the channels themselves in addition to the affinities of binding sites for agonists, antagonists and modulators and their coupling to channel opening.

V. EFFECTS OF POLYAMINES ON RECOMBINANT NMDA RECEPTORS

Spermine has stimulatory and inhibitory effects on recombinant NMDA receptors similar to effects seen at native NMDA receptors. The effects of spermine are controlled by the type of NR1 splice variant and the type of NR2 subunit present in the receptor complex. In this section the effects of spermine on recombinant NMDA receptors, their subunit-dependency, and their relationship to effects seen at native NMDA receptors are considered.

A. Homomeric NR1 Receptors

Durand et al[148] studied effects of spermine on homomeric NR1 receptors expressed in *Xenopus* oocytes and found that stimulation by spermine was dependent on the type of NR1 splice variant. Homomeric receptors expressed from variants such as NR1A and NR1E that lack the N-terminal insert were potentiated by spermine in the presence of a saturating concentration of glycine (glycine-independent stimulation). Stimulation was not seen with variants such as NR1B or NR1G that contain the N-terminal insert. In contrast there was no obvious relationship between spermine stimulation and the splicing of either C-terminal insert. All types of homomeric NR1 receptors exhibited a similar degree of voltage-dependent block by spermine regardless of the N- and C-terminal splicing patterns.[148] These results suggest that alternative splicing of the N-terminal insert controls sensitivity to glycine-independent stimulation by spermine, at least in homomeric NR1 receptors. This also applies to heteromeric NR1/NR2 receptors, with stimulation being seen at receptors that contain NR1A but not NR1B. However, in NR1/NR2 receptors the NR2 subunit exerts a dominant effect over properties seen with NR1A (see below).

The N-terminal insert (21 amino acids) in NR1 contains six basic amino acids (three lysine and three arginine residues) and three acidic amino acids (one glutamate and two aspartate residues). The insert thus bears a net positive charge. Site-directed mutagenesis to convert all six basic amino acids to alanine instead of lysine and arginine was found to restore spermine stimulation at homomeric NR1G receptors.[149] Because alanine is a neutral amino acid, the positive charge on the N-terminal insert is neutralized in the mutant NR1G subunit. The removal of positive charges may account for the restoration of spermine sensitivity,[149] although further studies with mutations containing different neutral and charged amino acids will be required to define this point. If the presence of positive charge normally prevents spermine stimulation, then the charged residues may interfere with binding of spermine to a site on the NMDA receptor. Alternatively, the charged N-terminal insert may itself function as a "spermine-like" moiety and interact with a spermine binding site on the NR1 protein.[149] In this case, spermine stimulation would not be seen because the spermine binding site would be constitutively occupied by the N-terminal insert portion of the subunit protein.

The N-terminal insert in NR1 also affects sensitivity to histamine and to Zn^{2+}. Potentiation by histamine is seen at homomeric NR1A receptors, which lack the N-terminal insert, and at some heteromeric NR1A/NR2 receptors but not at homomeric NR1B receptors, which contain the N-terminal insert, nor at heteromeric NR1B/NR2 receptors.[150] Zinc inhibits native NMDA receptors. The inhibitory effect is complex, being partly voltage-dependent, and may involve two separate Zn^{2+} binding sites.[151-153] A surprising finding reported for recombinant NMDA receptors was that Zn^{2+} can potentiate (as well as inhibit) responses of some homomeric NR1 receptors.[112] Potentiation is seen at concentrations of 0.1-1 μM Zn^{2+} whereas inhibition is seen with higher concentrations of Zn^{2+}. Similar to potentiation by spermine and histamine, potentiation by Zn^{2+} was only seen at homomeric receptors expressed from NR1 variants such as NR1A that lack the N-terminal insert.[112] Stimulatory effects of Zn^{2+} are not seen at heteromeric NR1/NR2 receptors. In summary, inclusion of the N-terminal insert in NR1 abolishes stimulation by spermine, histamine and Zn^{2+}. This could involve a common mechanism such as charge repulsion by the insert if the binding sites for spermine, histamine and Zn^{2+} are all located near the positively charged insert region of NR1.

B. Heteromeric NR1/NR2 Receptors

Although homomeric NR1 receptors have many of the properties of native NMDA receptors, they produce only very small whole-cell currents. In addition, responses of homomeric NR1 receptors have been observed in *Xenopus* oocytes but not in other heterologous expression systems. Heteromeric NR1/NR2 receptors generative very large responses, which makes them considerably easier to study than homomeric NR1 receptors. Furthermore, many native NMDA receptors are likely to be composed of combinations of NR1 and NR2 subunits. Most studies of recombinant NMDA receptors have focused on the properties of binary NR1/NR2 receptors such as NR1A/NR2A and NR1A/NR2B. It is possible that some native NMDA receptors contain two or more different types of NR2 subunits[128,129] or NR1 splice variants. It is impractical to study all possible combinations of NR1 and NR2 subunits (for example, there are 32 possible NR1/NR2 combinations at a 1:1 stochiometry and 48 possible NR1/NR2/NR2 combinations at a 1:1:1 stochiometry). However, studies with NR1/NR2 receptors have already provided valuable information concerning the influence of different NR2 subunits on functional and pharmacological properties of recombinant receptors that is likely to hold true for native NMDA receptors.

Glycine-independent and glycine-dependent stimulation and voltage-dependent inhibition by spermine are seen at homomeric NR1A receptors. However, these effects are differentially manifested in heteromeric NR1A/NR2 receptors containing different NR2 subunits. A fourth effect of spermine, a decrease in agonist affinity, was also uncovered in studies of NR1/NR2 receptors.

1. Glycine-Independent Stimulation and Voltage-Dependent Block

With saturating concentrations of glycine, glycine-independent stimulation and voltage-dependent inhibition can be studied at NR1A/NR2 receptors[139] (Figs. 6.9 and 6.10). At NR1A/NR2A receptors only voltage-dependent inhibition is seen in oocytes voltage-clamped at -100 to -25 mV, with a larger inhibition at more negative membrane potentials (Fig. 6.9). Oocytes are very big cells and it is difficult to voltage-clamp the membrane at positive potentials for a prolonged period because the holding current becomes large and unstable. However, voltage jumps or voltage ramps can be used to clamp oocytes briefly at very positive (and very negative) membrane potentials. Using voltage ramps, spermine was found to have no effect at NR1A/NR2A receptors at positive membrane potentials (up to +40 mV) (Fig. 6.10). Thus, the reason why spermine stimulation is not seen at NR1A/NR2A receptors is not because stimulation is masked by the voltage-dependent block, it is because the stimulation simply does not occur. At NR1A/NR2B receptors both voltage-dependent inhibition and glycine-independent stimulation are seen[139] (Figs. 6.9 and 6.10). At very negative membrane potentials the net effect of spermine is inhibitory. As the membrane potential is made more positive (and the inhibition is relieved) the stimulatory effect of spermine predominates and, with 100 μM spermine, pure stimulation is seen at positive membrane potentials

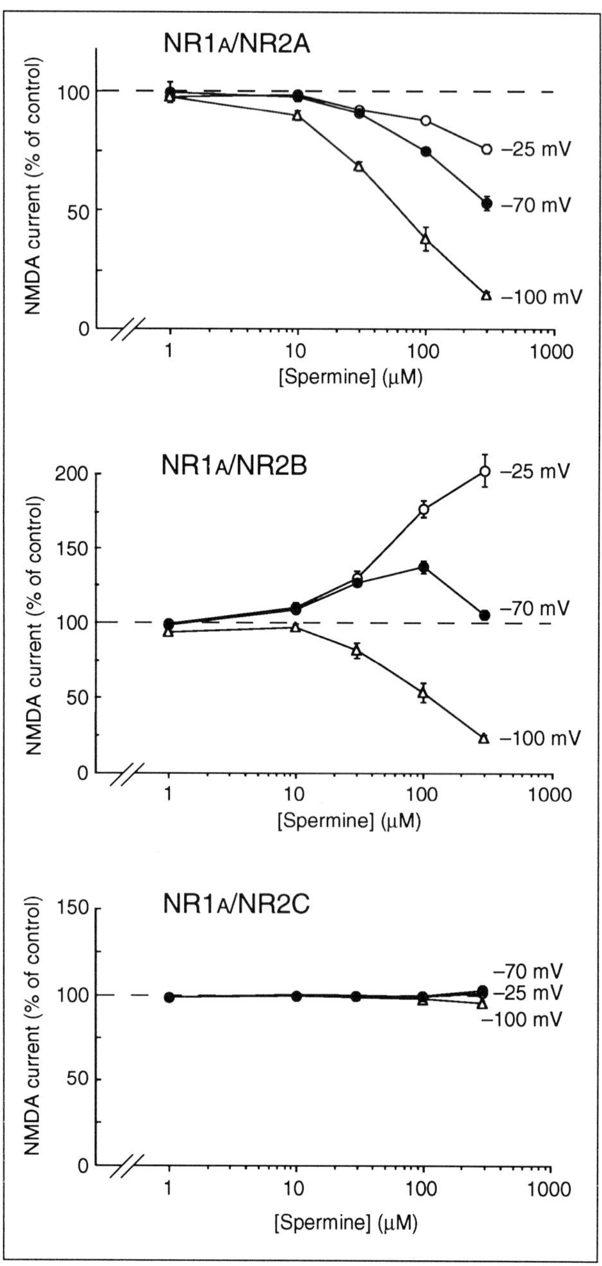

Fig. 6.9. Subunit-specific effects of spermine. The effects of spermine on responses to NMDA (100 μM with 10 μM glycine) were determined in oocytes expressing NR1A/NR2A, NR1A/NR2B, and NR1A/NR2C receptors and voltage-clamped at -25 to -100 mV. Reproduced, with permission, from K Williams et al, Mol Pharmacol 46: 531-541, 1994.

(Fig. 6.10). In contrast to NR1A/NR2A and NR1A/NR2B receptors, spermine has no effect on NR1A/NR2C receptors (Fig. 6.9) or NR1A/NR2D receptors.[139,154] Thus, neither glycine-independent stimulation nor voltage-dependent block occur at receptors containing NR2C or NR2D.

These results suggest that an NR1 variant, such as NR1A, that lacks the N-terminal insert is required for spermine stimulation, but that the manifestation of this effect is controlled by the type of NR2 subunit in heteromeric NR1A/NR2 receptors. If the binding site for spermine is located on the NR1 subunit, then its properties may be

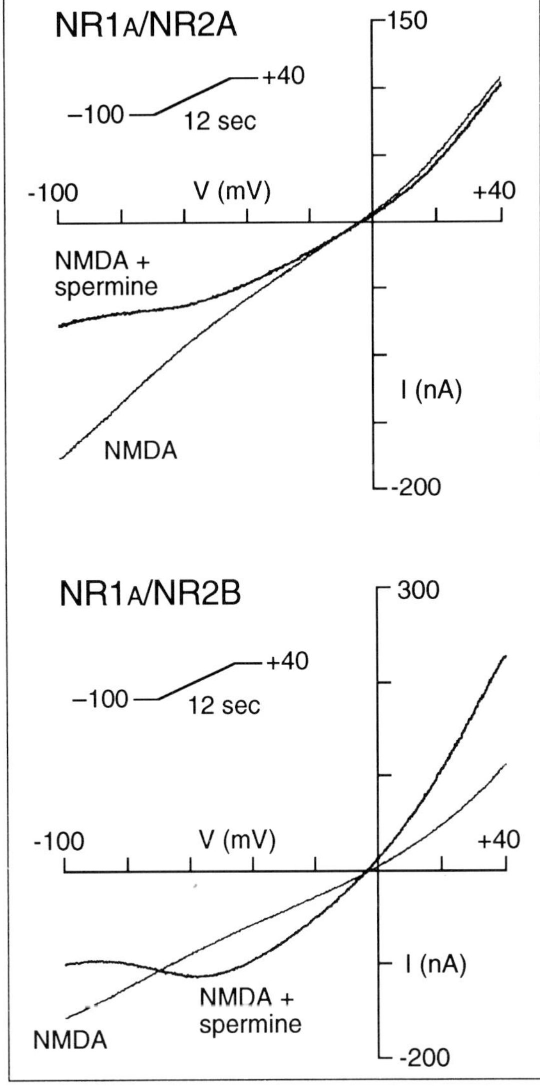

Fig. 6.10. Current-voltage (I-V) relationships. I-V curves for NMDA were measured by voltage ramps (−100 to +40 mV; 12 sec) in the absence and presence of 100 μM spermine in oocytes expressing NR1A/NR2A and NR1A/NR2B receptors. Reproduced, with permission, from K Williams et al. Mol Pharmacol 46: 531-541, 1994.

altered in heteromeric NR1/NR2 receptors resulting in a loss of glycine-independent stimulation in receptors containing NR2A, NR2C, and NR2D. This could involve an allosteric interaction that, for example, reduces the affinity of the binding site for spermine. Alternatively, the differential effects of spermine at various NR1/NR2 receptors may be due to intrinsic differences in receptor/channel properties, such as gating or desensitization, rather than to differences in the properties of the polyamine binding site. For example, spermine stimulation may occur only at NR1A/NR2B receptors because these receptors could adopt a particular gating mode that NR1A/NR2A, NR1A/NR2C, and NR1A/NR2D receptors cannot.

Ligand-binding assays have also been used to study the properties of recombinant NMDA receptors expressed in HEK 293 cells.[129,141,155-157] The effects of spermidine but not of spermine have been reported. Spermidine was found to potentiate the binding of ^{125}I-MK-801 to NR1A/NR2B but not to NR1A/NR2A receptors.[155,156] Again, this suggests that potentiation of MK-801 binding by polyamines is the biochemical correlate of glycine-independent stimulation measured electrophysiologically. Ligand binding assays with recombinant NMDA receptors should be of some use in dissecting the sites and mechanisms of action of polyamines.

The differences in sensitivity to voltage-dependent block by spermine at various NR1/NR2 receptors are reminiscent of differences in sensitivity to Mg^{2+}. Thus, voltage-dependent block by spermine and Mg^{2+} is weak or absent at NR1A/NR2C and NR1A/NR2D receptors compared with NR1A/NR2A and NR1A/NR2B receptors (see Fig. 6.12). A similar difference in sensitivity to block by the polyamine-containing spider toxin argiotoxin$_{636}$ has been reported for receptors containing NR2A, NR2B and NR2C.[158] Sensitivity to argiotoxin$_{636}$ was affected by an asparagine residue that is present in TMII of NR1, NR2A and NR2C subunits.[158] This is the same asparagine that controls Ca^{2+} permeability and sensitivity to Mg^{2+} block,[135-137] suggesting that the Mg^{2+} and argiotoxin binding sites may overlap. However, differences in the TMII amino acid sequences of NR2A and NR2C were not responsible for differences in the sensitivity of NR1/NR2A and NR1/NR2C receptors to argiotoxin$_{636}$. Thus, the regions of NR2 that control the affinity of NR1/NR2 channels for argiotoxins must lie outside the TMII region.[158] It will be interesting to carry out similar experiments to look at voltage-dependent block by spermine using mutant NR1 and NR2 subunits.

2. Glycine-Dependent Stimulation and Decrease in Agonist Affinity

The second form of spermine stimulation involves an increase in the affinity of NMDA receptors for glycine and is seen with sub-saturating concentrations of glycine. This form of stimulation (termed "glycine-dependent") occurs at homomeric NR1A receptors and at heteromeric NR1A/NR2A and NR1A/NR2B receptors (Fig. 6.11) but not at NR1A/NR2C or NR1A/NR2D receptors.[139,154] Thus, the specificity conferred by NR2 subunits is different from that of glycine-independent stimulation, which occurs only at NR1A/NR2B receptors (Fig. 6.12). Therefore, the two forms of stimulation may involve two distinct polyamine binding sites or be controlled by different structural features in the NR2 subunits. This is consistent with results from binding assays and with results of patch-clamp studies of native NMDA receptors in which glycine-dependent and glycine-independent stimulation were separable.[93]

The effects of spermine on recombinant NR1/NR2 receptors were initially characterized using high or saturating concentrations of NMDA or glutamate to activate the receptors.[139] Subsequently it was found that stimulation by spermine is dependent on the concentration of NMDA or glutamate.[159] For example, 100 μM spermine has virtually no effect on responses to 10 μM NMDA but potentiates responses to 100 μM NMDA at NR1A/NR2B receptors (Fig. 6.13A). This phenomenon is due to a decrease in the affinity of the receptor for NMDA

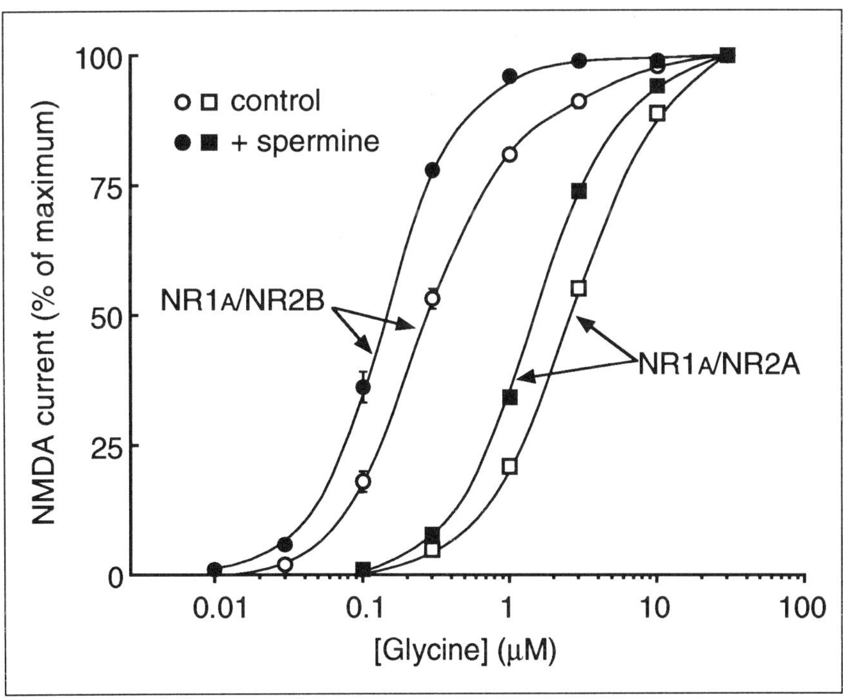

Fig. 6.11. Glycine-dependent stimulation by spermine. Concentration-response curves for glycine (all with 100 μM NMDA) were measured in the absence and presence of 100 μM spermine in oocytes expressing NR1A/NR2A and NR1A/NR2B receptors. Reproduced, with permission, from K Williams et al. Mol. Pharmacol 46:531-541, 1994.

in the presence of spermine.[159] At sub-saturating concentrations of NMDA, a decrease in affinity will reduce the size of the response and this effect balances stimulation by spermine resulting in no net effect of spermine (Fig. 6.13). The decrease in agonist affinity is seen at NR1A/NR2B but not at NR1A/NR2A, NR1A/NR2C or NR1A/NR2D receptors (Fig. 6.12). The effect may be mechanistically related to stimulation by spermine and may involve a common polyamine binding site (Fig. 6.12).

It is notable that potentiation of NMDA receptors by histamine is also dependent on agonist concentration. Potentiation by histamine, like that of spermine, is seen at NR1A/NR2B receptors activated with high but not low concentrations of NMDA and may involve a histamine-induced decrease in agonist affinity.[150] Stimulation by spermine and by histamine may have mechanistic similarities involving a net

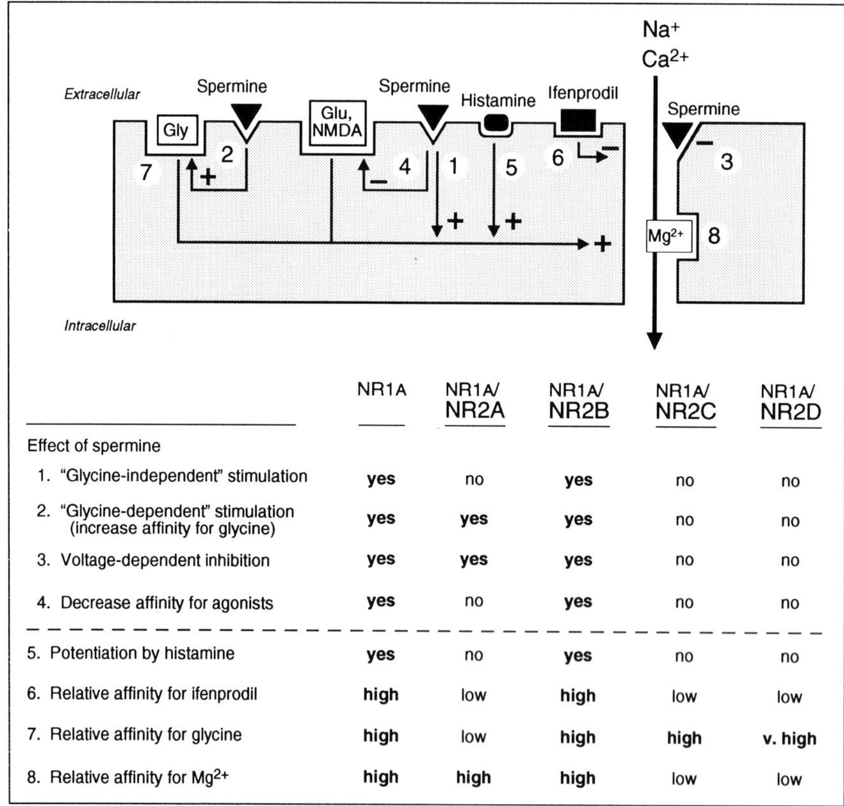

	NR1A	NR1A/ NR2A	NR1A/ NR2B	NR1A/ NR2C	NR1A/ NR2D
Effect of spermine					
1. "Glycine-independent" stimulation	**yes**	no	**yes**	no	no
2. "Glycine-dependent" stimulation (increase affinity for glycine)	**yes**	**yes**	**yes**	no	no
3. Voltage-dependent inhibition	**yes**	**yes**	**yes**	no	no
4. Decrease affinity for agonists	**yes**	no	**yes**	no	no
5. Potentiation by histamine	**yes**	no	**yes**	no	no
6. Relative affinity for ifenprodil	**high**	low	**high**	low	low
7. Relative affinity for glycine	**high**	low	**high**	**high**	**v. high**
8. Relative affinity for Mg^{2+}	**high**	**high**	**high**	low	low

Fig. 6.12. Subunit-specific modulation of recombinant NMDA receptors. A schematic model of the NMDA receptor is shown to illustrate the four macroscopic effects of spermine and the modulatory effects of histamine and ifenprodil. The effects of spermine (1-4), histamine (5), ifenprodil (6), and the relative affinities for glycine (7) and Mg^{2+} (8) are compared at homomeric NR1A receptors and heteromeric NR1A/NR2 receptors containing NR2A, NR2B, NR2C and NR2D.

increase in channel "activity" (frequency, bursting, or open time) with a concomitant decrease in apparent affinity for agonists.

C. Relationship to Native NMDA Receptors

The results that are summarized in Fig. 6.12 deal with the macroscopic effects of spermine on NMDA receptors. Very little is known about the mechanisms underlying these effects. For example, what changes in channel gating properties are responsible for glycine-independent stimulation? Why is stimulation seen at NR1A/NR2B receptors but not at receptors containing NR2A, NR2C and NR2D? It is presumed that stimulation by spermine involves an allosteric effect of spermine on the conformation of the receptor/channel protein that somehow increases the frequency of channel opening.[95] A consequence of this allosteric effect may be a change in the affinity of the binding site for glutamate. This would explain why a decrease in agonist affinity is seen at the same receptor subtypes (NR1A and NR1A/NR2B) at which spermine produces glycine-independent stimulation (Fig. 6.12). Similarly, a change in the conformation of the protein presumably underlies the increase in affinity for glycine, although this may involve a second polyamine binding site (Fig. 6.12). The ability of spermine to potentiate responses at NR1A/NR2B receptors but not at other NR1A/NR2 receptors may mean that NR1A/NR2B receptors can undergo a particular conformational change that the other receptors cannot, regardless of the presence of a polyamine binding site of those receptors. The same mechanism could hold true for potentiation by histamine, which acts at a discrete binding site but has the same subunit-specificity as spermine (Fig. 6.12). Future studies using mutant and chimeric NR1 and NR2 subunits will, no doubt, identify regions of the receptor subunits that control sensitivity to spermine and other modulators. This information, together with knowledge of the tertiary and quaternary structures of the receptor proteins and of the mechanisms involved in channel gating, should eventually make it possible to determine how spermine does (or does not) alter receptor activity.

The subunit-specific effects of spermine on recombinant NMDA receptors provide an explanation for some of the confusing and apparently contradictory effects of spermine seen at native NMDA receptors where stimulation by spermine varies widely between individual neurons. The regional and temporal expression of various NR2 subunits and NR1 splice variants in brain and in cultured neurons is probably responsible for the existence of receptors having different sensitivities to polyamines. In the cerebral cortex and hippocampus NR1 variants lacking the N-terminal insert (i.e. sensitive to polyamine stimulation) are the major NR1 species.[114,115] In these brain regions, expression of different NR2 subunits probably controls sensitivity to polyamines.[139] It is also possible that only NR1 variants with the N-terminal insert (i.e. not sensitive to polyamine stimulation) are expressed in subpopulations of neurons. In areas such as the cerebellum where there is a greater proportion of NR1 variants with the N-terminal insert, alternative splicing of NR1 and expression of NR2 subunits may both be major determinants of sensitivity to polyamines.

Studies with recombinant NMDA receptors have shed some light on the relationship between ifenprodil and polyamines. Ifenprodil is a

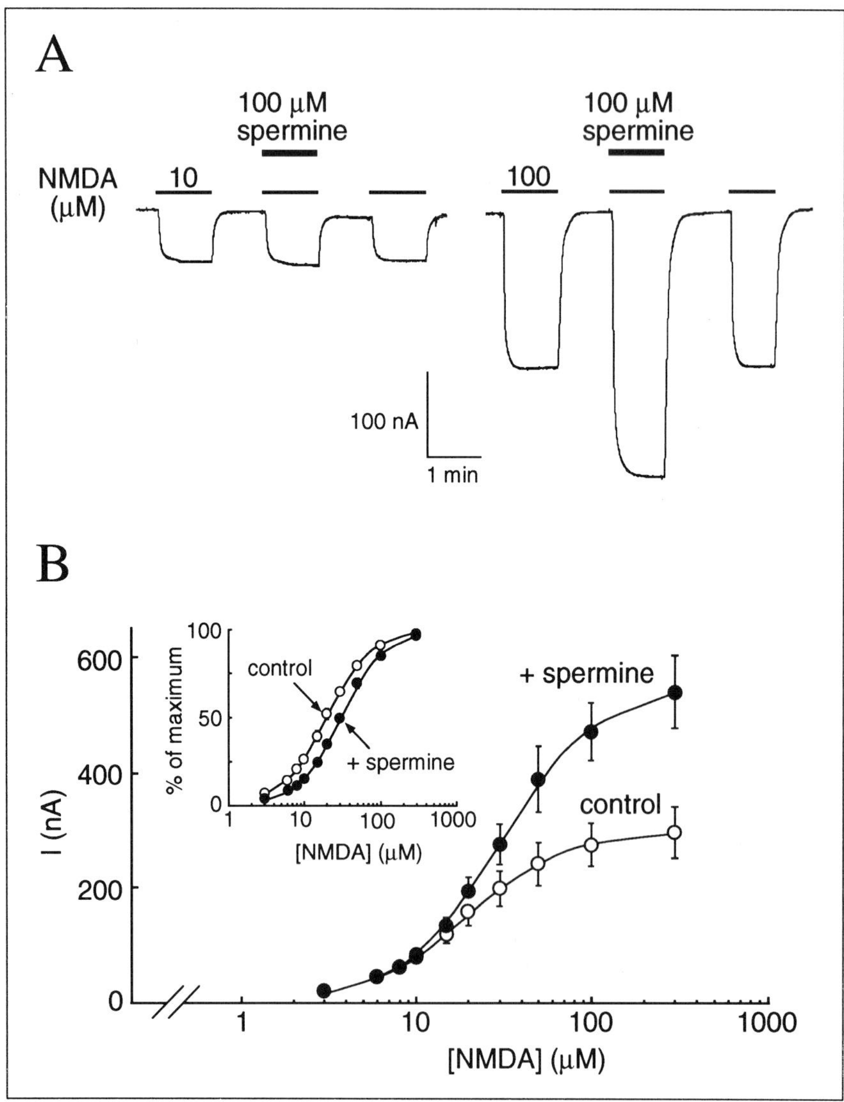

Fig. 6.13. Spermine decreases the affinity of NR1A/NR2B receptors for NMDA. A, inward currents induced by 10 and 100 μM NMDA in the absence and presence of 100 μM spermine in an oocyte expressing NR1A/NR2B receptors and voltage-clamped at −25 mV. B, Concentration-response curves for NMDA measured in the absence and presence of 100 μM spermine; I is the NMDA-induced current (nA). Inset: the same data are expressed as a percentage of the maximum response to NMDA or NMDA + spermine. Data in B are reproduced, with permission, from K Williams, Mol. Pharmacol. 46:161-168, 1994.

phenylethanolamine that is neuroprotective in vivo and acts as an antagonist at NMDA receptors.[160] Ifenprodil was proposed to act as a competitive antagonist at the stimulatory polyamine site.[87,161] However, such a mechanism seems unlikely because antagonist effects of ifenprodil are seen in the absence of extracellular polyamines.[140,162,163] Because ifenprodil is a high-affinity antagonist only at NR1A and NR1A/NR2B receptors and these receptors are sensitive to glycine-independent stimulation by spermine (Fig. 6.12), interactions between ifenprodil and polyamines at native NMDA receptors are probably coincidental and receptor subtype-dependent.

VI. IMPLICATIONS FOR THE IN VIVO BIOLOGY OF POLYAMINES

Polyamines are present in high concentrations (500-1500 nmoles/g) in the brain with a nonuniform regional distribution.[164,165] It is possible that neuronal polyamines are involved only in biochemical and cellular processes such as RNA and protein synthesis that are common to all types of cells. On the other hand, polyamines could play a role as neuromodulators in the brain (Fig. 6.14). There is as yet no direct evidence of a role for polyamines in synaptic transmission. Uptake and

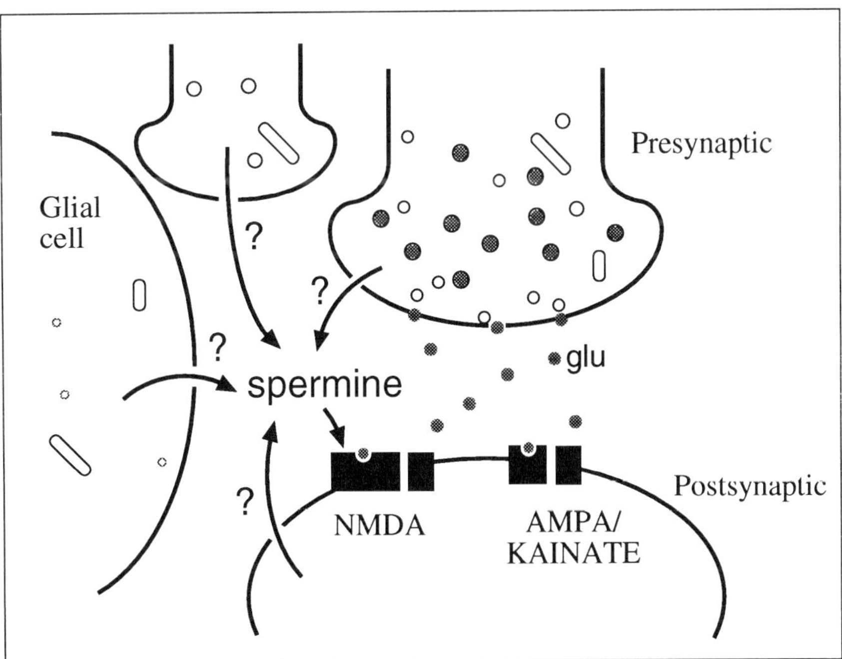

Fig. 6.14. Spermine at excitatory synapses. At glutamatergic synapses, spermine could be released from pre- or postsynaptic neurons, from glial cells, or from neighboring neurons to potentiate activity of NMDA receptors. This scheme is speculative; there is as yet no direct evidence for modulation of NMDA receptors by polyamines in vivo.

depolarization-induced release of polyamines from brain tissue has been reported[166-169] but the specificity of these effects is largely unknown. Polyamines could potentiate the activity of NMDA receptors if they are released from pre- or postsynaptic neurons, from glial cells, or from nearby neurons (Fig. 6.14), although it is unclear whether polyamines released from neurons or glial cells would reach extracellular concentrations (10-1000 μM) sufficient to affect NMDA receptors. Nonetheless, the NMDA receptor may represent the first identifiable target for polyamines at brain synapses.

The question of whether endogenous compounds that modulate NMDA receptors in vitro actually affect NMDA receptors in vivo is not unique to polyamines. Many of the question marks shown in Fig. 6.14 apply to other modulators such as histamine, neurosteroids, and Zn^{2+}. Histamine has a well-documented role as a neurotransmitter, and histaminergic innervation is widespread in the brain.[170] However, it is not known whether neuronally released histamine acts on NMDA receptors in vivo. In the case of Zn^{2+}, the cation is known to be present in high concentrations in some areas of the brain including hippocampal areas that are rich in NMDA receptors, and neuronal uptake and release of Zn^{2+} have been described[171-173] as has modulation of GABA-mediated inhibitory transmission by Zn^{2+}.[174] It is possible that a variety of endogenous ligands, including polyamines, influence synaptic transmission in the CNS by altering the properties of postsynaptic glutamate and GABA receptors. Such effects could be region-specific, receptor-specific, or even synapse-specific depending on the modulator in question. Endogenous ligands that modulate postsynaptic receptors may provide a "gain control" for synaptic transmission or may function as synaptic lubricants, smoothing the onset, magnitude and duration of synaptic responses. An experimental challenge is to define what role, if any, the growing numbers of modulators have in synaptic function.

Spermine has four macroscopic effects on NMDA receptors (Fig. 6.12). If endogenous spermine acts on NMDA receptors at intact synapses its net effect will depend on the concentrations of glutamate, glycine, and spermine, and the membrane potential, as well as the types of subunits present in the receptor complex. Given these variables, it is difficult to predict the effects of polyamines on NMDA receptors at intact synapses. However, the voltage-dependent block caused by 100 μM spermine is masked in the presence of physiological concentrations of Mg^{2+}.[139] Assuming that glycine is present at saturating concentrations in the synapse,[175] the predominant effects of spermine may involve glycine-independent stimulation and a decrease in the affinity for glutamate. Thus, at receptors containing the NR2B subunit, high micromolar concentrations of spermine would potentiate responses to high concentrations of glutamate. If, during synaptic transmission, the concentration of glutamate in the synaptic cleft is high but declines very rapidly then the duration of postsynaptic currents will be controlled by the kinetic and activation properties of NMDA receptors and the slow unbinding of glutamate from the receptors.[143-145] Under conditions where glutamate is present at high concentrations for a very brief time, an effect of spermine on the affinity of NMDA receptors for glutamate could be manifest as an increase in the rate of dissociation of glutamate from the receptor and subsequent shortening of the NMDA component of the postsynaptic potential. Such an effect, together with the stimulatory effect of spermine, could serve to alter the magnitude and duration of NMDA receptor-mediated responses. This could have an important influence on the induction of processes such as LTP and LTD that are dependent not only on the magnitude of NMDA responses but also on their time course. In this regard, an increase in the magnitude of NMDA receptor-mediated synaptic currents together with a decrease in the duration of the response (exactly the effects predicted for spermine) have been reported for histamine acting at excitatory synapses.[176] It will clearly be of interest to look at effects of spermine in the same experimental paradigm.

The NR2B subunit, which forms receptors that are sensitive to stimulation by spermine, is the predominant NR2 subunit in embryonic and neonatal brain.[122,123] Variants of NR1 that lack the N-terminal insert (i.e. sensitive to spermine stimulation) are the predominant forms of NR1 in neonatal brain.[114,115] Therefore, NMDA receptors in embryonic and neonatal brain may be particularly sensitive to the stimulatory effects of polyamines. This, of course, is the developmental period of cell proliferation and migration in the CNS and a period when levels of polyamines and activity of ornithine decarboxylase are higher than in adult animals.[177] It is possible that polyamines acting on NMDA receptors could influence neuronal growth, migration and synaptogenesis during development.[178-181]

Excessive activation of NMDA receptors leads to neurodegeneration. Spermine has been found to potentiate neurotoxic effects of glutamate on cultured neurons, an effect that may involve potentiation of NMDA receptor activity.[182] It is conceivable that endogenous polyamines could contribute to excitotoxicity in vivo and that blocking the stimulatory polyamine site on NMDA receptors after ischemia or hypoglycemia would reduce the neurotoxic potential of glutamate. It is very difficult to test these hypotheses without antagonists to selectively block the stimulatory polyamine site on NMDA receptors. A number of studies have described neurotoxic effects following injection of polyamines directly into the brain but the mechanism and specificity of such effects is entirely unknown; those studies are not discussed here.

Another area of polyamine biology that is relevant to NMDA receptors concerns the use of polyamine analogs as anti-tumor agents. A number of *bis*(ethyl)polyamines have been developed as potential antitumor agents.[183-186] Some *bis*(ethyl)polyamines such as BE4444 (Fig. 6.6) are potent antagonists at NMDA receptors.[98] The block seen with *bis*(ethyl)polyamines is unrelated to the stimulatory effect of spermine and represents a voltage-dependent channel block similar to that seen with spider toxins.[98] Highly charged organic cations such as BE4444 may not easily cross the blood-brain barrier under normal conditions and would, therefore, be expected to have little effect on NMDA receptors in the brain and spinal cord when administered systemically. However, it is conceivable that *bis*(ethyl)polyamines could antagonize receptors in the CNS under conditions where the integrity of the blood-brain barrier is compromised by, for example, brain tumors. *Bis*(ethyl)polyamines could therefore have a range of potentially serious side effects due to blockade of NMDA receptors.

VII. CONCLUSIONS

The use of cloned subunits is of considerable value in understanding the properties, diversity and regulation of NMDA receptors. Studies of these receptors at the molecular level are in the early stages and much remains to be uncovered. With regard to polyamines, the current status of knowledge about polyamine/NMDA receptor interactions is summarized below:

1) Spermine has four macroscopic effects on NMDA receptors that involve at least two distinct polyamine binding sites.
2) The type of NR1 splice variant and the type of NR2 subunit present in NMDA receptors control sensitivity to polyamines. In the hippocampus and cerebral cortex differential expression of NR2 subunits is probably the major factor that influences sensitivity to polyamines.
3) The structural features that control sensitivity of NMDA receptors to polyamines may also control sensitivity to histamine.
4) If endogenous polyamines act on some subtypes of NMDA receptors in the brain their major effect could be to increase the size and decrease the duration of synaptic responses mediated by NMDA receptors.

A more detailed understanding of the sites and mechanisms of action of spermine at NMDA receptors should come quite rapidly given the combined power of molecular biology and electrophysiology to study the receptors. Questions about the role of polyamines in synaptic function and whether endogenous polyamines affect NMDA receptors in vivo remain largely ignored. These are interesting and important questions that deserve some attention now that potential targets for polyamines have been identified. A major obstacle is the lack of selective antagonists to block stimulatory effects of polyamines on NMDA receptors.

Detailed and rigorous investigations of the localization of polyamines, of polyamine transport systems, and of the enzymes involved in polyamine biosynthesis and metabolism in the brain are also an essential step in determining what polyamines do in the nervous system. As other chapters in this volume indicate, a great deal of knowledge has accumulated in recent years concerning the biosynthesis and metabolism of polyamines in eukaryotes and prokaryotes, and many of the genes encoding key enzymes involved in polyamine metabolism have been cloned. Thus, a number of chemical and genetic tools are available to study the functions of polyamines in the brain and could be used to answer some of the questions illustrated in Fig. 6.14.

Polyamines may act on other ion channels in the nervous system in addition to glutamate receptors[4-6] A recently described example involves inward rectifier K^+ (IRK) channels.[5,187,188] Very low concentrations of spermine and spermidine (10-100 nM) were found to have dramatic effects on the gating properties of these channels in inside-out membrane patches studied by patch-clamp recording.[5] That is, polyamines act at the intracellular side on these channels to alter their function. Thus, intracellular polyamines may normally function to control gating and rectification of IRK channels.[5,187,188] If this is so, gating of IRK channels could be altered by conditions in which polyamine levels are disturbed or by intracellular polyamine analogs used in experimental or clinical settings to interfere with polyamine systems. These observations raise the question of whether polyamines may act at intracellular sites on other types of ion channels (including NMDA receptors!) and this is likely to become an area of intense investigation over the next few years.

References

1. Dudley HW, Rosenheim O. Notes on spermine. Biochem J 1925; 19:1034-1036.
2. Ransom RW, Stec NL. Cooperative modulation of [^{3}H]MK-801 binding to the *N*-methyl-D-aspartate receptor-ion channel complex by L-glutamate, glycine, and polyamines. J Neurochem 1988;51:830-836.
3. Williams K, Romano C, Molinoff PB. Effects of polyamines on the binding of [^{3}H]MK-801 to the *N*-methyl-D-aspartate receptor: pharmacological evidence for the existence of a polyamine recognition site. Mol Pharmacol 1989;36:575-581.
4. Scott RH, Sutton KG, Dolphin AC. Interactions of polyamines with neuronal ion channels. Trends Neurosci 1993;16:153-160.
5. Ficker E, Taglialatela M, Wilbe BA et al. Spermine and spermidine as gating molecules for inward rectifier K^+ channels. Science 1994;266:1068-1072.
6. Weiger T, Hermann A. Polyamines block Ca^{2+}-activated K^+ channels in pituitary tumor cells (GH3). J Membrane Biol 1994;140:133-142.
7. Collingridge GL, Lester RAJ. Excitatory amino acid receptors in the vertebrate central nervous system. Pharmacol Rev 1989;41:143-210.

8. Mayer ML, Westbrook GL. The physiology of excitatory amino acids in the vertebrate central nervous system. Prog Neurobiol 1987;28:197-276.
9. Hollmann M, Heinemann S. Cloned glutamate receptors. Annu Rev Neurosci 1994;17:31-108.
10. Nakanishi S. Molecular diversity of glutamate receptors and implications for brain function. Science 1992;258:597-603.
11. Seeburg PH. The molecular biology of mammalian glutamate receptor channels. Trends Neurosci 1993;16:359-365.
12. Watkins JC, Krogsgaard-Larsen P, Honoré T. Structure-activity relationships in the development of excitatory amino acid receptor agonists and competitive antagonists. Trends Pharmacol Sci 1990; 11:25-33.
13. McBain CJ, Mayer ML. *N*-Methyl-D-aspartic acid receptor structure and function. Physiol Rev 1994;74:723-760.
14. Jones KA, Baughman RW. Both NMDA and non-NMDA subtypes of glutamate receptors are concentrated at synapses on cerebral cortical neurons in culture. Neuron 1991;7:593-603.
15. Bekkers JM, Stevens CF. NMDA and non-NMDA receptors are co-localized at individual excitatory synapses in cultured rat hippocampus. Nature (Lond) 1989;341:230-233.
16. Bliss TVP, Collingridge GL. A synaptic model of memory: long-term potentiation in the hippocampus. Nature (Lond) 1993;361:31-39.
17. Choi DW, Rothman SM. The role of glutamate neurotoxicity in hypoxic-ischemic neuronal death. Annu Rev Neurosci 1990;13:171-182.
18. Choi DW. Glutamate neurotoxicity and diseases of the nervous system. Neuron 1988;1:623-634.
19. Collingridge GL, Singer W. Excitatory amino acid receptors and synaptic plasticity. Trends Pharmacol Sci 1990;11:290-296.
20. Meldrum B, Garthwaite J. Excitatory amino acid neurotoxicity and neurodegenerative disease. Trends Pharmacol Sci 1990;11:379-387.
21. Olney J. Excitotoxic amino acids and neuropsychiatric disorders. Annu Rev Pharmacol Toxicol 1990;30:47-71.
22. Shatz CJ. Impulse activity and the patterning of connections during CNS development. Neuron 1990;5:745-756.
23. Constantine-Paton M, Cline HT, Debski E. Patterned activity, synaptic convergence, and the NMDA receptor in developing visual pathways. Annu Rev Neurosci 1990;13:129-154.
24. Brose N, Gasic GP, Vetter DE et al. Protein chemical characterization and immunocytochemical localization of the NMDA receptor subunit NMDA R1. J Biol Chem 1993;268:22663-22671.
25. Petralia RS, Yokotani N, Wenthold RJ. Light and electron microscope distribution of the NMDA receptor subunit NMDAR1 in the rat nervous system using a selective anti-peptide antibody. J Neurosci 1994;14:667-696.
26. Nowak L, Bregestovski P, Ascher P et al. Magnesium gates glutamate-activated channels in mouse central neurons. Nature (Lond) 1984; 307:462-465.

27. Mayer ML, Westbrook GL, Guthrie. PB. Voltage-dependent block by Mg^{2+} of NMDA responses in spinal cord neurones. Nature (Lond) 1984;309:261-263.
28. Nicoll RA, Kauer JA, Malenka RC. The current excitement in long-term potentiation. Neuron 1988;1:97-103.
29. Artola A, Bröcher S, Singer W. Different voltage-dependent thresholds for inducing long-term depression and long-term potentiation in slices of rat visual cortex. Nature (Lond) 1990;347:69-72.
30. Dudek SM, Bear MF. Homosynaptic long-term depression in area CA1 of hippocampus and effects of *N*-methyl-D-aspartate receptor blockade. Proc Natl Acad Sci USA 1992;89:4363-4367.
31. Mulkey RM, Malenka RC. Mechanisms underlying induction of homosynaptic long-term depression in area CA1 of the hippocampus. Neuron 1992;9:967-975.
32. Linden DJ. Long-term synaptic depression in the mammalian brain. Neuron 1994;12:457-472.
33. Haley JE, Wilcox GL, Chapman PF. The role of nitric oxide in hippocampal long-term potentiation. Neuron 1992;8:211-216.
34. Schuman EM, Madison DV. A requirement for the intercellular messenger nitric oxide in long-term potentiation. Science 1991;254:1503-1506.
35. Lisman J. The CaM kinase II hypothesis for the storage of synaptic memory. Trends Neurosci 1994;17:406-412.
36. Kleinschmidt A, Bear MF, Singer W. Blockade of "NMDA" receptors disrupts experience-dependent plasticity of kitten striate cortex. Science 1987;238:355-358.
37. Schlaggar BL, Fox K, O'Leary DDM. Postsynaptic control of plasticity in developing somatosensory cortex. Nature (Lond) 1993;364:623-626.
38. Rabacchi S, Bailly Y, Delhaye-Bouchard N et al. Involvement of *N*-methyl-D-aspartate (NMDA) receptor in synapse elimination during cerebellar development. Science (Washington DC) 1992;256:1823-1825.
39. Choi DW. Glutamate neurotoxicity in cortical cell culture. J Neurosci 1987;7:357-368.
40. Dingledine R, McBain CJ, McNamara JO. Excitatory amino acid receptors in epilepsy. Trends Pharmacol Sci 1990;11:334-338.
41. Rogawski MA. The NMDA receptor, NMDA antagonists and epilepsy therapy. A status report. Drugs 1992;44:279-292.
42. Young AB, Greenamyre JT, Hollingsworth Z et al. NMDA receptor losses in putamen from patients with Huntington's disease. Science 1988; 241:981-983.
43. Turski L, Bressler K, Rettig K-J et al. Protection of substantia nigra from MPP^+ neurotoxicity by *N*-methyl-D-aspartate antagonists. Nature (Lond) 1991;349:414-418.
44. Spencer PS, Nunn PB, Hugon J et al. Guam amyotrophic lateral sclerosis-parkinsonism-dementia linked to a plant excitant neurotoxin. Science 1987;237:517-522.

45. Carlsson M, Carlsson A. Interactions between glutamatergic and monoaminergic systems within the basal ganglia-implications for schizophrenia and Parkinson's disease. Trends Neurosci 1990;13:272-276.
46. Williams K, Romano C, Dichter MA et al. Minireview: Modulation of the NMDA receptor by polyamines. Life Sci 1991;48:469-498.
47. Wang YT, Salter MW. Regulation of NMDA receptors by tyrosine kinases and phosphatases. Nature (Lond) 1994;369:233-235.
48. Urushihara H, Tohda M, Nomura Y. Selective potentiation of *N*-methyl-D-aspartate-induced current by protein kinase C in *Xenopus* oocytes injected with rat brain RNA. J Biol Chem 1992;267:11697-11700.
49. Kelso SR, Nelson TE, Leonard JP. Protein kinase C-mediated enhancement of NMDA currents by metabotropic glutamate receptors in *Xenopus* oocytes. J Physiol (Lond) 1992;449:705-718.
50. Chen L, Huang L-YM. Protein kinase C reduces Mg^{2+} block of NMDA-receptor channels as a mechanism of modulation. Nature (Lond) 1992;356:521-523.
51. Kleckner NW, Dingledine R. Requirement for glycine in activation of NMDA-receptors expressed in *Xenopus* oocytes. Science 1988;241:835-837.
52. Sills MA, Fagg G, Pozza M et al. [^{3}H]CGP 39653: a new *N*-methyl-D-aspartate antagonist radioligand with low nanomolar affinity in rat brain. Eur J Pharmacol 1991;192:19-24.
53. Murphy DE, Schneider J, Boehm C et al. Binding of [^{3}H]3-(2-carboxypiperazin-4-yl)propyl-1-phosphonic acid to rat brain membranes: a selective, high affinity ligand for *N*-methyl-D-aspartate receptors. J Pharmacol Exp Ther 1987;240:778-784.
54. Baron BM, Siegel BW, Slone AL et al. [^{3}H]5,7-Dichlorokynurenic acid, a novel radioligand labels NMDA receptor-associated glycine binding sites. Eur J Pharmcol (Mol Pharmacol Sec) 1991;206:149-154.
55. Snell LD, Morter RS, Johnson KM. Glycine potentiates *N*-methyl-D-aspartate-induced [^{3}H]TCP binding to rat cortical membranes. Neurosci Lett 1987;83:313-317.
56. Williams K, Hanna JL, Molinoff PB. Developmental changes in the sensitivity of the *N*-methyl-D-aspartate receptor to polyamines. Mol Pharmacol 1991;40:774-782.
57. Wong EHF, Knight AR, Woodruff GN. [^{3}H]MK-801 labels a site on the N-methyl-D-aspartate receptor channel complex in rat brain membranes. J Neurochem 1988;50:274-281.
58. Reynolds IJ, Murphy SN, Miller RJ. ^{3}H-labeled MK-801 binding to the excitatory amino acid receptor complex from rat brain is enhanced by glycine. Proc Natl Acad Sci USA 1987;84:7744-7748.
59. Reynolds IJ, Miller RJ. Multiple sites for the regulation of the *N*-methyl-D-aspartate receptor. Mol Pharmacol 1988;33:581-584.
60. Mayer ML, Miller RJ. Excitatory amino acid receptors, second messengers and regulation of intracellular Ca^{2+} in mammalian neurons. Trends Pharmacol Sci 1990;11:254-260.
61. Williams K, Russell SL, Shen YM et al. Developmental switch in the expression of NMDA receptors occurs in vivo and in vitro. Neuron

1993;10:267-278.

62. Zhong J, Russell SL, Pritchett DB et al. Expression of mRNAs encoding subunits of the *N*-methyl-D-aspartate receptor in cultured cortical neurons. Mol Pharmacol 1994;45:846-853.
63. Snutch TP. The use of *Xenopus* oocytes to probe synaptic communication. Trends Neurosci 1988;11:250-256.
64. Verdoorn TA, Kleckner NW, Dingledine R. Rat brain *N*-methyl-D-aspartate receptors expressed in *Xenopus* oocytes. Science 1987;238:1114-1116.
65. Verdoorn TA, Kleckner NW, Dingledine R. *N*-methyl-D-aspartate/glycine and quisqualate/kainate receptors expressed in *Xenopus* oocytes :antagonist pharmacology. Mol Pharmacol 1989;35:360-368.
66. Verdoorn TA, Dingledine R. Excitatory amino acid receptors expressed in *Xenopus* oocytes: agonist pharmacology. Mol Pharmacol 1988; 24:298-307.
67. Leonard JP, Kelso SR. Apparent desensitization of NMDA responses in *Xenopus* oocytes involves calcium-dependent chloride current. Neuron 1990;2:53-60.
68. McGurk JF, Bennett MVL, Zukin RS. Polyamines potentiate responses of *N*-methyl-D-aspartate receptors expressed in *Xenopus* oocytes. Proc Natl Acad Sci (USA) 1990;87:9971-9974.
69. Kleckner NW, Dingledine RJ. Regulation of hippocampal NMDA receptors by magnesium and glycine during development. Mol Brain Res 1991;11:151-159.
70. Anis N, Sherby S, Goodnow RJ et al. Structure-activity relationships of philanthotoxin analogs and polyamines on *N*-methyl-D-aspartate and nicotinic acetylcholine receptors. J Pharmacol Exp Ther 1990;254:764-773.
71. Nussenzveig IZ, Sircar R, Wong M-L et al. Polyamine effects upon *N*-methyl-D-aspartate receptor functioning: differential alteration by glutamate and glycine site antagonists. Brain Res 1991;561:285-291.
72. Reynolds IJ, Miller, R. J. Ifenprodil is a novel type of *N*-methyl-D-aspartate receptor antagonist: interaction with polyamines. Mol Pharmacol 1989;36:758-765.
73. Reynolds IJ. Arcaine uncovers dual interactions of polyamines with the N-methyl-D-aspartate receptor. J Pharmacol Exp Ther 1990; 255:1001-1007.
74. Sacaan AI, Johnson KM. Characterization of the stimulatory and inhibitory effects of polyamines on [^{3}H]N-(1-[thienyl]cyclohexyl)piperidine binding to the *N*-methyl-D-aspartate receptor ionophore complex. Mol Pharmacol 1990;37:572-577.
75. Yoneda Y, Ogita K, Enomoto R. Characterization of spermidine-dependent [^{3}H](+)-5-methyl-10,11-dihydro-5H-dibenzo[a,d]cyclohepten-5,10-imine (MK-801) binding in brain synaptic membranes treated with triton X-100. J Pharmacol Exp Ther 1991;256:1161-1172.
76. Romano C, Williams K. Modulation of NMDA receptors by polyamines. In: C. Carter, ed. The Neuropharmacology of Polyamines. London: Academic Press, 1994:81-106.

77. Romano C, Williams K, DePriest S et al. Effects of mono-, di-, and triamines on the *N*-methyl-D-aspartate receptor complex: a model of the polyamine recognition site. Mol Pharmacol 1992; 41:785-792.
78. Williams K, Dawson VL, Romano C et al. Characterization of polyamines having agonist, antagonist, and inverse agonist effects at the polyamine recognition site of the NMDA receptor. Neuron 1990;5:199-208.
79. Bonhaus DW, McNamara JO. *N*-methyl-D-aspartate receptor regulation of uncompetitive antagonist binding in rat brain membranes: kinetic analysis. Mol Pharmacol 1988;34:250-255.
80. Kloog Y, Haring R, Sokolovsky M. Kinetic characterization of the phencyclidine-*N*-methyl-D-aspartate receptor interaction: evidence for a steric blockade of the channel. Biochemistry 1988;27:843-848.
81. Romano C, Williams K, Molinoff PB. Polyamines modulate the binding of [^{3}H]MK-801 to the solubilized *N*-methyl-D-aspartate receptor. J Neurochem 1991;57:811-818.
82. Ogita K, Yoneda Y. Solubilization of spermidine-sensitive (+)-[^{3}H]5-methyl-10,11-dihydro-5H-dibenzo[a,d]cyclohepten-5,10-imine ([^{3}H]MK-801) binding activity from rat brain. J Neurochem 1990;55:1515-1520.
83. Bakker MHM, McKernan RM, Wong EHF et al. [^{3}H]MK-801 binding to *N*-methyl-D-aspartate receptors solubilized from rat brain: effects of glycine site ligands, polyamines, ifenprodil, and desipramine. J Neurochem 1991;57:39-45.
84. Reynolds IJ. [^{3}H]CGP 39653 binding to the agonist site of the *N*-methyl-D-aspartate receptor is modulated by Mg^{2+} and polyamines independently of the arcaine-sensitive polyamine site. 1994;62:54-62.
85. Ransom RW, Deschenes NL. Polyamines regulate glycine interaction with the *N*-methyl-D-aspartate receptor. Synapse 1990;5:294-298.
86. Sacaan AI, Johnson KM. Spermine enhances binding to the glycine site associated with the *N*-methyl-D-aspartate receptor complex. Mol Pharmacol 1989;36:836-839.
87. Carter CJ, Lloyd KG, Zivkovic B et al. Ifenprodil and SL 82.0715 as cerebral antiischemic agents. III. Evidence for antagonistic effects at the polyamine modulatory site within the *N*-methyl-D-aspartate receptor complex. J Pharmacol Exp Ther 1990;253:475-482.
88. Pullan LM, Britt M, Chapdelaine MJ et al. Stereoselectivity for the (R)-enantiomer of HA-966 (1-hydroxy-3-aminopyrrolidone-2) at the glycine site of the *N*-methyl-D-aspartate receptor complex. J Neurochem 1990;55:1346-1351.
89. Williams K, Pullan LM, Romano C et al. An antagonist/partial agonist at the polyamine recognition site of the *N*-methyl-D-aspartate receptor that alters the properties of the glutamate recognition site. J Pharmacol Exp Ther 1992;262:539-544.
90. Reynolds IJ. Arcaine is a competitive antagonist of the polyamine site on the NMDA receptor. Eur J Pharmacol 1990;177:215-216.
91. Donevan SD, Jones SM, Rogawski MA. Arcaine blocks *N*-methyl-D-as-

partate receptor responses by an open-channel mechanism: whole-cell and single-channel recording studies in cultured hippocampal neurons. Mol Pharmacol 1992;41:727-735.

92. Rock DM, Macdonald RL. The polyamine diaminodecane (DA-10) produces a voltage-dependent flickery block of single NMDA receptor channels. Neurosci Lett 1992;144:111-115.
93. Benveniste M, Mayer ML. Multiple effects of spermine on *N*-methyl-D-aspartatic acid receptor responses of rat cultured hippocampal neurons. J Physiol (Lond) 1993;464:131-163.
94. Lerma J. Spermine regulates *N*-methyl-D-aspartate receptor desensitization. Neuron 1992;8:343-352.
95. Rock DM, MacDonald RL. The polyamine spermine has multiple actions on *N*-methyl-D-aspartate receptor single-channel currents in cultured cortical neurons. Mol Pharmacol 1992;41:83-88.
96. Araneda RC, Zukin RS, Bennett MVL. Effects of polyamines on NMDA-induced currents in rat hippocampal neurons: a whole-cell and single-channel study. Neurosci Lett 1993;152:107-112.
97. Rock DM, Macdonald RL. Spermine and related polyamines produce a voltage-dependent reduction of *N*-methyl-D-aspartate receptor single-channel conductance. Mol Pharmacol 1992;42:157-164.
98. Igarashi K, Williams K. Antagonist properties of polyamines and *bis*(ethyl)polyamines at *N*-methyl-D-aspartate receptors. J Pharmacol Exp Ther 1995;272:1101-1109.
99. Usherwood PNR, Blagbrough IS. Spider toxins affecting glutamate receptors: Polyamines in therapeutic neurochemistry. Pharm Ther 1991; 52:245-268.
100. Jackson H, Usherwood PNR. Spider toxins as tools for dissecting elements of excitatory amino acid transmission. Trends Neurosci 1988; 11:278-283.
101. Jackson H, Parks T. Spider toxins: recent applications in neurobiology. Annu Rev Neurosci 1989;12:405-414.
102. Williams K. Effects of Agelenopsis aperta toxins on the *N*-methyl-D-aspartate receptor: polyamine-like and high-affinity antagonist actions. J Pharmacol Exp Ther 1993;266:231-236.
103. Priestley T, Woodruff GN, Kemp JA. Antagonism of responses to excitatory amino acids on rat cortical neurones by the spider toxin, argiotoxin$_{636}$. Br J Pharmacol 1989;97:1315-1323.
104. Parks TN, Mueller AL, Artman LD et al. Arylamine toxins from funnel-web spider (Agelenopsis aperta) venom antagonize *N*-methyl-D-aspartate receptor function in mammalian brain. J Biol Chem 1991;266:21523-21529.
105. Ragsdale D, Gant DB, Anis NA et al. Inhibition of rat brain glutamate receptors by philanthotoxin. J Pharmacol Exp Ther 1989;251:156-163.
106. Moriyoshi K, Masu M, Ishii T et al. Molecular cloning and characterization of the rat NMDA receptor. Nature (Lond) 1991;354:31-37.
107. Nakanishi N, Axel R, Shneider NA. Alternative splicing generates functionally distinct *N*-methyl-D-aspartate receptors. Proc Natl Acad Sci (USA)

1992;89:8552-8556.

108. Yamazaki M, Mori H, Araki K et al. Cloning, expression and modulation of a mouse NMDA receptor subunit. FEBS Lett 1992;300:39-45.
109. Betz H. Ligand-gated ion channels in the brain: the amino acid receptor superfamily. Neuron 1990;5:383-392.
110. Tingley WG, Roche KW, Thompson AK et al. Regulation of NMDA receptor phosphorylation by alternative splicing of the C-terminal domain. Nature (Lond) 1993;364:70-73.
111. Kuryatov A, Laube B, Betz H et al. Mutational analysis of the glycine-binding site of the NMDA receptor: structural similarity with bacterial amino acid-binding proteins. Neuron 1994;12:1291-1300.
112. Hollmann M, Boulter J, Maron C et al. Zinc potentiates agonist-induced currents at certain splice variants of the NMDA receptor. Neuron 1993;10:943-954.
113. Sugihara H, Moriyoshi K, Ishii T et al. Structures and properties of seven isoforms of the NMDA receptor generated by alternative splicing. Biochem Biophys Res Comm 1992;185:826-832.
114. Laurie DJ, Seeburg PH. Regional and developmental heterogeneity in splicing of the rat brain NMDAR1 mRNA. J Neurosci 1994;14:3180-3194.
115. Zhong J, Carrozza DP, Williams K et al. Expression of mRNAs encoding subunits of the NMDA receptor in developing rat brain. J Neurochem 1995;64:531-539.
116. Monyer H, Sprengel R, Schoepfer R et al. Heteromeric NMDA receptors: molecular and functional distinction of subtypes. Science 1992;256:1217-1221.
117. Ishii T, Moriyoshi K, Sugihara H et al. Molecular characterization of the family of the *N*-methyl-D-aspartate receptor subunits. J Biol Chem 1993;268:2836-2843.
118. Kutsuwada T, Kashiwabuchi N, Mori H et al. Molecular diversity of the NMDA receptor channel. Nature (Lond) 1992;358:36-41.
119. Meguro H, Mori H, Araki K et al. Functional characterization of a heteromeric NMDA receptor channel expressed from cloned cDNAs. Nature (Lond) 1992;357:70-74.
120. Ikeda K, Nagasawa M, Mori H et al. Cloning and expression of the ε4 subunit of the NMDA receptor channel. FEBS Lett 1992;313:34-38.
121. Watanabe M, Inoue Y, Sakimura K et al. Distinct distributions of five *N*-methyl-D-aspartate receptor channel subunit mRNAs in the forebrain. J Comp Neurol 1993;338:377-390.
122. Monyer H, Burnashev N, Laurie DJ et al. Developmental expression in the rat brain and functional properties of four NMDA receptors. Neuron 1994;12:529-540.
123. Watanabe M, Inoue Y, Sakimura K et al. Developmental changes in distribution of NMDA receptor channel subunit mRNAs. NeuroReport 1992;3:1138-1140.
124. Sheng M, Cummings J, Roldan LA et al. Changing subunit composition

of heteromeric NMDA receptors during development of rat cortex. Nature (Lond) 1994;368:144-147.

125. Standaert DG, Testa CM, Young AB et al. Organization of N-methyl-D-aspartate glutamate receptor gene expression in the basal ganglia of the rat. J Comp Neurol 1994;343:1-16.
126. Watanabe M, Mishina M, Inoue Y. Distinct spatiotemporal expressions of five NMDA receptor channel subunit mRNAs in the cerebellum. J Comp Neurol 1994;343:513-519.
127. Stern P, Cik M, Colquhoun D et al. Single channel properties of cloned NMDA receptors in a human cell line: comparison with results from Xenopus oocytes. J Physiol (Lond) 1994;476:391-397.
128. Wafford KA, Bain CJ, LeBourdelles B et al. Preferential co-assembly of recombinant NMDA receptors composed of three different subunits. NeuroReport 1993;4:1347-1349.
129. Chazot PL, Coleman SK, Cik M et al. Molecular characterization of *N*-methyl-D-aspartate receptors expressed in mammalian cells yields evidence for the coexistence of three subunit types within a discrete receptor molecule. J Biol Chem 1994;269:24403-24409.
130. Benveniste M, Mayer ML. Kinetic analysis of antagonist action at *N*-methyl-D-aspartic acid receptors. Two binding sites each for glutamate and glycine. Biophys J 1991;59:560-573.
131. Patneau DK, Mayer ML. Structure-activity relationships for amino acid transmitter candidates acting at N-methyl-D-aspartate and quisqualate receptors. J Neurosci 1990;10:2385-2399.
132. Sommer B, Köhler M, Sprengel R et al. RNA editing in brain controls a determinant of ion flow in glutamate-operated channels of the CNS. Cell 1991;67:11-19.
133. Verdoorn TA, Burnashev N, Monyer H et al. Structural determinants of ion flow through recombinant glutamate receptor channels. Science 1991;252:1715-1718.
134. Hume RI, Dingledine R. Identification of a site in glutamate receptor subunits that controls calcium permeability. Science 1991;253:1028-1031.
135. Sakurada K, Masu M, Nakanishi S. Alteration of Ca^{2+} permeability and sensitivity to Mg^{2+} and channel blockers by a single amino acid substitution in the *N*-methyl-D-aspartate receptor. J Biol Chem 1993;268:410-415.
136. Mori H, Masaki H, Yamakura T et al. Identification by mutagenesis of a Mg^{2+}-block site of the NMDA receptor channel. Nature (Lond) 1992;358:673-675.
137. Burnashev N, Schoepfer R, Monyer H et al. Control by asparagine residues of calcium permeability and magnesium blockade in the NMDA receptor. Science 1992;257:1415-1419.
138. Yamakura T, Mori H, Masaki H et al. Different sensitivities of NMDA receptor channel subtypes to non-competitive antagonists. NeuroReport 1993;4:687-690.
139. Williams K, Zappia AM, Pritchett DB et al. Sensitivity of the *N*-methyl-D-aspartate receptor to polyamines is controlled by NR2 subunits. Mol Pharmacol 1994;45:803-809.

140. Williams K. Ifenprodil discriminates subtypes of the *N*-methyl-D-aspartate receptor: selectivity and mechanisms at recombinant heteromeric receptors. Mol Pharmacol 1993;44:851-859.
141. Laurie DJ, Seeburg PH. Ligand affinities at recombinant *N*-methyl-D-aspartate receptors depend on subunit composition. Eur J Pharmacol (Mol Pharmacol Sec) 1994;268:335-345.
142. Buller AL, Larson HC, Schneider BE et al. The molecular basis of NMDA receptor subtypes: native receptor diversity is predicted by subunit composition. J Neurosci 1994;14:5471-5484.
143. Clements JD, Lester RAJ, Tong G et al. The time course of glutamate in the synaptic cleft. Science 1992;258:1489-1501.
144. Hestrin S, Sah P, Nicoll RA. Mechanisms generating the time course of dual component excitatory synaptic currents recorded in hippocampal slices. Neuron 1990;5:247-253.
145. Lester RAJ, Clements JD, Westbrook GL et al. Channel kinetics determine the time course of NMDA receptor-mediated synaptic currents. Nature (Lond) 1990;346:565-567.
146. Stern P, Béhé P, Schoepfer R et al. Single-channel conductances of NMDA receptors expressed from cloned cDNAs: comparison with native receptors. Proc R Soc Lond B 1992;250:271-277.
147. Tsuzuki K, Mochizuki S, Iino M et al. Ion permeation properties of the cloned mouse ε2/ζ1 NMDA receptor channel. Mol Brain Res 1994;26:37-46.
148. Durand GM, Bennett MVL, Zukin RS. Splice variants of the *N*-methyl-D-aspartate receptor NR1 identify domains involved in regulation by polyamines and protein kinase C. Proc Natl Acad Sci (USA) 1993;90:6731-6736.
149. Zheng X, Zhang L, Durand GM et al. Mutagenesis rescues spermine and Zn^{2+} potentiation of recombinant NMDA receptors. Neuron 1994;12:811-818.
150. Williams K. Subunit-specific potentiation of recombinant *N*-methyl-D-aspartate receptors by histamine. Mol Pharmacol 1994;46:531-541.
151. Peters S, Koh J, Choi. DW. Zinc selectively blocks the action of N-methyl-D-aspartate on cortical neurons. Science 1987;236:589-593.
152. Christine CW, Choi DW. Effect of zinc on NMDA receptor-mediated channel currents in cortical neurons. J Neurosci 1990;10:108-116.
153. Westbrook GL, Mayer ML. Micromolar concentrations of Zn^{2+} antagonize NMDA and GABA responses of hippocampal neurons. Nature (Lond) 1987;328:640-643.
154. Williams K. Pharmacological properties of recombinant *N*-methyl-D-aspartate (NMDA) receptors containing the ε4 (NR2D) subunit. Neurosci Lett 1995;184:181-184.
155. Lynch DR, Anegawa NJ, Verdoorn T et al. *N*-methyl-D-aspartate receptors: different subunit requirements for binding of glutamate antagonists, glycine antagonists, and channel-blocking agents. Mol Pharmacol 1994;45:540-545.

156. Lynch DR, Lawrence JJ, Lenz S et al. Pharmacological characterization of heterodimeric NMDA receptors composed of NR1a and-2B subunits: differences with receptors formed from NR1a and-2A. J Neurochem 1995;64:1462-1468.
157. Cik M, Chazot PL, Stephenson FA. Optimal expression of cloned NMDAR1/NMDAR2A heteromeric glutamate receptors: a biochemical characterization. Biochem J 1993;296:877-883.
158. Raditsch M, Ruppersberg JP, Kuner T et al. Subunit-specific block of cloned NMDA receptors by argiotoxin$_{636}$. FEBS Lett 1993;324:63-66.
159. Williams K. Mechanisms influencing stimulatory effects of spermine at recombinant *N*-methyl-D-aspartate receptors. Mol Pharmacol 1994;46:161-168.
160. Carter C, Benavides J, Legendre P et al. Ifenprodil and SL 82.0715 as cerebral anti-ischemic agents. II. Evidence for *N*-methyl-D-aspartate receptor antagonist properties. J Pharmacol Exp Ther 1988;247:1222-1232.
161. Carter C, Rivy J-P, Scatton B. Ifenprodil and SL 82. 0715 are antagonists at the polyamine site of the *N*-methyl-D-aspartate (NMDA) receptor. Eur J Pharmacol 1989;164:611-612.
162. Church J, Fletcher EJ, Baxter K et al. Blockade by ifenprodil of high voltage-activated Ca^{2+} channels in rat and mouse cultured hippocampal pyramidal neurones: comparison with *N*-methyl-D-aspartate receptor antagonist actions. Br J Pharmacol 1994;113:499-507.
163. Legendre P, Westbrook GL. Ifenprodil blocks *N*-methyl-D-aspartate receptors by a two-component mechanism. Mol Pharmacol 1991; 40:289-298.
164. Seiler N. Polyamine metabolism and function in brain. Neurochem Int 1981;3:95-110.
165. Seiler N. Formation, catabolism and properties of the natural polyamines. In: C. Carter, ed. The Neuropharmacology of Polyamines. London: Academic Press, 1994:1-36
166. Fage D, Voltz C, Scatton B et al. Selective release of spermine and spermidine from the striatum by *N*-methyl-D-aspartate receptor activation in vivo. J Neurochem 1992;58:2170-2175.
167. Harman RJ, Shaw GG. High-affinity uptake of spermine by slices of rat cerebral cortex. J Neurochem 1981;36:1609-1615.
168. Harman RJ, Shaw GG. The spontaneous and evoked release of spermine from rat brain in vitro. Br J Pharmacol 1981;73:165-174.
169. Russell DH, Gfeller E, Marton LJ et al. Distribution of putrescine, spermidine, and spermine in rhesus monkey brain: decrease in spermidine and spermine concentrations in motor cortex after electrical stimulation. J Neurobiol 1974; 5:349-354.
170. Schwartz J-C, Arrang J-M, Garbarg M et al. Histaminergic transmission in the mammalian brain. Physiol Rev 1991;71:1-51.
171. Assaf SY, Chung S-H. Release of endogenous Zn^{2+} from brain tissue during activity. Nature (Lond) 1984;308:734-736.
172. Crawford IL, Connor JD. Zinc in maturing rat brain: hippocampal concentration and localization. J Neurochem 1972;19:1451-1458.

173. Howell GA, Welch MG, Frederickson CJ. Stimulation-induced uptake and release of zinc in hippocampal slices. Nature (Lond) 1984;308:736-738.
174. Xie X, Smart TG. A physiological role for endogenous zinc in rat hippocampal synaptic transmission. Nature (Lond) 1991;349:521-524.
175. Kemp JA, Leeson PD. The glycine site of the NMDA receptor—five years on. Trends Pharmacol Sci 1993;14:20-25.
176. Bekkers JM. Enhancement by histamine of NMDA-mediated synaptic transmission in the hippocampus. Science 1993; 261:104-106.
177. Seiler N. Polyamines. In: A. Lajtha, ed. Handbook of Neurochemistry. London: Plenum Publishing Corp, 1982:223-255
178. Scheetz AJ, Constantine-Paton M. Modulation of NMDA receptor function: implications for vertebrate neural development. FASEB J 1994;8:745-752.
179. Rossi DJ, Slater NT. The developmental onset of NMDA receptor-channel activity during neuronal migration. Neuropharmacology 1993;32:1239-1248.
180. LoTurco JJ, Blanton MG, Kriegstein AR. Initial expression and endogenous activation of NMDA channels in early neocortical development. J Neurosci 1991;11:792-799.
181. Komuro H, Rakic P. Modulation of neuronal migration by NMDA receptors. Science 1993;260:95-97.
182. Lombardi G, Szekely AM, Bristol LA et al. Induction of ornithine decarboxylase by *N*-methyl-D-aspartate receptor activation is unrelated to potentiation of glutamate excitotoxicity by polyamines in cerebellar granule neurons. J Neurochem 1993;60:1317-1324.
183. Porter C, Cavanaugh PFJ, Stolowich N et al. Biological properties of N^4- and N^1,N^8-spermidine derivatives. Cancer Res 1985;45:2050-2057.
184. He Y, Suzuki T, Kashiwagi K et al. Correlation between the inhibition of cell growth by *bis*(ethyl)polyamine analogs and the decrease in the function of mitochondria. Eur J Biochem 1994;221:391-398.
185. Bernacki RJ, Bergeron RJ, Porter CW. Antitumor activity of N,N-*bis*(ethyl)spermine homologues against human MALME-3 melanoma xenografts. Cancer Res 1992;52:2424-2430.
186. Basu HS, Pellarin M, Feuerstein BG et al. Interaction of a polyamine analog, 1,19-*bis*-(ethylamino)-5,10,15-triazanonadecane (BE-4-4-4-4), with DNA and effect on growth, survival, and polyamine levels in seven human brain tumor cell lines. Cancer Res 1993;53:3948-3955.
187. Lopatin AN, Makhina EN, Nichols CG. Potassium channel block by cytoplasmic polyamines as the mechanism of intrinsic rectification. Nature (Lond) 1994;372:366-369.
188. Fakler B, Brändle U, Glowatzki E et al. Strong voltage-dependent inward rectification of inward rectifier K^+ channels is caused by intracellular spermine. Cell 1995;80:149-154.

CHAPTER 7

S-Adenosylmethionine Decarboxylase and Spermidine/Spermine-N^1-Acetyltransferase-Emerging Targets for Rational Inhibitor Design

Patrick M. Woster

I. INTRODUCTION

The polyamine pathway represents a logical target for chemotherapeutic intervention, since depletion of polyamines results in the disruption of a variety of cellular functions, and may in specific cases result in cytotoxicity.[1] Inhibitors of the polyamine pathway, therefore, have traditionally been developed as potential antitumor and/or antiparasitic agents. Such inhibitors also play a critical role as research tools to elucidate the cellular functions of the naturally occurring polyamines, especially if these agents are specific for a single enzyme in the pathway. To date, a large portion of the biochemical and pharmacological studies aimed at understanding the cellular roles of polyamine metabolism, and most of the studies involving inhibitors of polyamine biosynthesis, have focused on the enzymes comprising the forward biosynthetic pathway, and most notably on ornithine decarboxylase and S-adenosylmethionine decarboxylase. A number of authoritative reviews

Polyamines: Regulation and Molecular Interaction, edited by Robert Casero. © 1995 R.G. Landes Company.

have appeared in the literature which detail the results of these studies.[2-9] Over the past few years, there has been an upsurge in interest in the cellular role of the enzymes mediating the polyamine back-conversion pathway,[10] which has been fueled by the recent discovery of the *bis*(alkyl)polyamines, a novel class of polyamine analogs which exhibit promising antitumor effects in vitro and in vivo. Recent evidence suggests that *bis*(alkyl)polyamine analogs may become an important new class of antitumor agent. The biochemistry, function and molecular biology of SSAT has recently been described in a recent review.[11]

A large number of polyamine biosynthesis inhibitors have been developed over the last 20 years, and agents are now available which are targeted to each of the enzymes in the forward biosynthetic pathway. The mechanism based ornithine decarboxylase (ODC) inhibitor α-diflluoromethylornithine (DFMO, also known as eflornithine), which was synthesized at Merrill Dow in 1978,[12] has become a research tool which has provided a tremendous impetus for studying the effects of interruption of the polyamine pathway. Although a number of additional ODC inhibitors have been developed,[3] none of them have become as scientifically or therapeutically important as DFMO, which is currently used clinically for the treatment of parasitic infections such as trypanosomiasis. As a result, there are few recent examples of novel synthetic inhibitors for ODC, even though it serves as a control point for the polyamine biosynthetic pathway. Only a handful of inhibitors for the constituative enzymes spermidine synthase and spermine synthase have been developed, presumably because these enzymes do not play a regulatory role in the cell. The most promising of the aminopropyltransferase inhibitors are the multisubstrate adduct analogs known as AdoDATO[13] and AdoDATAD,[14] which act as extremely potent and remarkably specific inhibitors for spermidine and spermine synthase, respectively. Since the introduction of these two analogs, no additional inhibitors for the aminopropyltransferases have been described. In recent years, the development of synthetic inhibitors for the polyamine pathway has been directed towards S-adenosylmethionine decarboxylase (AdoMet-DC), and more recently towards spermidine/spermine-N^1-acetyltransferase (SSAT). AdoMet-DC, like ODC, acts as a controlling enzyme in the forward polyamine pathway, and as such has become an attractive target for the development of novel synthetic inhibitors. More recently, it has been shown that SSAT, the rate-limiting step in the polyamine back-conversion pathway, plays a critical role in the regulation of intracellular polyamine levels. Agents which interact with SSAT appear to act as cell-type specific cytotoxic agents, and as promising antitumor agents both in vitro and in vivo. This chapter will focus on AdoMet-DC and SSAT as targets for the rational development of chemotherapeutic agents, and will review the progress made in the development of synthetic inhibitors to date.

II. S-ADENOSYLMETHIONINE DECARBOXYLASE

A. Structure, Function and Mechanism

The required aminopropyl donor for the aminopropyltransferases (spermidine synthase and spermine synthase) is decarboxylated S-adenosylmethionine (dc-AdoMet), produced from S-adenosylmethionine (AdoMet) by the action of AdoMet-DC. AdoMet-DC, like ODC, is a highly regulated enzyme in mammalian cells, and serves as a regulatory point in the forward biosynthetic pathway.[2,4] AdoMet-DC is also an inducible enzyme, and responds dramatically to either polyamine depletion, or to elevation of spermidine or spermine. However, unlike ODC, all known forms of AdoMet-DC belong to a class of enzymes that do not require a pyridoxal phosphate cofactor, but rather contain a covalently linked pyruvate cofactor at the amino terminus of the active site-containing subunit.[14] The *Escherichia coli* form of the enzyme, first purified by Tabor,[16,17] is now thought to exist as an octamer consisting of 4 α subunits (molecular weight 19,000, and containing one pyruvoyl moiety each) and 4 β subunits (molecular weight 14,000).[18] In addition, AdoMet-DC from *E. coli* requires a divalent cation such as Mg^{+2} for activity.[16] The mammalian form of AdoMet-DC was first isolated from rat prostate[19-21] and is a pyruvoyl enzyme which requires putrescine rather than a divalent cation for activation.[20] The proenzyme form of rat prostate AdoMet-DC (38,000 M_r) is processed into two subunits of 32,000 and 6000 M_r, and the active enzyme contains two copies of each subunit ($\alpha_2\beta_2$).[22] Mammalian AdoMet-DC has been expressed in *E. coli*, and the specific residues mediating putrescine stimulation of activity and processing have now been identified.[23] During the processing reaction, the pyruvate cofactor is generated on the larger subunit. Pegg and coworkers have recently purified the human form of AdoMet-DC to homogeneity.[24] In this study, human AdoMet-DC was expressed in high yield in *E. coli* using the pIN-III(lpp^{P-5}) expression vector, and then purified using MGBG-Sepharose affinity chromatography. The human enzyme, in terms of sequence and subunit structure, is very similar to the rat prostate form, and exhibits only 9 conservative changes in 334 amino acids. The proenzyme form of human AdoMet-DC (38,000 M_r) is autocatalytically processed into two subunits of 30,700 (pyruvate containing) and 7700 M_r, and the active enzyme contains two copies of each subunit ($\alpha_2\beta_2$). Site-directed mutagenesis studies have now been used to identify a putrescine binding site which mediates both enzyme activation and stimulation of processing.[25] Despite its similarity to the mammalian isozyme, the human form of AdoMet-DC is quite different from the *E. coli* form, both in subunit structure, and in primary sequence, showing less than 10% sequence homology.[21] In addition, a trypanosomal form of AdoMet-DC has been identified, and appears to exhibit significant differences in structure from both the human and bacterial forms of the enzyme.[26]

Thus, the differences in structure between the various forms of AdoMet-DC suggest that it may be possible to design isozyme-specific inhibitors for use as chemotherapeutic agents.

The key to designing effective inhibitors for any of the known forms of AdoMet-DC lies in exploiting the catalytic mechanism for the decarboxylation of AdoMet, which is shown in Scheme 7.1. An enzyme-substrate complex results when the substrate, AdoMet, forms an imine linkage with the terminal pyruvate of the active site-containing subunit. The subsequent derboxylation step the produces an enolate, which rearranges via Route A to form a decarboxylated imine species. Hydrolysis of this imine then releases dc-AdoMet and regenerates the enzyme. Anton and Kutny have demonstrated that a substrate-dependent inactivation of AdoMet-DC from *E. coli* occurs approximately once every 6000-7000 turnovers, as shown in Scheme 7.1.[27] During this inactivation, incorrect protonation of the enolate intermediate results in the formation of an isomerized imine species (Route B). Upon hydrolysis, an inactive form of AdoMet-DC is generated in which the terminal pyruvate has been transaminated to an alanine. The by-product of this reaction is a thioadenosine-containing aldehyde, which is further metabolized to methylthioadenosine (MTA) and the electrophilic species acrolein. Diaz and Anton have shown that this electrophilic species is responsible for alkylating the residue derived from cysteine-140 of the proenzyme form of *E. coli* AdoMet-DC.[28] This residue lies in a short sequence which is conserved between the bacterial and rat liver forms of the enzyme, suggesting that it participates in the catalytic mechanism.

B. Non-nucleoside Inhibitors of AdoMet-DC

The finding that known antiparasitic agents such as pentamidine and berenil act as inhibitors of mammalian and parasitic AdoMet-DC[29] has spurred a renewed interest in the development of non-nucleoside inhibitors of the enzyme. The antileukemic agent methylglyoxal *bis*(guanylhydrazone) (MGBG) is a potent competitive inhibitor of the putrescine-activated mammalian enzyme, with a K_i value of less than 1 μM.[30] However, MGBG is of limited use as a chemotherapeutic agent due to a wide variety of other effects on cells (induction of severe mitochondrial damage, interference with polyamine transport, induction of SSAT, etc.).[4] Thus, MGBG has not been particularly useful in the clinical setting, but has been used extensively as a research tool. A number of early attempts to modify the structure of MGBG resulted in active antitumor agents, but all of these MGBG analogs were as toxic or nonspecific as MGBG itself.[31-34] Interestingly, one of the newest developments in the area of AdoMet-DC inhibitors has been the discovery of a promising new series of MGBG analogs by Stanek et al.[35] All of the analogs showing significant activity are conformationally restricted derivatives of MGBG in which the backbone assumes a partially

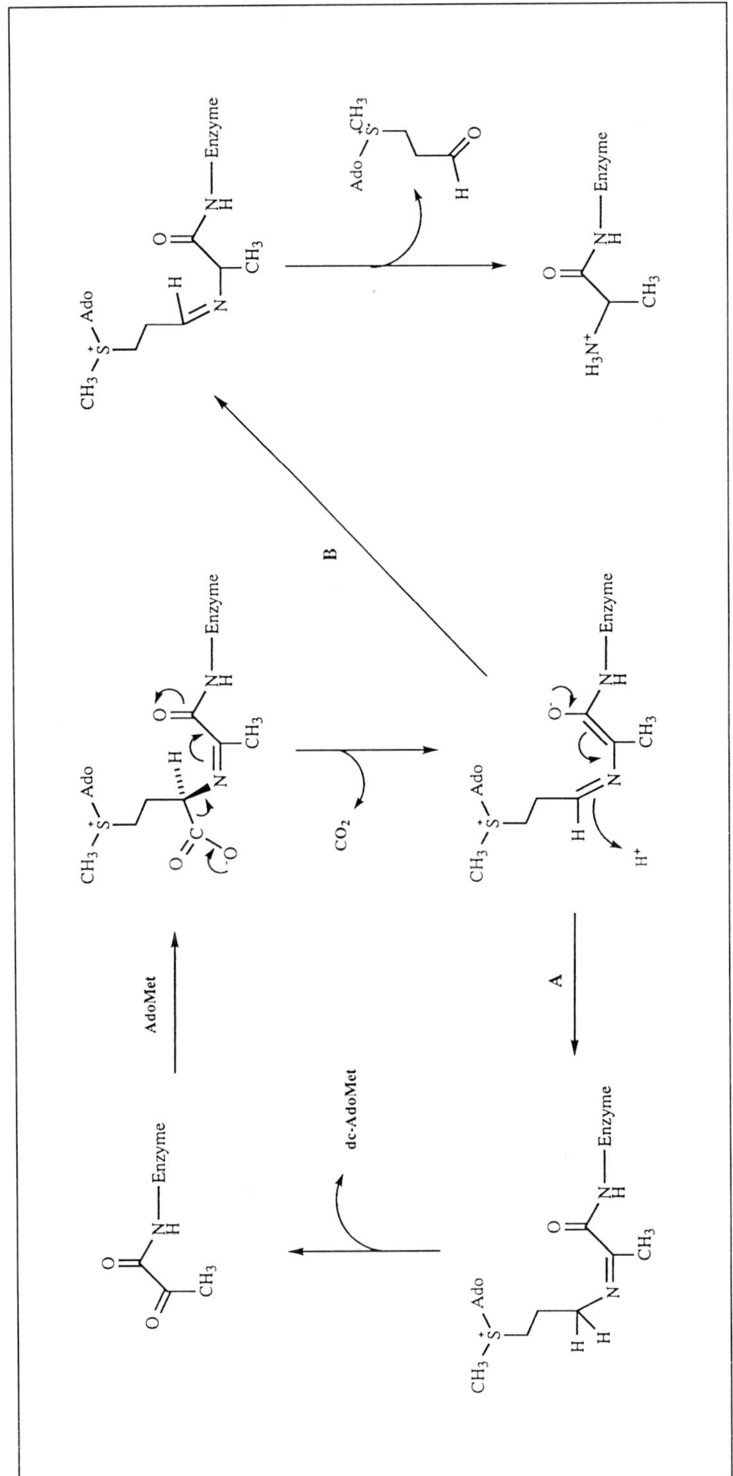

Scheme 7.1. Mechanism for enzymatic decarboxylation of AdoMet and for substrate/product inactivation of AdoMet-DC.

fixed, all *trans* conformation. Two of the most promising of these analogs are CGP 39937 and CGP 33829, shown in Figure 7.1, which inhibit AdoMet-DC with IC_{50} values of 6 and 36 nM, respectively, and show good antitumor activity in vitro.[36] Along these same lines, the conformationally restricted analog 4-amidinoindan-1-one 2'-amidinohydrazone shown in Figure 7.1 was synthesized and tested as an inhibitor of rat liver AdoMet-DC.[37] This analog is a remarkably potent, selective inhibitor of AdoMet-DC (IC_{50} = 5 nM), and also inhibits the growth of cultured T24 human bladder cancer cells with an IC_{50} of 0.71 μM. This compound and its congeners are currently being evaluated as antitumor agents. In addition, there is preliminary evidence that they may act as effective antiprotozoal agents, and may also be effective against *Pneumocystis carinii.*

C. Nucleoside-based Inhibitors of AdoMet-DC

Early attempts to develop nucleoside analogs as inhibitors of AdoMet-DC were concentrated on the synthesis of substrate and product analogs.[38-40] Although some of these analogs showed competitive, and in some cases noncompetitive inhibitory activity, none of them have proven to be useful clinically, and therefore none have been developed further. Interestingly, the α-difluoromethyl analog of AdoMet (analogous to DFMO) has been synthesized,[38] but has little useful activity against AdoMet-DC. This observation is presumably due to the chemical

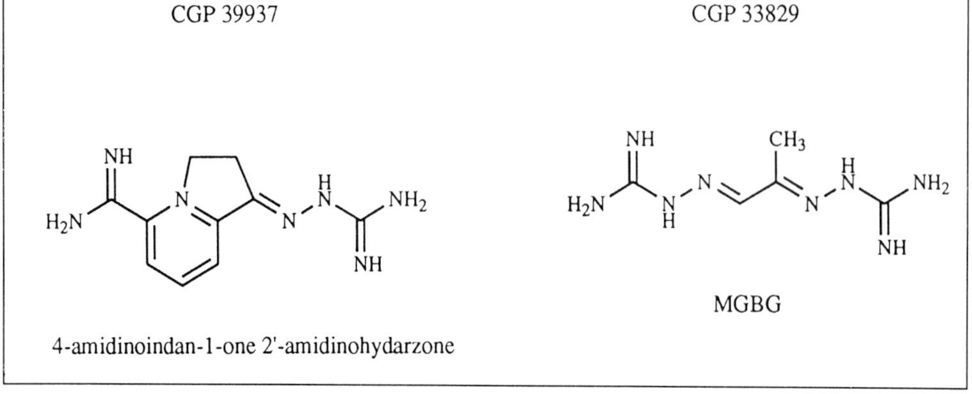

CGP 39937

CGP 33829

4-amidinoindan-1-one 2'-amidinohydarzone

MGBG

Fig. 7.1. Analogs of MGBG which reversibly inhibit AdoMet-DC.

instability of the analog, a trait of adenosylsulfonium compounds in general. The benefit of the early studies involving the synthesis of substrate and product analogs lies in the elucidation of critical structure activity relationships. Specifically, it was determined from these studies that both AdoMet and dc-AdoMet analogs could inhibit the enzyme, and that the methylsufonium center of AdoMet could be replaced by a charged nitrogen species. These data have proven critical for the development of second generation AdoMet-DC inhibitors.

A number of structural analogs of AdoMet have been developed as potential inhibitors of AdoMet-DC, in which a nucleophilic amine surrogate has been appended to the molecule. S-(5'-Deoxy-5'-adenosyl)methylthioethylhydroxylamine (AMA) has been shown to be an irreversible inhibitor of AdoMet-DC in L1210 cells with an IC_{50} concentration of 100 µm.[41-43] This compound acts presumably by forming a stable oxime linkage, rather than the usual imine, at the terminal pyruvate of AdoMet-DC, thus inactivating the enzyme. AMA has been reported to inhibit AdoMet-DC from rat liver at concentrations as low as 3 nM, however, this compound has not been studied adequately in purified enzyme preparations, and no kinetic inhibitor constants are available. Secrist has described the synthesis of a series of nucleophilic AdoMet analogs which act as potent inhibitors of AdoMet-DC.[44] The two most promising inhibitors in this series were 5'-deoxy-5'-[(3-hydrazinopropyl)methylamino]adenosine (MHZPA, Fig. 7.2) and 5'-deoxy-5'-[(3-hydrazinopropyl)methylamino]adenosine (MAOEA, Fig. 7.2). These derivatives are potent, irreversible inhibitors of the enzyme, with IC_{50} values of 7.5 and 40 nM against AdoMet-DC isolated from MRC5 cells, and I_{50} values of 400 and 70 nM against rat prostate AdoMet-DC,[45] respectively. The analogs act as effective growth inhibitors in L1210 cells, causing an elevation of putrescine and depletion of spermidine; however, these analogs appear to be rapidly metabolized in these cells, a fact which may limit their usefulness in vivo.[45] In a

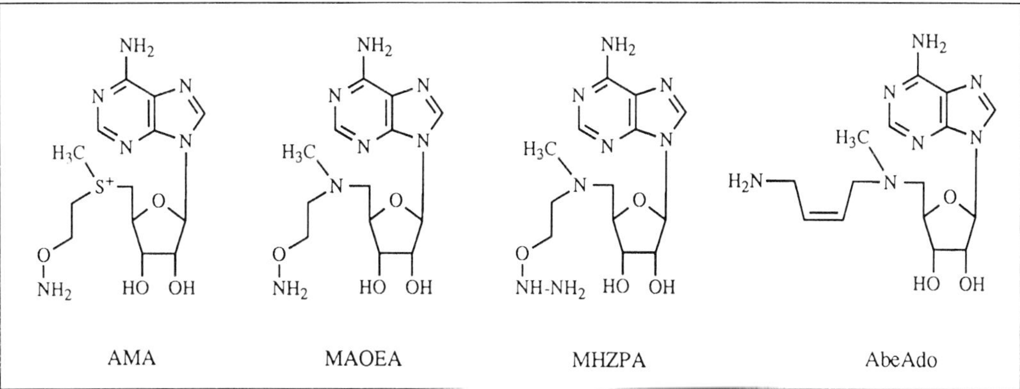

Fig. 7.2. Structures of nucleoside-based, irreversible inactivators of AdoMet-DC.

recent study, the adduct between MHZPA and the terminal pyruvate of human AdoMet-DC was characterized by isolating it from a LYS-C digest of the enzyme-inhibitor complex, confirming that MHZPA forms a stable hydrazone with the enzyme as predicted.[24]

Danzin has described the synthesis and enzymatic evaluation of 5'-{[(Z)-4-amino-2-butenyl]methylamino}-5'-deoxyadenosine (AbeAdo, Fig. 7.2), a potent, irreversible inhibitor of AdoMet-DC.[46] This analog, which contains a nitrogen in place of the natural methylsulfonium center, is the first inactivator of AdoMet-DC which can be considered "enzyme activated", since it was designed to rearrange to a latent electrophile in the catalytic site of AdoMet-DC, as shown in Scheme 7.2. In theory, formation of an imine linkage between the primary amine of AbeAdo and the pyruvate cofactor of AdoMet-DC would activate the proton α to the imine. Deprotonation by general base catalysis (Route A) would then lead to the formation of a latent electrophilic Michael acceptor following elimination of methylaminoadenosine (MAA). Attack at this electrophilic center by a nucleophilic amino acid residue could then be envisioned, resulting in irreversible inactivation of the enzyme. As predicted, AbeAdo produces a rapid time-dependent loss of enzyme activity, with a K_i of 0.3 μM and a K_{inact} of 3.6 min^{-1} against the *E. coli* form of the enzyme. The analog is also a potent inactivator of the rat liver form of AdoMet-DC (K_i = 0.56 μM, turnover number =1.5), and produces a long lasting, dose-dependent decrease in AdoMet-DC activity in vivo.[47] The Z-isomer of AbeAdo is 100 times more potent against rat liver AdoMet-DC,[47] and 1000-times more potent against *E. coli* AdoMet-DC,[46] than the corresponding E-isomer. Interestingly, there is now evidence that AbeAdo inactivates *E. coli* and human AdoMet-DC by different mechanisms. Inactivation of bacterial AdoMet-DC by AbeAdo is accompanied by a quasi-stoichiometric production of the by product, MAA.[46] However, inactivation of human AdoMet-DC by AbeAdo results in transamination of the terminal pyruvate moiety to an alanine, consistent with the putative mechanism shown in Scheme 7.2, Route B.[24] The transamination most likely proceeds by incorrect protonation of an enolate resulting from formation of the AbeAdo/AdoMet-DC adduct. Hydrolysis of this intermediate then produces the resulting alanine residue, and an aldehyde-containing aminoadenosine analog. The data available to date concerning the inhibition of AdoMet-DC by AbeAdo suggest that there are fundamental differences between the active sites of bacterial and human AdoMet-DC and support the contention that it may be possible to develop isozyme-specific inhibitors of the various forms of the enzyme.

Wu and Woster have recently reported the synthesis and biological evaluation of S-(5'-deoxy-5'-adenosyl)-1-amino-4-methylthio-2-cyclopentene (AdoMac, Fig. 7.3), an enzyme activated, irreversible inhibitor of AdoMet-DC.[48] AdoMac, first isolated as a mixture of

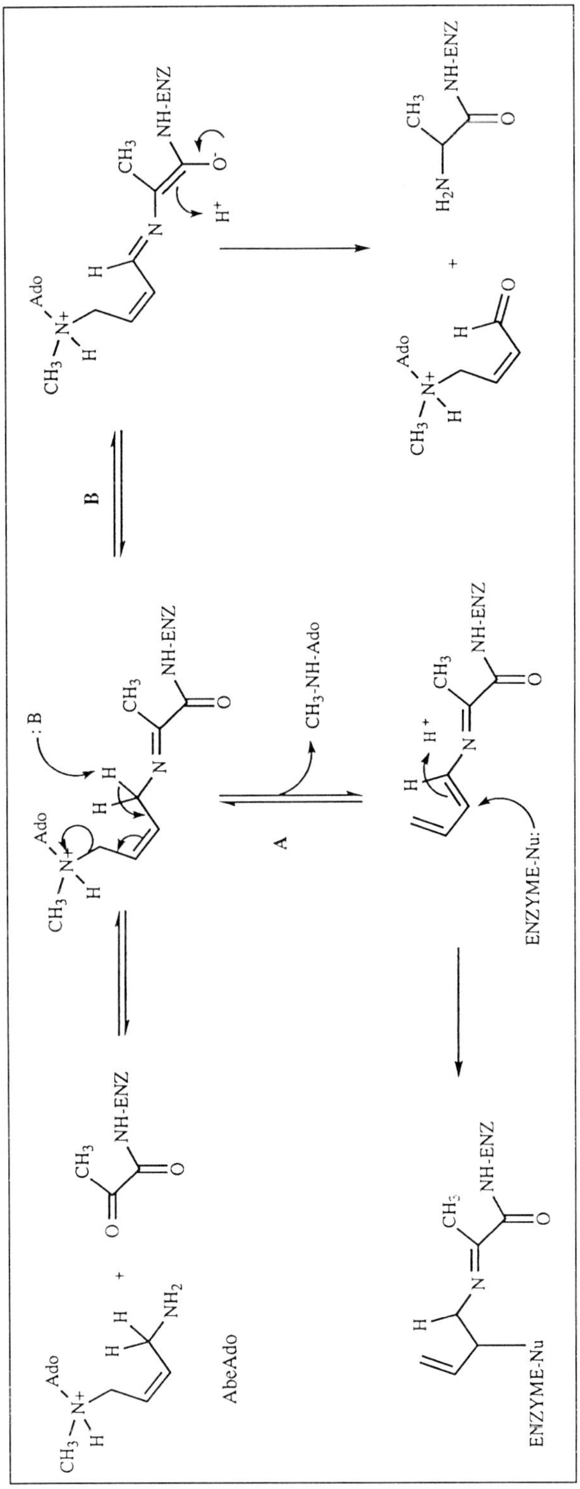

Scheme 7.2. Putative mechanisms for inactivation of AdoMet-DC by the irreversible inhibitor AbeAdo.

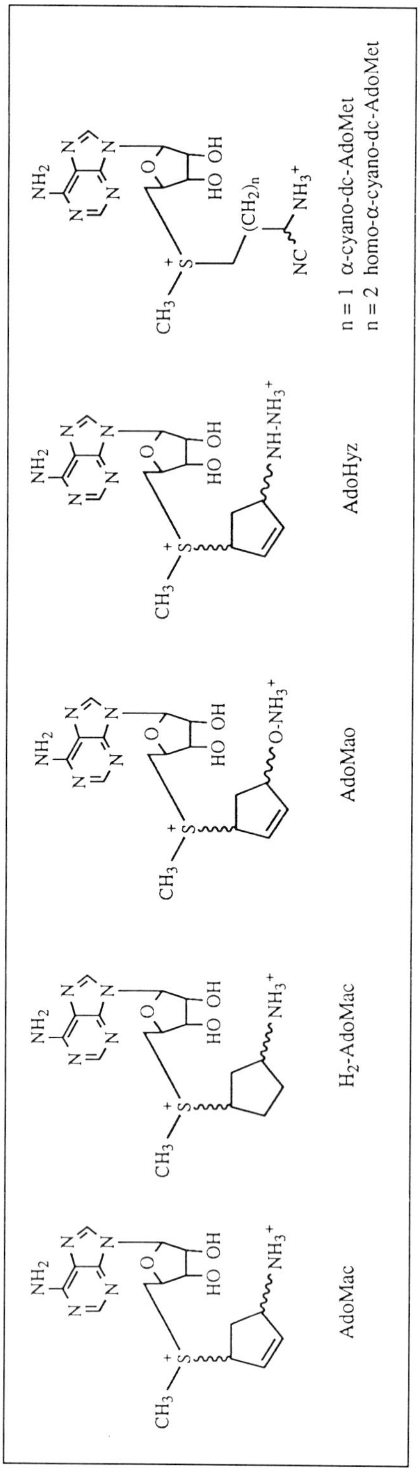

AdoMac

H_2-AdoMac

AdoMao

AdoHyz

n = 1 α-cyano-dc-AdoMet
n = 2 homo-α-cyano-dc-AdoMet

Fig. 7.3. Structures of nucleoside-based reversible and irreversible inhibitors of AdoMet-DC..

diastereomers, inhibits AdoMet-DC from *E. coli* with a K_i of 18 μM, and a k_{inact} of 0.133 min^{-1}. This inactivation is thought to proceed via the mechanism shown in Scheme 7.3. Formation of an imine linkage between AdoMac and the pyruvate cofactor of AdoMet-DC would result in activation of the cyclopentene proton α to the imine linkage. Abstraction of this proton would then result in the formation of a latent electrophile, which could react with a nucleophilic amino acid residue within the catalytic site, irreversibly inactivating the enzyme. Inhibition of AdoMet-DC by AdoMac meets all of the criteria ascribed to irreversible, enzyme activated inactivation. Enzyme activity could not be restored following prolonged dialysis of the enzyme/inhibitor complex, and the enzyme could be protected from inactivation by AdoMac when preincubated with MGBG. In addition, the expected by-product, MTA, was detected by HPLC during the enzymatic reaction.

The inhibitor AdoMac has the potential to exist in four distinct diastereomeric forms resulting from chirality at the 1 and 4 positions of the cyclopentene ring. In order to assess the configurational dependence of the binding of AdoMac to AdoMet-DC, and of the subsequent inactivation process, the pure diastereomers were individually synthesized and evaluated as inhibitors of the *E. coli* form of AdoMet-DC.[49] The K_i values observed, which range between 4 and 40 μM (Table 7.1), demonstrate that the binding of AdoMac to

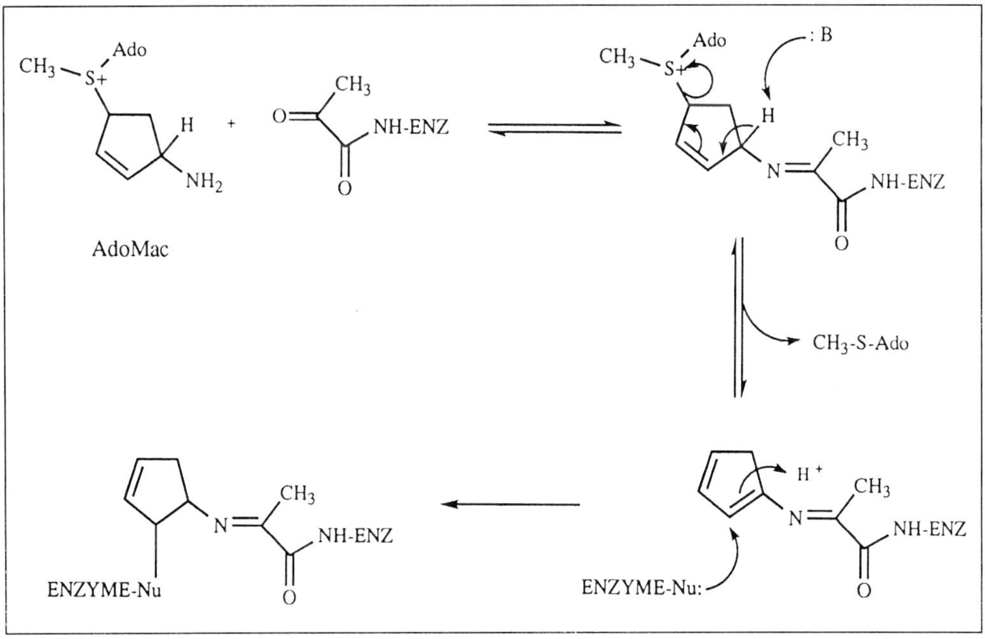

Scheme 7.3. Putative mechanism for the inactivation of AdoMet-DC by the irreversible inhibitor AdoMac. Reprinted with permission from Bioch Pharmacol 1995; 49:1125-1133.

AdoMet-DC is significantly configuration dependent. The *cis*-diastereomers are significantly more potent than the trans forms, with the *cis*-1S,4R-AdoMac being most potent. However, the k_{inact} values for the four diastereomers are remarkably constant, varying only between 0.064 and 0.099 min^{-1}. Following reversible formation of the imine linkage, each pure diastereomer of AdoMac appears to inactivate AdoMet-DC at the same rate, suggesting that this inactivation may proceed through a common intermediate or set of intermediates. If this hypothesis is correct, the configuration-dependent formation of the Schiff's base between AdoMet-DC and a given diastereomer of AdoMac may result in the observed differences in K_i. In the case of each diastereomer, HPLC experiments revealed the time-dependent appearance of a peak which co-eluted with MTA, and the corresponding disappearance of the peak co-eluting with AdoMac, suggesting that MTA was generated from AdoMac in the enzymatic reaction as predicted. Removal of the driving force for the elimination of methylthioadenosine resulted in a reversibly binding inhibitor. Thus, the unmethylated thioether analog corresponding to AdoMac (nor-AdoMac), and the corresponding dihydro derivative (H_2-AdoMac, Fig. 7.3), reversibly inhibit the enzyme. As expected, compounds *cis*-1R,4S-nor-AdoMac and *cis*-1S,4R-H_2-AdoMac act as weak competitive inhibitors of AdoMet-DC, with K_i values of 293 and 93 μM, respectively, as shown in Table 7.1. In addition, the inhibition observed in the presence of these analogs was not time-dependent, and no generation of MTA or any other metabolite related to AdoMac was detected. As was mentioned above, it now appears plausible that AbeAdo may inactivate the rat liver and Escherichia coli forms of AdoMet-DC by different mechanisms, since the by-product MAA is detected only following inactivation of the bacterial enzyme. In contrast, our preliminary studies show that the generation of MTA

Table 7.1. Comparison of the kinetic parameters for the inactivation of human and* Escherichia coli *AdoMet-DC by the pure diastereomeric forms of AdoMac,* cis-*1S, 4R-H_2-AdoMac and* cis-*1R,4S-nor-AdoMac

Inhibitor	K_i (human)	k_{inact} (human)	K_i (*E. coli*)	k_{inact} (*E. coli*)	Time Dependence	MTA Generation
cis-1R,4S-AdoMac	11 μM	0.073 min^{-1}	8 μM	0.099 min^{-1}	yes	yes
cis-1S,4R-AdoMac	17 μM	0.068 min^{-1}	4 μM	0.064 min^{-1}	yes	yes
trans-1S,4S-AdoMac	53 μM	0.088 min^{-1}	24 μM	0.068 min^{-1}	yes	yes
trans-1R,4R-AdoMac	63 μM	0.082 min^{-1}	40 μM	0.079 min^{-1}	yes	yes
cis-1S,4R-H_2-AdoMac	72 μM	–	93 μM	–	no	no
cis-1R,4S-nor-AdoMac	307 μM	–	293 μM	–	no	no

Reprinted with permission from Bioch Pharmacol 1995; 49:1125-1133.

following inactivation of AdoMet-DC by AdoMac can be detected using both the bacterial and human forms of the enzyme (see below). In terms of the inactivation of AdoMet-DC, the implications of this observation are as yet unknown, especially in light of the dramatically different amino acid composition of the *E. coli* and human forms of the enzyme.[21] However, since each pure diastereomer of AdoMac represents a distinct conformational mimic exhibiting restricted sidechain rotation, the data suggests that these and related analogs may be useful as conformational probes for the catalytic sites of the various isozymes of AdoMet-DC.

The inactivation of human AdoMet-DC by the pure diastereomers of AdoMac has recently been studied.[50] As was observed for the inactivation of *E. coli* AdoMetDC, each pure diastereomer of AdoMac inactivated the human enzyme in a time- and concentration-dependent manner. The K_i values for the four diastereomeric forms of AdoMac ranged between 11 and 63 μM, while the corresponding k_{inact} values did not vary significantly, and were similar to those observed for the inactivation of the bacterial enzyme, as shown in Table 7.1. Again, HPLC experiments demonstrated that MTA was generated from AdoMac in the enzymatic reaction as predicted. Further, the value of the partition ratio (i.e., the ratio of k_{cat}/k_{inact}) for *cis*-1R,4S-AdoMac, the most potent diastereomeric inactivator of human AdoMet-DC, was evaluated indirectly by the titration method, and was determined to be 8.84. This observation then allowed for the calculation of the k_{cat} value for *cis*-1R,4S-AdoMac, which was found to be 0.645 min^{-1}. As expected, *cis*-1S,4R-H_2-AdoMac reversibly inhibits human AdoMet-DC with a K_i value of 72 μM (Table 7.1). Likewise, *cis*-1R,4S-nor-AdoMac acts as a weak competitive inhibitor of human AdoMet-DC, with a K_i value of 307 μM (Table 7.1). The inhibition observed in the presence of H_2-AdoMac and nor-AdoMac was not time-dependent, and no generation of MTA or any other related metabolite was detected in the enzymatic reaction mixture following exposure to these analogs.

As shown in Table 7.1, both the human and *E. coli* forms of AdoMet-DC are able to discriminate between the four diastereomers of AdoMac, and in each case the *cis* diastereomers (with respect to the cyclopentene ring) are significantly more potent than the trans. However, human AdoMet-DC prefers the *cis*-1R,4S-diastereomer, while the bacterial enzyme preferentially binds to *cis*-1S,4R-AdoMac. When subjected to computer-assisted molecular mechanics analysis, these two molecules possess significantly different least-energy conformations, as shown in Figure 7.4. Computer simulation suggests that there are significant conformational differences between the *cis*-1R-4S- and *cis*-1S,4R-diastereomers of AdoMac, which may account for the observed variation in the K_i values between the human and bacterial forms of the enzyme. In addition, the present findings suggest the possibility that AdoMac inhibits the bacterial and human forms of the enzyme by the same mechanism, and possibly through a common intermediate, since the k_{inact} values are strikingly similar for each determination. However, additional studies are required to determine the validity of these hypotheses. The data presented in Table 7.1 supports the contention that there are distinct differences in the conformational requirements of the catalytic sites of human and *E. coli* AdoMet-DC. These differences can potentially be exploited to design analogs which are specific for a given form of the enzyme.

Differences in the catalytic sites of the isozymic forms of AdoMet-DC can also be exploited using AdoMet analogs which are not conformationally restricted. Wu et al have recently described the synthesis of (5'-deoxy-5'-S-adenosyl)-2-amino-4-methylsulfonio-2-pentanenitrile (α-cyano-dc-AdoMet, Fig. 7.3), as well as an analog which contains one additional carbon in the sidechain, (5'-deoxy-5'-S-adenosyl)-2-amino-4-methylsulfonio-2-pentanenitrile (homo-α-cyano-dc-AdoMet, Fig. 7.3).[51] As expected, α-cyano-dc-AdoMet acted as an enzyme activated, irreversible inhibitor of AdoMet-DC from *E. coli*, with an IC_{50} value of 9 μM, and a K_i value of 31 μM. The unnatural analog, homo-α-cyano-dc-AdoMet, was significantly less potent, exhibiting an IC_{50} value of 50 μM. Unexpectedly, the human form of AdoMet-DC showed a reverse preference for these two analogs, since homo-α-cyano-dc-AdoMet (K_i 7 μM) was considerably more active as an inhibitor than α-cyano-dc-AdoMet (K_i 247 μM). These data further support the hypothesis that there are significant structural differences between the catalytic sites of the human and *E. coli* forms of AdoMet-DC.

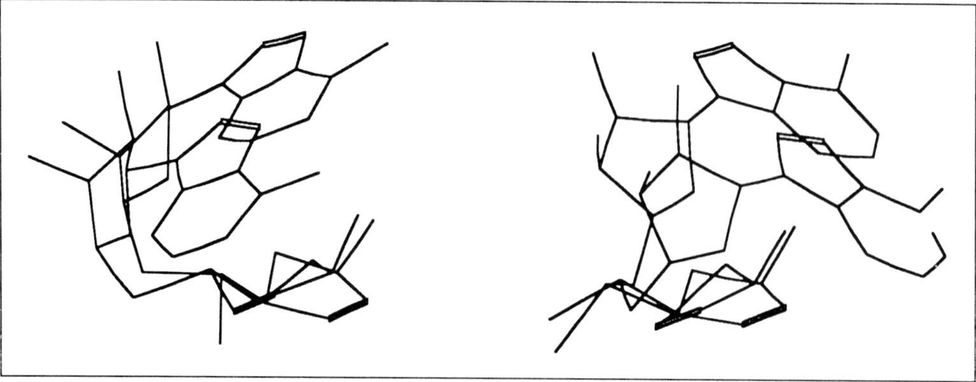

Fig. 7.4. Results for RMS fitting experiments for the least energy conformers of the cis-*1R,4S-* and cis-*1S,4R-diasteromers of AdoMac. Hydrogens have been removed from the molecules for clarity. Left—normal view. Right—orthogonal view. Reprinted with permission from Bioch Pharmacol 1995; 49:1125-1133.*

D. AdoMet-DC Inhibitors as Antiparasitic Agents

Efforts to develop an inhibitor of AdoMet-DC which is an effective antitumor agent have so far met with relatively little success. Probably the best hope in this area lies in the continuing development of novel MGBG analogs, which appear to have excellent antitumor activity in vitro. As has been alluded to, MGBG analogs such as CGP 39937 and CGP 33829, shown in Figure 7.1, are also under investigation as antiparasitic agents, and although there are no specific studies available in the literature, preliminary indications are that these analogs are worth pursuing in this regard. Nucleoside analogs which inhibit AdoMet-DC are less likely to act as effective antitumor agents, since in most cases they are unable to penetrate into mammalian cells in sufficient concentration to effect cell growth. However, there are a number of cases where adenosine analogs which are effective AdoMet-DC inhibitors have minimal activity in mammalian cells, but show good activity against parasitic organisms such as *Trypanosoma brucei brucei*. For example, 5'-deoxy-5'-[(2-hydrazinoethyl)methylamino]adenosine (MHZEA) and the previously described AdoMet analog MAOEA are particularly good trypanocides in vitro, exhibiting IC_{50} values of 40 nM and 1.3 μM, respectively.[52] In addition, Guo et al has recently described a series of conformationally restricted AdoMet analogs which are weak reversible inhibitors of AdoMet-DC, and have minimal effects in mammalian cells.[53] However, these analogs show good activity against cultured bloodforms of *T. b. brucei* in vitro. The ability of these analogs to inhibit trypanosomal growth most likely depends on their ability to enter trypanosomal cells via a recently discovered adenosine transporter.[54]

Although the inhibitor AbeAdo is a potent inactivator of AdoMet-DC, it has only modest effects on mammalian tumor cell growth in vitro.[55] In addition, the in vivo effects of the analog as an antitumor agent are disappointing.[56] The most promising potential use of AbeAdo appears to be as an antitrypanosomal agent. In fact, this agent cures *T. b. brucei* and multidrug resistant *T. b. rhodesiense* infections in mice, and is 100 times as potent as DFMO in this regard.[57] Interestingly, these effects appear to be due to a dramatic, parasite-specific rise in AdoMet, rather than to effects on the polyamine pathway. The uptake of AbeAdo into parasitic cells is time and temperature dependent, and is also saturable, suggesting that the analog is indeed a substrate for the trypanosomal adenosine transporter.[58] Despite the promise of AbeAdo as an antitrypanosomal agent, it is not currently being developed for clinical use.

The exploitation of conformational differences between isozymes of AdoMet-DC would be of particular value in the case of parasitic organisms such as *T. brucei*, which possess yet another distinct form of putrescine-activated AdoMet-DC.[26] To this end, the AdoMet analog S-(5'-deoxy-5'-adenosyl)-1-aminoxy-4-methylsulfonio-2-cyclopentene

(AdoMao, Fig. 7.3) was synthesized in two of its four possible diastereomeric forms (unpublished results). AdoMao was designed to inactivate AdoMet-DC by forming a stable oxime linkage with the terminal pyruvate of the enzyme. The trans-1R,4R- and *trans*-1S,4S-diastereomers of AdoMao, as well as the corresponding diastereomers of the unmethylated precursor molecule nor-AdoMao, acted as time-dependent, irreversible inhibitors of AdoMet-DC from *E. coli*, exhibiting remarkably constant K_i values ranging between 20.6 and 23.7 μM. These analogs also inhibited the human form of AdoMet-DC, although this form of the enzyme was able to discriminate between AdoMao (K_i values of 21.2 μM for the *trans*-1R,4R form and 19.6 μM for the *trans*-1S,4S form) and nor-AdoMao (K_i values of 95.2 μM for the *trans*-1R,4R form and 30.9 μM for the *trans*-1S,4S form). The observed differences in the inactivation of bacterial and human AdoMet-DC further support the contention that it is possible to design isozyme-specific inhibitors of AdoMet-DC. The *trans* diastereomers of AdoMao and nor-AdoMao were next evaluated for their ability to inhibit trypanosomal growth in vitro against cultured *T. b. brucei* bloodforms. All four of these analogs were effective growth inhibitors, with IC_{50} values ranging between 0.9 and 10.1 μM. The two most effective analogs, *trans*-1S,4S-AdoMao (IC_{50} 0.9 μM) and *trans*-1S,4S-AdoMao (IC_{50} 3.0 μM) were also effective against two clinical isolates of the pathogenic organism *T. b. rhodesiense*, KETRI 243 and KETRI 269. The most promising analog in all respects was *trans*-1S,4S-AdoMao, which was subsequently found to have minimal effects on cell growth, AdoMet-DC activity and intracellular polyamine levels in the human promyelocytic leukemia cell line HL60. Thus, the S-adenosylmethionine analog *trans*-1S,4S-AdoMao acts as an effective inhibitor of AdoMet-DC and appears to serve as a parasite-specific trypanocidal agent in vitro. The related analog S-(5'-deoxy-5'-adenosyl)-1-aminoxy-4-methylsulfonio-2-cyclopentene (AdoHyz, Fig. 7.3), which contains a hydrazino- rather than an aminoxy-amine surrogate, is also a conformation-dependent inhibitor of trypanosomal growth, but is much less potent in this regard (unpublished results). Studies involving the kinetics of the inhibition of human and bacterial AdoMet-DC produced by AdoHyz are currently ongoing.

III. SPERMIDINE/SPERMINE N^1-ACETYLTRANSFERASE*

In addition to the controlling enzymes of the forward pathway, ODC and AdoMet-DC, cellular polyamine content is modulated by a pair of acetyltransferases. Spermidine in the cell nucleus is acetylated on the four carbon end by spermidine N^8-acetyltransferase, reducing the number of positive charges on the molecule and possibly altering its ability to stabilize DNA.[59,60] A specific cytoplasmic deacetylase, N^8-acetylspermidine deacetylase, can then reverse this enzymatic

**A portion of this chapter was adapted from a previously published article (Curr Opin Invest Drugs 1993; 2:1291-1299.*

acetylation. Specific inhibitors for N^8-acetylspermidine deacetylase have been synthesized[59,61] and are being used to investigate the function of the N^8-acetylation/deacetylation pathway. Cytoplasmic spermidine and spermine serve as substrates for spermidine/spermine-N^1-acetyltransferase (SSAT), resulting in acetylation on the three carbon end of each molecule, as shown in Figure 7.5. This enzyme is the first and rate limiting step in the catabolism of the polyamines spermidine and spermine.[11] SSAT is characterized as being highly substrate specific, rapidly inducible, and having a very short biological half-life. It has also been shown in several cellular and in vivo systems to be inducible by a variety of stimuli including growth factors, toxic insults, the natural polyamines and polyamine analogs. Acetylated spermidine or spermine resulting from catalysis by SSAT then acts as a substrate for polyamine oxidase (PAO), which catalyzes the formation of 3-acetamidopropionaldehyde and either putrescine or spermidine, respectively. Thus SSAT and PAO together serve as a reverse route facilitating the interconversion of cellular polyamines. Current information indicates that the only fates available to N^1-acetylated polyamines are PAO-mediated oxidation or export from the cell. It is important to note that the combination of the highly regulated catabolic enzyme SSAT, coupled with the finely controlled synthetic enzymes ODC and AdoMetDC, allow the cell considerable control of intracellular polyamine concentrations.

A. Kinetics and Substrate Specificity

Under normal conditions, SSAT is present in very low levels in mammalian cells, but can be induced many fold by a variety of agents, a fact which led to its discovery in rat liver following treatment with carbon tetrachloride.[62] The structural requirements for interaction with the SSAT active site have not been fully investigated, but some structure/affinity data are available. The natural substrates for the enzyme, spermidine and spermine (Fig. 7.6), have K_m values of 130 and 34 μM, while norspermidine and norspermine exhibit K_m values of 9 and 10 μM, respectively.[63] N^1-Acetylspermine is also a substrate (K_m = 51 μM), while N^1-acetylspermidine is not, since it does not possess the requisite free aminopropyl moiety. 1,3-Diaminopropane acts as a substrate with a K_m of 460 μM, while putrescine, 1,5-diaminopentane (cadaverine), homospermidine and histones do not.[62,63] A series of α-methyl-[64] and α-gem-dimethyl[65] substituted polyamines have been synthesized, but in neither case acted as substrates for SSAT. These observations underscore the fact that a molecule must possess a terminal aminopropyl moiety in order to act as a substrate for SSAT. The reaction catalyzed by SSAT has been characterized as an ordered bimolecular reaction in which spermidine or spermine binds before acetylCoA, and the corresponding N^1-acetylpolyamine is the final product to be released.[63] The reaction likely proceeds through a binary complex between the polyamine

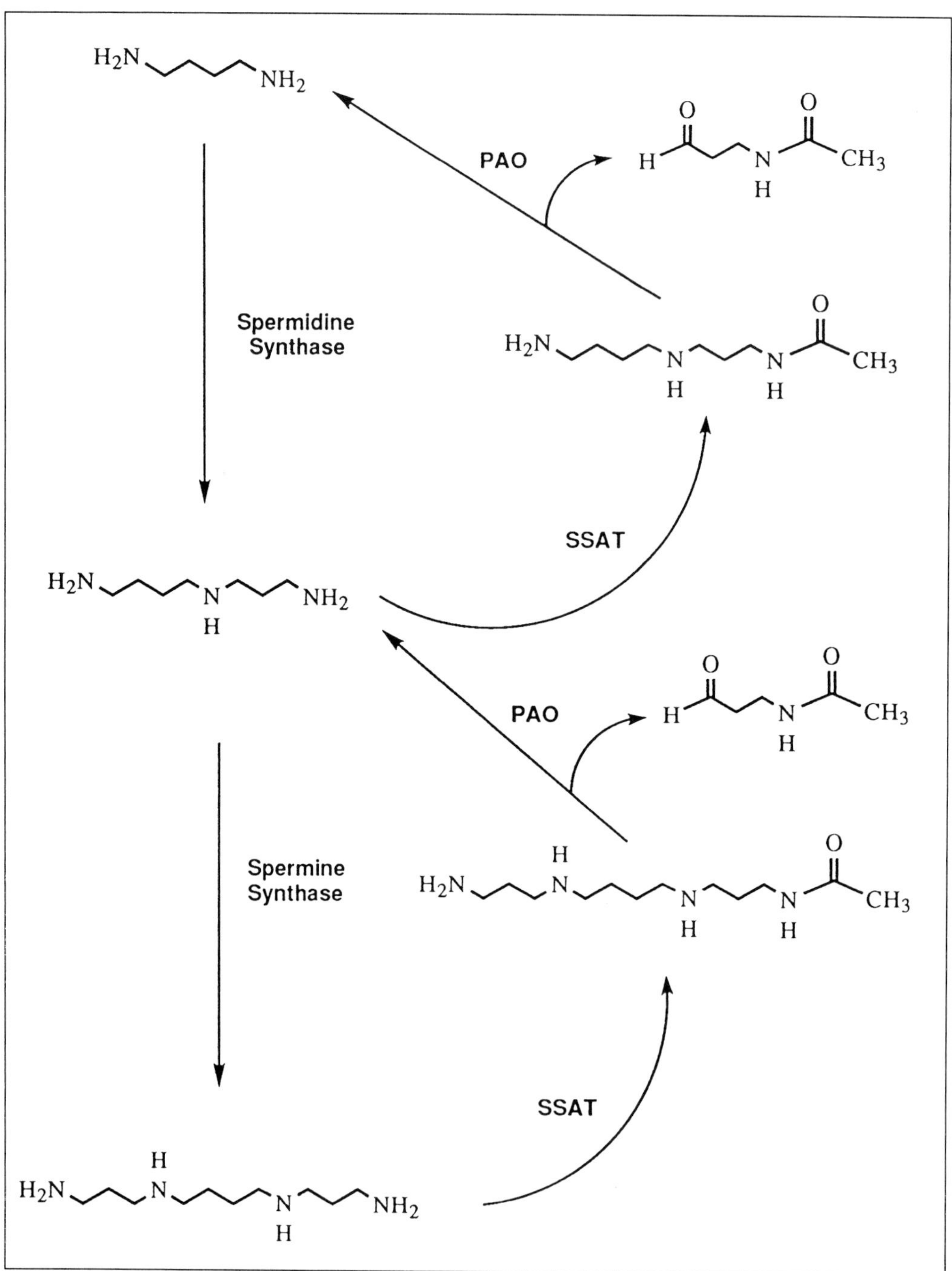

Fig. 7.5. The catabolism of mammalian polyamines by SSAT and polyamine oxidase.

and acetyl CoA, since it is strongly inhibited by polyamine/acetyl CoA-based multisubstrate adduct inhibitors.[66] The potency of these inhibitors, which varies with the polyamine portion of the multisubstrate adduct, closely follows the substrate affinities mentioned above, and thus distinguishes SSAT from enzymes which are known to acetylate histones.[67]

In addition to the studies mentioned above, a number of structure/function relationships have been inferred based on the ability of substituted polyamine analogs to inhibit cell growth in vitro and in vivo. In these studies, however, except in the case of the *bis*(ethyl)polyamines, a direct correlation between modulation of cellular SSAT and antiproliferative effects has not been established. It has been suggested that bis-monoalkylation of both terminal nitrogens of spermidine or spermine is optimal for antiproliferative activity, and that alkylation at an internal nitrogen reduces in vitro activity.[68] Elongation of the central alkyl chain of spermine or bis-alkylated spermine to C-8 produces analogs which have dramatic effects on in vitro[69] and in vivo[70] tumor cell growth, levels of intracellular polyamines[69,71] and DNA conformation and aggregation.[69] The terminal *bis*(benzyl) substituted C-8 spermine homologues have been shown to down-regulate ODC and AdoMet-DC in rat hepatoma cells, resulting in depletion of cellular polyamines and complete abolition of cell division.[71] However, this

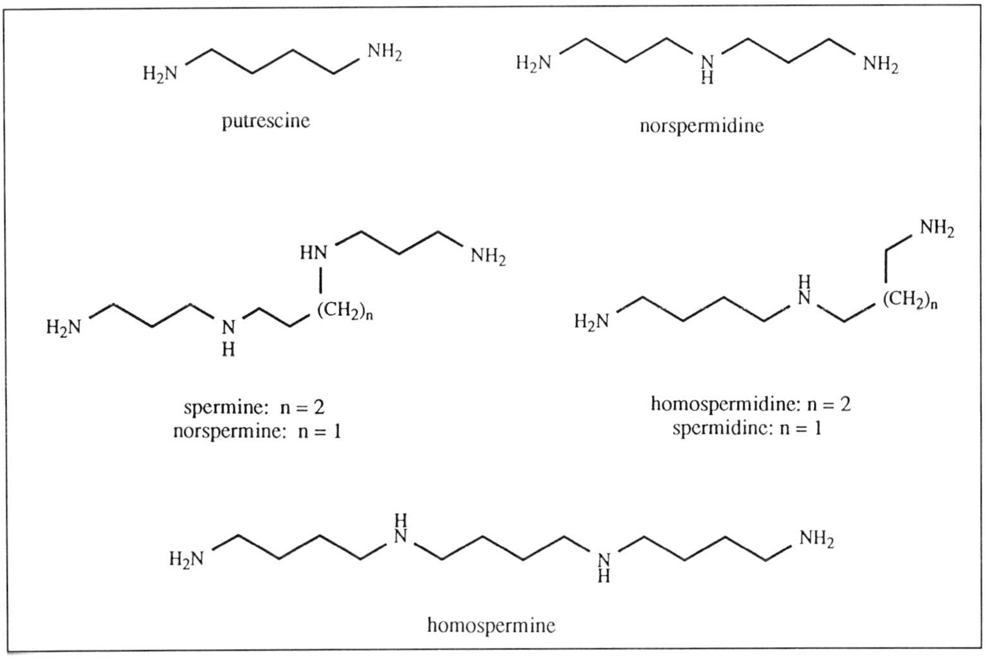

Fig. 7.6. Structures of the naturally occurring polyamines and their nor- or homo- analogs.

effect appears to be related to metabolism of the *bis*(benzyl) derivative to the corresponding *bis*-primary amine. The *bis*(ethyl)polyamines, and in particular *bis*(ethyl)norspermine, have excellent affinity for SSAT and serve as competitive inhibitors, but can dramatically induce the enzyme to levels that are as much as 2000% above baseline.[72] This effect, combined with the down regulation of polyamine biosynthesis by *bis*(ethyl)polyamine analogs,[73,74] leads to a depletion of cellular polyamines and cytotoxicity in some cell lines. The cellular and antiproliferative effects of these compounds will be discussed below.

B. Development of Synthetic Polyamine Analogs

The observation that polyamines and related compounds could induce SSAT in vitro, as well as deplete intracellular polyamines,[75] coupled with the finding that polyamine analogs such as norspermidine showed antiproliferative activity in vitro,[76] led to the hypothesis that synthetic polyamine analogs could be developed as potential antineoplastic agents. In this regard, the ideal agent would be a substrate for the polyamine cellular transport system, and would retain the ability to down-regulate polyamine biosynthetic enzymes, but would not support the cellular functions of the natural polyamines. A number of analogs have now appeared in the literature which, in preliminary studies, appear to satisfy these requirements. The antiproliferative activity of some of these analogs has been linked to their ability to induce SSAT and down-regulate polyamine biosynthesis. The remaining agents show similar biological activity in vitro, but their relationship to SSAT has not been established. The most successful of these agents to date have been the *bis*(alkyl)polamines, which in recent studies have shown great promise in vitro and in vivo as antitumor agents.

A series of bisbenzyl polyamine analogs has been described by Bitonti et al[77] which exhibit potent effects on cell growth and intracellular polyamine levels in cultured rat hepatoma (HTC) cells.[78] Originally developed for their marked antimalarial activity,[79] these analogs, and in particular MDL 27695 (Fig. 7.7), appear to act by regulation of the polyamine biosynthetic pathway. Thus in HTC cells, treatment with a 1 μM concentration of MDL 27695 led to a rapid loss of ODC and AdoMet-DC activity, and a complete cessation of cell growth, although cellular polyamine levels did not change dramatically. These effects appear to depend on the ability of the compound to be metabolized to the corresponding diamine derivative MDL 26752 (Fig. 7.7), since the effects of MDL 27695 were completely blocked by the addition of a polyamine oxidase inhibitor. Indeed, MDL 26752 inhibited polyamine biosynthesis and cell growth more rapidly and to a greater extent than MDL 27695 when added directly to cell culture. Surprisingly, MDL 26752 appears to be inactive as an antimalarial agent.[79] The ability of these compounds to inhibit or induce SSAT has not been determined.

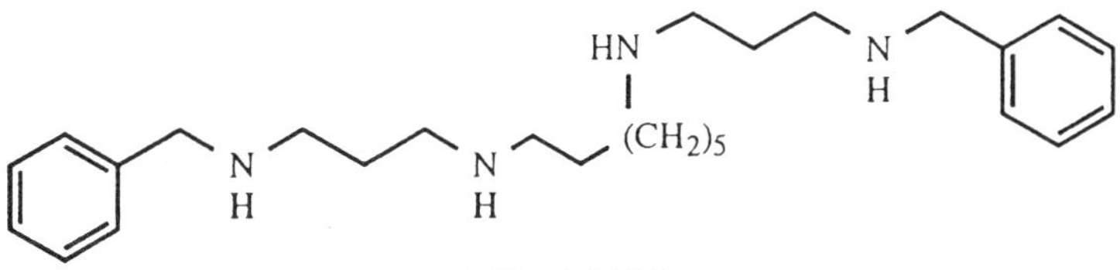

MDL 27695

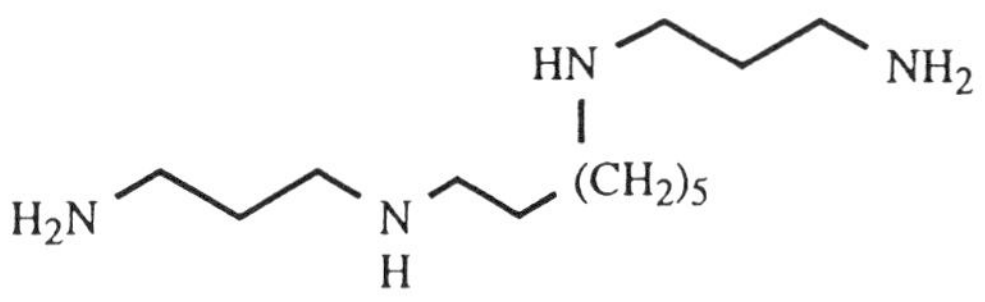

MDL 26752

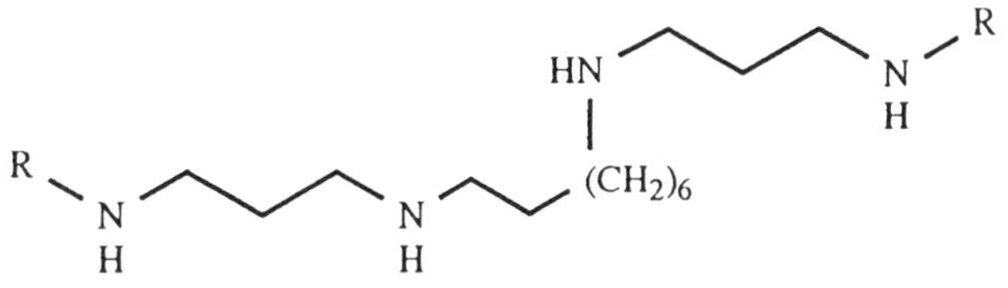

General structure of tetraamines based on 1,8-diaminooctane

N^1,N^{11}-*bis*(ethyl)norspermine

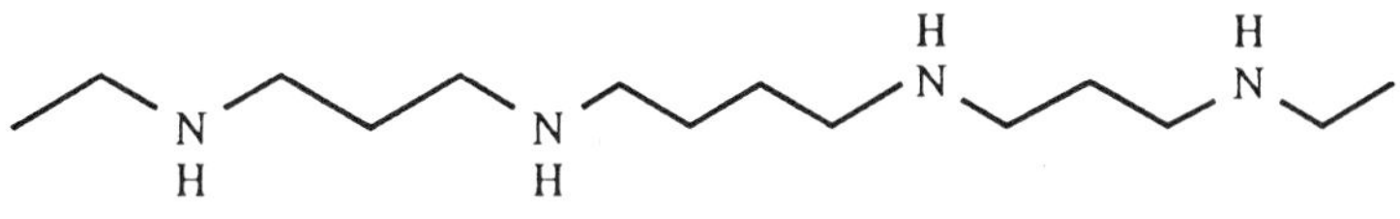

N^1,N^{12}-*bis*(ethyl)spermine

Fig. 7.7 Various tetra-amine analogs of the natural polyamines

Edwards et al has synthesized a series of di- and triamines related to spermidine, and a series of tetra-amines derived from 1,8-diaminooctane, and these analogs have been evaluated for antitumor activity in an L1210 cultured cell system.[70] In the series of di- and triamines, substitution of alkyl groups at the terminal nitrogens, or replacing the central nitrogen with heteroatoms, failed to produce spermidine analogs with antitumor effects superior to norspermidine. However, tetramines with the general structure shown in Fig. 7.7 generally showed improved antitumor activity. The most active compound in the series (R = H) increased survival time in male mice inoculated with L1210 leukemia from 7.7 days to 16.2 days. Coadministration of spermidine was shown to reverse the antitumor activity of this compound, presumably due to a competition for the polyamine transport system. In addition, coadministration of a polyamine oxidase inhibitor potentiated the observed antitumor activity, suggesting that these analogs may be metabolized by polyamine oxidase. Analogs in which R = CH_3 or R = CH_2CH_3 also showed activity in the L1210 model, but substitution of larger alkyl groups resulted in a reduction of activity. Polyamine levels did not appear to be dramatically reduced following treatment with the tetra-amine analogs, and the ability of the compounds to interact with SSAT was not established. In a related study, these tetra-amines, the aforementioned bisbenzyl analogs related to MDL 27695, and an additional series of substituted tetra-amines were evaluated for the ability to inhibit proliferation of HeLa cells.[80] The ability of each compound to displace ethidium bromide from calf thymus DNA was also determined, although no correlation between the DNA binding properties and antitumor activity was detected. Not surprisingly, the bisbenzyl polyamine analog MDL 27695 and the tetra-amine shown above (R = H) were active antiproliferative compounds, exhibiting IC_{50} values of 5 and 50 μM, respectively.

C. The bis(ethyl)polyamines

Bergeron et al have recently reported the synthesis and preliminary biological evaluation of a series of terminally alkylated polyamine analogs with promising antitumor activity.[68,81] The most active analog in the series, N^1,N^{12}-*bis*(ethyl)spermine (BESpm), was cytostatic against cultured L1210 cells, with an IC_{50} value of less than 1 μM at 96 hours. All of the spermine analogs described were more potent than the previously evaluated spermidine analogs,[76] and caused a depletion of cellular polyamines and down-regulation of ODC and AdoMet-DC. BESpm also caused an increase in life span in excess of 200% in male DBA/2 mice which had been inoculated with L1210 leukemia cells. In terms of structure/activity relationships, ethyl substitution at a terminal nitrogen was optimal for activity, followed in order by propyl and methyl substitution. However, alkyl substitution at an internal nitrogen resulted in a dramatic decrease in activity.

The continuing interest in the *bis*(alkyl)polyamine analogs is based on findings demonstrating that natural polyamines utilize several feedback mechanisms which regulate their synthesis.[8,82,83] The N,N'-*bis*(ethyl)-polyamines are readily transported into mammalian cells, apparently by the same pathway as the natural polyamines.[8,83] Treatment of most mammalian cell types with these analogs leads to a reduction of all three natural polyamines (Spd, Spm, Put), decreases in ODC activity and AdoMetDC activity, and ultimately in cytostasis.[11,68,83,84] Subsequent to these effects, the *bis*(ethyl)polyamines are cytotoxic to several representative nonsmall cell lung cancers which are typically refractory to any nonsurgical intervention. In fact, *bis*(ethyl)spermidine (BESpd) and BESpm both elicit a cytotoxic response in the human large cell lung tumor cell line H157, which has been shown to be refractory to the effects of the ODC inhibitor α-difluoromethylornithine.[85] There is now substantial data which indicate that the superinduction of the polyamine catabolic enzyme SSAT is associated with these observed cytotoxic responses.[11,82,86]

The induction of SSAT in two human lung cancer cell lines which respond differently to treatment with inhibitors of polyamine biosynthesis has recently been studied.[72,87] In these cell lines, the level of induction of SSAT appears to correlate inversely with the degree of resistance to cytotoxicity following treatment with the polyamine analog BESpm.[72] Rate of growth and cellular polyamine content in the human small cell lung carcinoma (SCLC) line NCI H82 are minimally affected by BESpm, which appears to down-regulate polyamine biosynthesis by the same mechanism as the natural polyamines.[88] By contrast, BESpm was found to be markedly cytotoxic at a concentration of 10 μM in the large cell lung carcinoma (LCLC) line NCI H157, accompanied by nearly complete depletion of all intracellular polyamines and a decrease in ornithine decarboxylase (ODC) activity to undetectable levels. In the responsive LCLC line, BESpm was found to induce SSAT in a time- and dose-dependent manner to maximum levels greater than 1000-fold above baseline. Conversely, in the unresponsive SCLC cell line, minimal (less than 7-fold) induction of SSAT was observed, regardless of the time period or the concentration of BESpm employed. These results suggest that the differential induction of SSAT may play a role in determining cell specific sensitivity to polyamine antimetabolites. It has further been determined that in cell lines representing the major forms of lung cancer, SSAT induction correlates positively with cytotoxic response following treatment with BESpm. In addition, SSAT induction appears to occur at the level of increased steady-state mRNA and new protein synthesis.[89] The correlation between high induction of the SSAT activity and cytotoxicity has now been demonstrated for other examples of human malignancies including human adenocarcinoma and melanoma.[90-93]

The *bis(*ethyl)polyamines, and in particular, N^1,N^{11}-bis(ethyl)-norspermine (BENSpm, see Fig. 7.7) have recently been evaluated for antitumor activity in a series of preclinical in vivo trials.[90-93] The *bis*(ethyl)polyamine analogs were evaluated in human MALME-3 melanoma xenografts and found to exhibit potent antitumor activity in this model.[90] The most potent analog, BENSpm, was determined to be the least toxic and best tolerated analog in the series. BENSpm was also effective in the PANUT-3 human melanoma xenograft model.[91] In both cases, treatment with BENSpm was accompanied by superinduction of SSAT and reduction of polyamine pools, although a decrease in ODC and AdoMet-DC activity was not detected. In an in vivo human pancreatic adenocarcinoma model, BENSpm was one of the most effective agents yet tested, having greater activity than the conventional agent 5-fluorouracil (alone or in combination with adriamycin and mitomycin C).[92] In this model, a striking induction of SSAT was detected, as was a reduction in ODC and AdoMet-DC activity. However, the depletion of polyamines was minimal, suggesting that the observed antitumor activity may in this case result from a mechanism other than polyamine depletion.[93] As a result of the studies outlined above, BENSpm is currently in Phase I clinical trials.

Saab et al have recently reported the synthesis and preliminary evaluation of two unsymmetrically substituted polyamine analogs, N^1-ethyl-N^{11}-propargyl-4,8-diazaundecane (PENSpm) and N^1-ethyl-N^{11}-(cyclopropyl)methyl-4,8-diazaundecane (CPENSpm), shown in Figure 7.8.[94] These analogs were found to be effective inhibitors of SSAT, exhibiting similar potency to the known inhibitor BESpm. In 96 hour dose-response experiments, the unsymmetrically substituted analogs, like BESpm, showed a cell-type specific selective cytotoxicity. Treatment of H157 cells with either PENSpm or CPENSpm resulted in cytotoxicity, with IC_{50} values of 0.79 and 0.57 μM, respectively. Both compounds were more potent against the H157 cell line than the known inhibitor BESpm. By contrast, compounds PENSpm and CPENSpm are minimally cytostatic in the H82 SCLC cell line, exhibiting IC_{50} values of 4.74 and 0.84 μM, respectively. This observation is similar to the results of studies involving BESpm, in which no significant cytotoxicity was observed at concentrations of up to 100 μM in the H82 cell line.[87]

PENSpm and CPENSpm were evaluated in vitro for their ability to induce SSAT in both H157 LCLC cells and H82 SCLC cells, and were found to be highly effective inducers of the human enzyme in the responsive H157 LCLC cell line, producing SSAT levels comparable to or greater than those observed following treatment with BESpm. Neither compound caused significant induction of SSAT in the treated H82 SCLC cell line, an observation which is also consistent with previous studies involving BESpm.[87] These findings support the assertion that there is a positive correlation between analog-initiated superinduction of SSAT and the development of cytotoxicity.

The promising antitumor activity of the norspermine analogs PENSpm and CPENSpm led to the synthesis of additional congeners in the series, all of which are shown in Figure 7.8. All of the analogs tested retain the remarkable cell-type specificity observed with CPENSpm, and are rapidly cytotoxic to H157 cells in culture, while having a minimal effect on the H82 small cell line. Compounds with smaller cycloalkylmethyl ring substituents (CPENSpm, CBENSpm) appear to effect H157 and H82 cultured lung carcinoma cells in exactly the same manner as

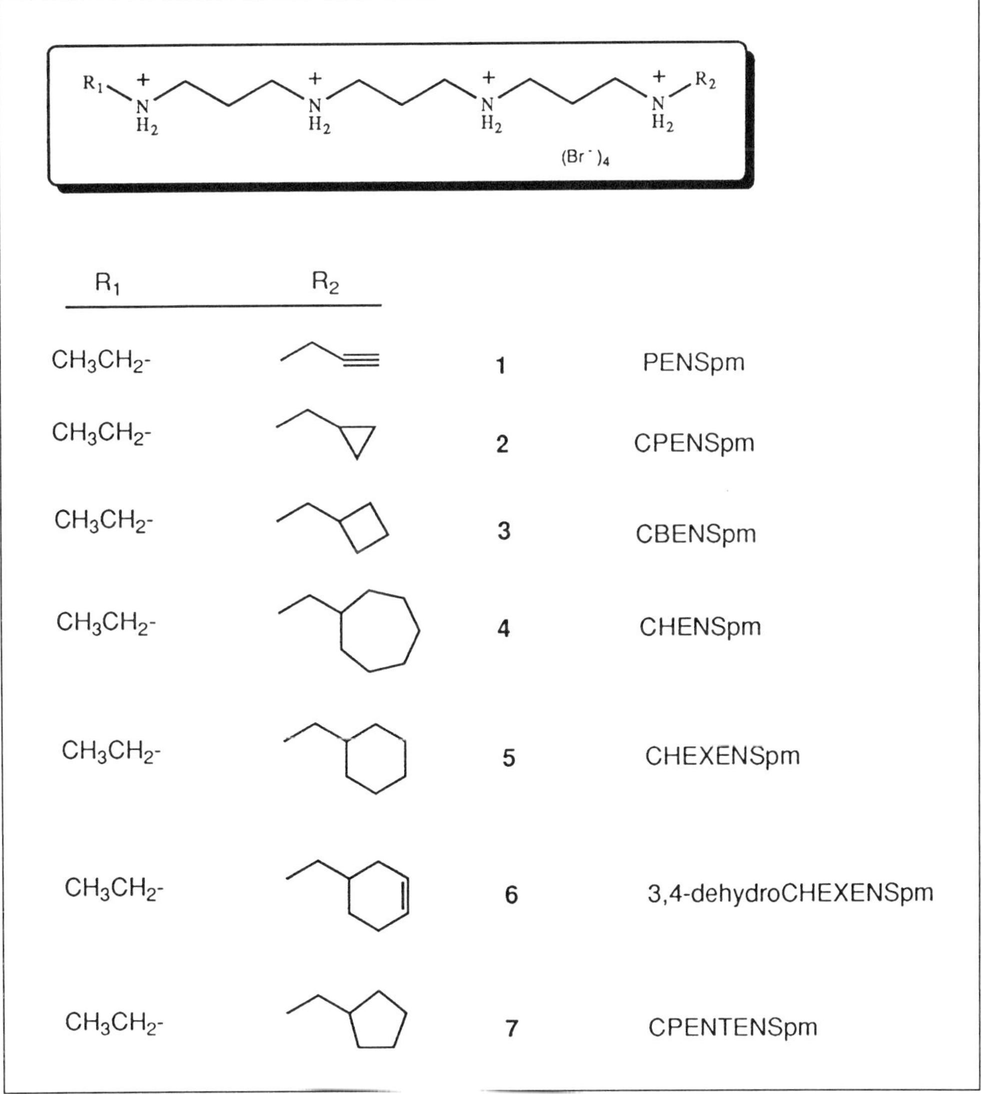

Fig. 7.8. Unsymetrically substituted polyamine analogs.

BENSpm, causing a dramatic induction of SSAT, down-regulation of ODC and AdoMet-DC, polyamine depletion and a rapid cytotoxicity at concentrations as low as 5 µM.[95] However, analogs with larger ring substituents, most notably CHENSpm (Fig. 7.8), appear to induce cytotoxicity by a different mechanism (unpublished observations). Following 24 hour treatment with 10 µM CHENSpm, there is little induction of SSAT, and polyamine levels are minimally affected in the H157 cell line. Despite this apparent lack of effect on the polyamine system, CHENSpm is currently the most cytotoxic alkylpolyamine produced to date, and is rapidly cytotoxic in H157 cells (and in other tumor cell lines) at concentrations as low as 1 µM. These observations are in sharp contrast to the cellular events that take place following treatment with 10 µM BENSpm, as shown in Table 7.2, and suggest that CHENSpm is causing a cytotoxic response by a distinctly different mechanism than BENSpm or CPENSpm. Interestingly, CHENSpm retains the dramatic cell-type specificity seen with CPENSpm, since it is cytostatic to the H82 cell line in culture. Experiments are now underway to elucidate the mechanism by which CHENSpm produces cytotoxicity in H157 and other cell lines.

Table 7.2. Polyamine and SSAT levels in cultured NCI H157 non-SCLC cells following 24 hour treatment with BENSpm and CHENSpm

Treatment	PUT	SPD	SPM	SSAT activity (pmol/mg/min)	Fold increase
Control	2.00	9.83	14.92	52	0.00
10 µM BENSpm	0.00	0.00	0.51	148141	2849
10 µM CHENSpm	4.28	6.44	15.97	781	15.01

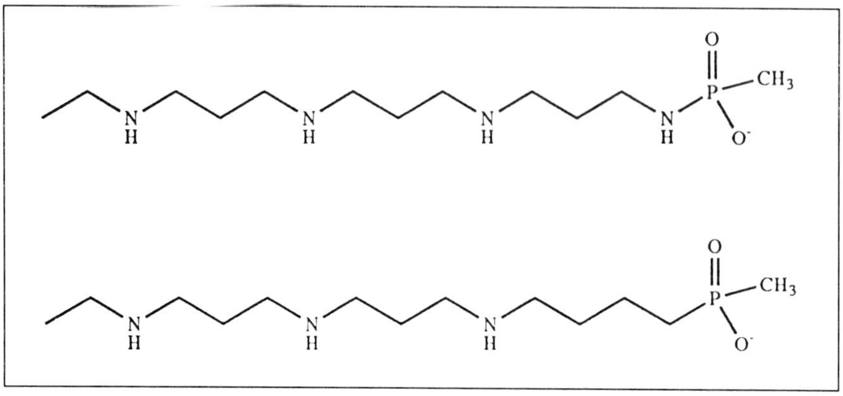

Fig. 7.9. Transition state analog inhibitors of SSAT.

Woster and coworkers have recently synthesized two phosphate-containing transition state analogs (Fig. 7.9) for the SSAT reaction, and these analogs have been evaluated as inhibitors of purified human SSAT (unpublished observations). The phosphonamidate shown in Figure 7.9 does not appear to inhibit SSAT in an in vitro assay. This lack of inhibition is presumably due to the lability of the phosphonamidate bond, and preliminary studies indicate that the analog is hydrolyzed to afford monoethylnorspermine, which is subsequently a substrate for SSAT. The lack of chemical stability observed with the phosphonamidate was overcome by synthesizing the corresponding phosphinate analog, as shown in Figure 7.9. This analog is an effective inhibitor of purified SSAT, exhibiting an IC_{50} concentration of 100 μM following direct addition of the analog, and 50 μM when assayed following a 10 or 30 minute preincubation. The phosphinate transition state analog may become an important research tool, especially if it inhibits SSAT without significant induction, since it could be used to study the effects of selective depletion of acetylated polyamines in the cell. In addition, its potential value as a chemotherapeutic agent is under investigation.

The discovery that SSAT participates in the development of cytotoxicity in some tumor cell lines has opened up a new avenue for research in the polyamine field. The synthetic analogs mentioned in this chapter should prove invaluable for determining the precise role of SSAT in normal cellular metabolism, and for elucidating the mechanism by which the induction of SSAT produces an antiproliferative effect. It is also clear that substituted polyamine analogs, and particularly those which are structurally related to spermine, have the potential to become novel antineoplastic agents. Studies are now being conducted which have the potential to address these important questions, and to produce additional analogs for development as antitumor agents.

References

1. Sunkara PS, Baylin SB and Luk GD. Inhibitors of polyamine biosynthesis: cellular and in vivo effects on tumor proliferation In: PP McCann AE Pegg and A Sjoerdsma, eds., Inhibition of Polyamine Metabolism. Biological Significance and Basis for New Therapies. Academic Press NY, 1987:121-140.
2. Coward JK. Polyamine metabolism—recent developments and implications for the design of new chemotherapeutic agents. Ann Rep Med Chem 1982;17: 253-259.
3. McCann PP and Pegg AE. Ornithine decarboxylase as an enzyme target for therapy. Pharmac Ther 1992;54:195-215.
4. Pegg AE. Recent advances in the biochemistry of polyamines in eukaryotes. Biochem J 1986;234:249-262.
5. Pegg AE. Polyamine metabolism and its importance in neoplastic growth and as a target for chemotherapy. Cancer Res 1988;48:759-774.

6. Pegg AE and McCann PP. S-Adenosylmethionine decarboxylase as an enzyme target for therapy. Pharmac Ther 1992;56:359-377.
7. Pegg AE and McCann PP. Polyamine metabolism and function. Am J Physiol 1982;243:C212-C221.
8. Porter CW and Sufrin J. Interference with polyamine biosynthesis and/or function by analogs of polyamines or methionine as a potential anticancer chemotherapeutic strategy. Anticancer Res 1986;6:525-542.
9. Tabor CW and Tabor H. Polyamines. Ann Rev Biochem 1984; 53:749-790.
10. Seiler N. Functions of polyamine acetylation. Can J Physiol Pharmacol 1987;65:2024-2035.
11. Casero JR RA and Pegg AE. Spermidine/spermine N^1-acetyltransferase - the turning point in polyamine metabolism. FASEB J 1993;7:653-661.
12. Metcalf BW, Bey P, Danzin C, et al. J Am Chem Soc 1978;100:2551-2553.
13. Tang K-C, Mariuzza R and Coward JK. Synthesis and evaluation of some stable multisubstrate adducts as specific inhibitors of spermidine synthase. J Med Chem 1981;24:1277-1284.
14. Woster PM, Black AY, Duff KJ, et al. Synthesis and biological evaluation of S-adenosyl-1,12-diamino-3-thio-9-azadodecane a multisubstrate adduct inhibitor of spermine synthase. J Med Chem 1989;32:1300-1307.
15. Recsei PA and Snell EE. Pyruvoyl enzymes. Ann Rev Bioch 1984; 53:357-387.
16. Wickner RB. Tabor CW and Tabor H. Purification of adenosylmethionine decarboxylase from *Escherichia coli* W. Evidence for covalently bound pyruvate. J Biol Chem 1970;245:2132-2139.
17. Markham GD. Tabor CW and Tabor H. S-Adenosylmethionine decarboxylase of *Escherichia coli.* Studies on the covalently linked pyruvate required for activity. J Biol Chem 1982;257:12063-12068.
18. Anton DL and Kutny R. *Escherichia coli* S-Adenosylmethionine decarboxylase: subunit structure reductive amination and NH_2-terminal sequences. J Biol Chem 1987; 262:2817-2822.
19. Pegg AE and Williams-Ashman HG. On the role of S-adenosylmethionine in the biosynthesis of spermidine by rat prostate. J Biol Chem 1969;244:682-693.
20. Shirahata A and Pegg AE. Regulation of S-adenosylmethionine decarboxylase activity in rat liver and prostate. J Biol Chem 1985;260:9583-9588.
21. Pajunen A, Crozat A, Janne O, et al. Structure and regulation of mammalian S-adenosylmethionine decarboxylase. J Biol Chem 1988; 263:17040-17049.
22. Stanley BA, Pegg AE and Holm I. Site of pyruvate formation and processing of mammalian S-adenosylmethionine decarboxylase proenzyme. J Biol Chem 1989;264:21073-21079.
23. Stanley BA, Shantz LM and Pegg AE. Expression of mammalian S-adenosylmethionine decarboxylase in *Escherichia coli.* Determination of sites for putrescine activation of activity and processing. J Biol Chem

1994;269:7901-7907.

24. Shantz LM, Stanley BA, Secrist III JA and Pegg AE. Purification of human S-adenosylmethionine decarboxylase expressed in *Escherichia coli* and the use of this protein to investigate the mechanism of inhibition by 5'-deoxy-5'-[(3hydrazinopropyl)methylamino]-adenosine and 5'{[(Z)-4-amino-2-butenyl]methylamino}-5'-deoxyadenosine. Biochemistry 1992;31:6848-6855.
25. Stanley BA and Pegg AE. Amino acid residues necessary for putrescine stimulation of human S-adenosylmethionine decarboxylase proenzyme processing and catalytic activity. J Biol Chem 1991;266:18502-18506.
26. Tekwani BL, Bacchi CJ and Pegg AE. Putrescine activated S-adenosylmethionine decarboxylase from *Trypanosoma brucei brucei.* Mol Cell Bio 1992;117:53-61.
27. Anton DL and Kutny R. Mechanism of substrate inactivation of *Escherichia coli* S-adenosylmethionine decarboxylase. Biochemistry 1987;26:6444-6447.
28. Diaz E and Anton DL. Alkylation of an active-site cysteinyl residue during substrate-dependent inactivation of *Escherichia coli* S-adenosylmethionine decarboxylase. Biochemistry 1991;30:4078-4081.
29. Karvonen E, Kauppinen L, Partanen T and Poso H. Irreversible inhibition of putrescine-stimulated S-adenosylmethionine decarboxylase by berenil and pentamidine. Biochem J 1985;231:165-169.
30. Williams-Ashman HG and Pegg AE. In: DR Morris and LJ Marton, eds. Polyamines in Biology and Medicine. NY: Dekker, 1981:43-73 .
31. Hibasami H, Tsukada T, Maekawa S, et al. Antitumor effect of methylglyoxal bis(3-aminopropylamidinohydrazone). Cancer Chemother Pharmac 1988;22:187-190.
32. Hibasami H, Maekawa S, Murata T and Nakashima K. Antitumor effect of a new multienzyme inhibitor of the polyamine synthetic pathway methylglyoxal bis(cyclopentylamidinohydrazone) against human and mouse leukemia cells. Cancer Res 1989;49:2065-2068.
33. Elo H, Mutikainen I, Alhonen-Hongisto L, Laine R and Janne J. Diethylglyoxal bis(guanylhydrazone): a novel highly potent inhibitor of S-adenosylmethionine decarboxylase with promising properties for potential chemotherapeutic use. Cancer Lett 1988;41:21-30.
34. Nakashima K, Hibasami H, Tsukuda T and Maekawa S. Methylglyoxal bis(guanylhydrazone) analogs: multifunctional inhibitors for polyamine enzymes. Eur J Med Chem 1987;22:553-558.
35. Stanek J, Caravatti G, Capraro HG, et al. S-Adenosylmethionine decarboxylase inhibitors. New aryl and heteroaryl analogs of methylglyoxal bis(guanylhydrazone). J Med Chem 1993;36:46-54.
36. Regenass U, Caravatti G, Mett H, et al. New S-adenosylmethionine decarboxylase inhibitors with potent antitumor activity. Cancer Res 1992;52:4712-4718.
37. Stanek J, Caravatti G, Frei J, et al. 4-Amidinoindan-1-one 2'-amidinohydrazone. A new potent and selective inhibitor of S-adenosylmethionine decarboxylase. J Med Chem 1993;36:2168-2171.
38. Bey P, Vevert JP, VanDorsselaer V and Kolb M. J Org Chem

1979;44:2732.

39. Pankaskie M and Abdel-Monem MM. Inhibitors of polyamine biosynthesis. 8. Irreversible inhibition of mammalian S-adenosylmethionine decarboxylase by substrate analogs. J Med Chem 1980;23:121-127.
40. Kolb M, Danzin C, Barth J and Claverie N. Synthesis and biochemical properties of chemically stable product analogs of the reaction catalyzed by S-adenosylmethionine decarboxylase. J Med Chem 1982;25:550-556.
41. Artamonova EY, Zavalova LL, Khomutov RM and Khomutov AR. Irreversible inhibition of S-adenosylmethionine decarboxylase by hydroxylamine-containing analogs of decarboxylated S-adenosylmethionine. Bioorg Khim 1986;12:206-212.
42. Khomutov AR, Kritsky AM, Artamonova EY and Khomutov RM. Polyfunctional O-substituted hydroxylamines. Modification of nucleic acids and inhibition of SAM-decarboxylase. Nucleosides and Nucleotides 1987;6:359-362.
43. Kramer DL, Khomutov RM, Bukin YV, et al. Cellular characterization of a new irreversible ianhibitor of S-adenosylmethionine decarboxylase and its use in determining the relative abilities of individual polyamines to sustain growth and viability of L1210 cells. Biochem J 1989;259:325-331.
44. Secrist III JA. New substrate analogs as inhibitors of S-adenosylmethionine decarboxylase. Nucleosides and Nucleotides 1987;6:73-83.
45. Pegg AE, Jones DB and Secrist III JA. Effect of inhibitors of S-adenosylmethionine decarboxylase on polyamine content and growth of L1210 cells. Biochemistry 1988;27:1408-1415.
46. Casara P, Marchal P, Wagner J and Danzin C. 5'{[(Z)-4-amino-2-butenyl]methylamino}-5'-deoxyadenosine. A potent enzyme activated irreversible inhibitor of S-adenosyl-l-methionine decarboxylase from *Escherichia coli*. J Am Chem Soc 1989;111:9111-9113.
47. Danzin C, Marchal P and Casara P. Irreversible inhibition of rat S-adenosylmethionine decarboxylase by 5'{[(Z)-4-amino-2-butenyl]methylamino}-5'-deoxyadenosine. Biochem Pharmacol 1990; 40:1499-1503.
48. Wu YQ and Woster PM. S-(5'-Deoxy-5'-adenosyl)-1-amino-4-methylthio-2-cyclopentene (AdoMac). A potent and irreversible inhibitor of S-adenosylmethionine decarboxylase. J Med Chem 1992;35:3196-3201.
49. Wu YQ and Woster PM. Resolution of the pure diastereomeric forms of S-(5'-deoxy-5'-adenosyl)-1-ammonio-4-methylsulfonio-2-cyclopentene and their evaluation as irreversible inhibitors of S-adenosylmethionine decarboxylase from *Escherichia coli*. Bio Med Chem 1993;1:349-360.
50. Wu YQ and Woster PM. Irreversible inhibition of human S-adenosylmethionine decarboxylase by the pure diastereomeric forms of AdoMac S-(5'-deoxy-5'-adenosyl)-1-ammonio-4-methylsulfonio-2-cyclopentene. Biochem Pharmacol 1995;49:1125-1133.
51. Wu YQ, Lawrence T, Guo JQ and Woster PM. α-Cyano-substituted analogs of S-adenosylmethionine as enzyme activated irreversible inhibitors of S-adenosylmethionine decarboxylase. Bioorg Med Chem Lett 1993;

3:2811-2816.

52. Tekwani BL, Bacchi CL, Secrist III JA and Pegg AE. Irreversible inhibition of S-adenosylmethionine decarboxylase of Trypanosoma brucei brucei by S-adenosylmethionine analogs. Biochem Pharmacol 1992;44:905-911.
53. Guo JQ, Wu YQ, Douglas KA, et al. Restricted rotation analog inhibitors of S-adenosylmethionine. Synthesis and evaluation of their selective toxicity in *Trypanosoma bruceii bruceii*. Bioorg Med Chem Lett 1993;3:147-152.
54. Carter NS and Fairlamb AH. Arsenical-resistant trypanosomes lack an unusual adenosine transporter. Nature 1993;361:173-175.
55. Byers TL, Ganem B and Pegg AE. Cytostasis induced in L1210 murine leukemia cells by the S-adenosylmethionine decarboxylase inhibitor 5'{[(Z)-4-amino-2-butenyl]methylamino}-5'-deoxyadenosine may be due to hypusine depletion. Biochem J 1992;287:717-724.
56. Seiler N, Sarhan S, Mamont P, Casara P and Danzin C. Some biological consequences of S-adenosylmethionine decarboxylase inhibition by MDL 73811. Life Chem Rep 1991;9:151-162.
57. Byers TL, Bush TL, McCann PP and Bitonti AJ. Antitrypanosomal effects of polyamine biosynthesis inhibitors correlate with increases in Trypanosoma brucei brucei S-adenosyl-L-methionine. Biochem J 1991;274:527-533.
58. Byers TL, Casara P and Bitonti AJ. Uptake of the antitrypanosomal drug 5'{[(Z)-4-amino-2-butenyl]methylamino}-5'-deoxyadenosine (MDL 73811) by the purine transport system of *Trypanosoma brucei brucei*. Biochem J 1992;283:755-758.
59. Dredar SA, Blankenship JW, Marchant PE, et al. Design and synthesis of inhibitors of N^8-acetylspermidine deacetylase. J Med Chem 1989;32:984-989.
60. Marchant P, Dredar S, Manneh V, et al. A selective inhibitor of N^8-acetylspermidine deacetylation in mice and HeLa cells without effects on histone deacetylation. Arch Bioch Bioph 1989;15:128-136.
61. Huang TL, Dredar SA, Manneh VA, et al. Inhibition of N^8-acetylspermidine deacetylase by active-site-directed metal coordinating inhibitors. J Med Chem 1992;35:2414-2418.
62. Matsui I, Wiegand L and Pegg AE. Properties of spermidine N-acetyltransferase from livers of rats treated with carbon tetrachloride and its role in the conversion of spermidine into putrescine. J Biol Chem 1981;256:2454-2459.
63. Della Ragione F and Pegg AE. Studies of the specificity and kinetics of rat liver spermidine/spermine N^1-acetyltransferase. Biochem J 1983; 213:701-706.
64. Lakanen JR, Coward JK and Pegg AE. α-Methyl polyamines. Metabolically stable spermidine and spermine mimics capable of supporting growth in cells depleted of polyamines. J Med Chem 1992;35:724-734.
65. Nagarajan S, Ganem B and Pegg AE. Studies of non-metabolizable polyamines that support growth of SV-3T3 cells depleted of natural polyamines by exposure to α-difluoromethylornithine. Biochem J

1988;254:373-378.

66. Erwin BG, Persson L and Pegg AE. Differential inhibition of histone and polyamine acetylases by multisubstrate analogs. Biochemistry 1984; 23:4250-4255.
67. Cullis PM, Wolfenden R, Cousens LS and Alberts BM. Inhibition of histone acetylation by N-[2-(S-coenzyme A)acetyl]spermidine amide a multisubstrate analog. J Biol Chem 1982;257:12165-12169.
68. Bergeron RJ, Neims AH, McManis JS, et al. Synthetic polyamine analogs as antineoplastics. J Med Chem 1988;31:1183-1190.
69. Basu HS, Feuerstein BG, Deen DF, et al. Correlation between the effects of polyamine analogs on and conformation and cell growth. Cancer Research 1989;49:5591-5597.
70. Edwards ML, Prakash NJ, Stemerick DM, et al. Polyamine analogs with antitumor activity. J Med Chem 1990;33:1369-1375.
71. Bitonti AJ, Bush TJ and McCann PP. Regulation of polyamine biosynthesis in rat hepatoma (HTC) cells by a bisbenzyl polyamine analog. Bioch J 1989;257:769-774.
72. Casero JR RA, Celano P, Ervin SJ, et al. Differential induction of spermidine/spermine-N^1-acetyltransferase in human lung cancer cells by the bis(ethyl)polyamine analogs. Cancer Res 1989;49:3829-3833.
73. Porter CW, McManis J, Casero JR RA and Bergeron RJ. The relative abilities of the bis(ethyl) derivatives of putrescine spermidine and spermine to regulate polyamine biosynthesis and inhibit cell growth. Cancer Res 1987;47:2821-2825.
74. Porter CW, Pegg AE, Ganis B, Madhabala R and Bergeron RJ. Combined regulation of ornithine and S-adenosylmethionine decarboxylases by the spermine analog N^1,N^{12}-bis(ethyl)spermine. Biochem J 1990;268:207-212.
75. Erwin BG and Pegg AE. Regulation of spermidine/spermine N^1-acetyltransferase in L6 cells by polyamines and related compounds. Biochem J 1986;238:581-587.
76. Porter CW, Berger FG, Pegg AE, Ganis B and Bergeron RJ. Regulation of ornithine decarboxylase activity by spermidine and the spermidine analog N^1,N^8-bis(ethyl)spermidine. Bioch J 1987;242:433-440.
77. Bitonti AJ, Dumont JA, Bush TL, et al. Bis(benzyl)polyamine analogs inhibit the growth of chloroquine-resistant human malaria parasites (*Plasmodium falciparum*) in vitro and in combination with α-difluoromethylornithine cure murine malaria. Proc Natl Acad Sci USA 1989;86:651-656.
78. Bitonti AJ, Bush TL and McCann PP. Regulation of polyamine biosynthesis in rat hepatoma (HTC) cells by a bisbenzyl polyamine analog Biochem J 1989;257:769-774.
79. Edwards ML, Stemerick DM, Bitonti AJ, et al. Antimalarial polyamine analogs. J Med Chem 1991;34:569-574.
80. Edwards ML, Snyder RD and Stemerick DM. Synthesis and DNA-binding properties of polyamine analogs. J Med Chem 1991;34:2414-2420.
81. Bergeron RJ, Garlich JR and Stolowich NJ. Reagents for the stepwise functionalization of spermidine homospermidine and bis(3-amino-

propyl)amine. J Org Chem 1984;49:2997-3001.
82. Casero RA JR and Baylin SB. Concepts for deriving specific inhibitors of polyamine biosynthesis - human lung cancer cells as a model system In: MP Palfreman PP McCann W Lovenberg JG Temple and A Sjoerdsma, eds. Enzymes as Targets for Drug Design. San Diego: Academic Press, 1989:185-200.
83. Porter CW, Ganis B, Vinson T, et al. A comparison and characterization of growth inhibition by α-difluoromethylornithine (DFMO) an inhibitor of ornithine decarboxylase and N^1,N^8-bis(ethyl)spermidine (BES) an apparent regulator of the enzyme. Cancer Res 1986;46:6279-6285.
84. Bergeron RJ, Hawthorne TR, Vinson JRT, et al. Role of the methylene backbone in the antiproliferative activity of polyamine analogs in L1210 cells. Cancer Res 1989;49:2959-2964.
85. Casero RA, Go B, Theiss HW, et al. Cytotoxic response of the relatively difluoromethylornithine-resistant human lung tumor cell line NCI H157 to the polyamine analog N^1,N^8-bis(ethyl)spermidine. Cancer Res 1987;47:3964-3967.
86. Casero RA, Celano P, Ervin SJ, et al. High specific induction of spermidine/spermine-N^1-acetyltransferase in a human large cell lung carcinoma. Biochem J 1990;270:615-620.
87. Casero RA, Ervin SJ, Celano P, et al. Differential response to treatment with bis(ethyl)polyamine analogs between human small cell lung carcinoma and undifferentiated large cell lung carcinoma in culture. Cancer Res 1989;49:639-643.
88. Porter CW, McManis J, Casero JR RA and Bergeron RJ. The relative abilities of the bis(ethyl) derivatives of putrescine, spermidine and spermine to regulate polyamine biosynthesis and inhibit cell growth. Cancer Res 1987;47:2821-2825.
89. Casero RA, Mank AR, Xiao L, et al. Steady-state messenger RNA and activity correlates with sensitivity to N^1,N^{12}-bis(ethyl)spermine in human cell lines representing the major forms of lung cancer. Cancer Res 1992;52:5359-5363.
90. Bernacki RJ, Bergeron RJ and Porter CR. Antitumor activity of N,N'-bis(ethyl)spermine homologues against human MALME-3 melanoma xenografts. Cancer Research 1992;52:2424-2430.
91. Porter CW, Bernacki RJ, Miller J and Bergeron RJ. Antitumor activity of N^1,N^{11}-bis(ethyl)norspermine against human melanoma xenografts and possible biochemical correlates of drug action. Cancer Research 1993;53:581-586.
92. Chang BK, Bergeron RJ, Porter CW and Liang Y. Antitumor effects of N-alkylated polyamine analogs in human pancreatic adenocarcinoma models. Cancer Chemother Pharmacol 1992;30:179-182.
93. Chang BK, Bergeron RJ, Porter CW, et al. Regulatory and antiproliferative effects of N-alkylated polyamine analogs in human and hamster pancreatic adenocarcinoma cell lines. Cancer Chemother Pharmacol 1992; 30:183-188.

94. Saab NH, West EE, Bieszk NC, et al. Synthesis and evaluation of unsymmetrically substituted polyamine analogs as modulators of human spermidine/spermine-N[1]-acetyltransferase (SSAT) and as potential antitumor agents. J Med Chem 1993;36:2998-3004.
95. Casero RA, Mank AR, Saab NH, et al. Growth and biochemical effects of asymmetrically substituted polyamine analogs in human lung tumor cells. Cancer Chemother Pharmacol 1995;36:69-74.

CHAPTER 8

The Polyamine Metabolic Pathway as a Target for Chemoprevention

Craig M. Berchtold, Thomas W. Kensler, Robert A. Casero, Jr.

I. INTRODUCTION

A functional polyamine pathway is considered essential for the process of cell proliferation in both normal and neoplastic cells and tissues.[1-4] Depletion of intracellular polyamines generally results in cytostasis. However there are cell type-specific instances where cytotoxicity results from polyamine depletion.[2,5] Therefore the polyamine pathway has been a target for the rational development of therapeutic agents in the treatment of cancer. More recently, there has been increased interest in polyamine metabolism in the field of chemoprevention. A common goal in the chemoprevention field is to reduce the incidence of cancer in a population that is highly susceptible to the disease because of familial background, previous cancer treatment, or increased exposure to an etiologic agent.[6-9] The chemoprevention trials of 2-difluoromethylornithine (DFM0), an inhibitor of ornithine decarboxylase (ODC), represents a significant advancement toward this end.[10] In most cases, the targeted population for chemoprevention is asymptomatic and any deleterious effects of chemical intervention is unacceptable.

The spectrum of action of chemopreventive agents encompasses many biochemical processes and may be grouped into two main categories: antimutagens and antiproliferatives such as the polyamine pathway inhibitor, DFMO.[11] A primary focus in chemoprevention is to retard or inhibit the transformation of benign intraepithelial neoplasia to a

Polyamines: Regulation and Molecular Interaction, edited by Robert Casero. © 1995 R.G. Landes Company.

clinically less treatable invasive carcinoma. The dynamic target tissue pertinent for chemical intervention, intraepithelial neoplasia, was proposed by Boone et al[11] (Fig. 8.1). In theory, the actions of chemopreventive agents could target the many basic biochemical activities regulating the fidelity of DNA and the proliferation of preneoplastic and neoplastic tissue before the onset of anaplasia.[11]

The mouse skin carcinogenesis model, on which the theory of intraepithelial neoplasia was founded, closely mimics the histological changes that take place in human carcinogenesis.[9] The chemical carcinogenesis stages of initiation, promotion, and progression were derived from this fundamental rodent paradigm, which has been instrumental in the assessment of putative chemopreventive agents. The multistage nature of chemical carcinogenesis in mouse skin has also been the progenitor to similar studies in the rodent mammary, urinary tract, colon and liver systems. Thus, the application of experimental chemical carcinogenesis protocols to internal organ systems has helped to foster clinical chemopreventive investigations involving tissue specific cancers.[7]

Early chemotherapy trials with inhibitors to the regulatory enzymes of the polyamine metabolic pathway revealed problems with phamacological interventions against a cycle with multiple compensatory mechanisms for the maintenance of polyamine homeostasis.[10,12] The molecular processes of the pathway that maintain physiological levels of intracellular polyamine concentrations are: 1) the extremely short half-life and rapid de novo synthesis of the regulatory biosynthetic enzymes, ODC and S-adenosylmethionine decarboxylase (AdoMetDC); 2) the back conversion pathway regulated by spermidine/spermine N^1-acetyltransferase (SSAT), which can recycle spermidine and putrescine from spermine and spermidine, respectively; and 3) the highly inducible polyamine transporter, which controls polyamine uptake into the cell. To circumvent these problems in the deregulation of the polyamine metabolic pathway, polyamine analogs were synthesized by Bergeron and collegues[13] to compete with the cellular uptake of natural polyamines and to down-regulate the biosynthetic enzymes of the pathway.[10,13] Several spermine analogs are currently involved in Phase I clinical trials and may be useful chemopreventive agents.

The products of the polyamine metabolic pathway, spermidine and spermine, are small polycationic, aliphatic, amines that bind electrostatically to the negatively charged macromolecules.[14] Presumably, the biochemical interactions of the polyamines can facilitate a change in the function of cellular molecular constituents by altering the charge and/ or conformation. Thus, a slight modification in the structure and charge of the polyamine target macromolecules may result in changes in the ordered proliferative response in cells and tissues. Although, the precise functions of the polyamines are unknown, potential molecular targets of the polyamines are being intensely examined (see Basu and Marton, this monograph).[15]

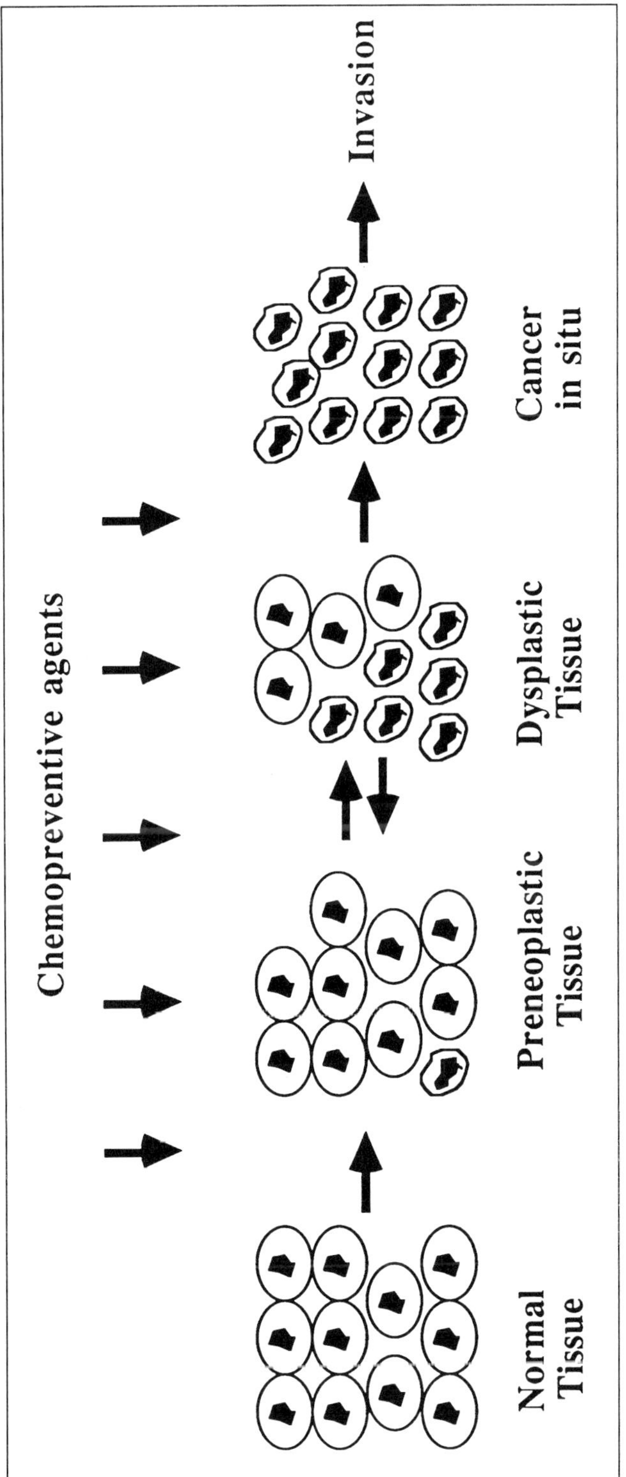

Fig. 8.1. Intra-epithelial neoplasia as a target of chemopreventive agents (adapted from Reference 11).

II. POLYAMINE METABOLIC PATHWAY

A. Components of the Pathway

The polyamine pathway has been extensively reviewed elsewhere[1,2,12,16,17] and is only outlined here briefly. In mammalian cells L-ornithine, a product of arginase activity in the urea cycle, is the substrate for the first regulatory enzyme of the polyamine metabolic pathway (See Morris, this monograph). ODC, a pyridoxal 5'-phosphate requiring enzyme, decarboxylates L-ornithine forming putrescine. The diamine is converted to the diamine spermidine and the tetra-amine spermine by the successive actions of spermidine and spermine synthases. Each of these two synthases sequentially transfers a single aminopropyl group to the putrescine molecule to form spermidine and spermine, respectively. The action of the two synthases is limited by the bioavailability of the aminopropyl donor decarboxylated S-adenosylmethionine (dcAdoMet). Thus, the activity of S-adenosylmethionine decarboxylase (AdoMetDC) becomes a rate limiting step in the biosynthesis of polyamines (See Stanley, this monograph). Although the biosynthetic pathway is essentially irreversible, spermine can be catabolized to putrescine by the activity of the two step catabolic pathway which is rate limited by the first enzymatic step, SSAT (See Xiao and Casero, this monograph). SSAT transfers the acetyl group from acetyl Co enzyme A to the N^1-position of spermidine or spermine. The acetylated spermine and spermidine, are catabolized by the FAD-requiring polyamine oxidase to spermidine and putrescine, respectively. This highly regulated and responsive pathway allows the fine attenuation of intracellular polyamine concentrations.

B. ODC in Transformation

ODC is extensively regulated at multiple levels during both the initiation and termination of growth stimuli. The activity of this enzyme is maintained at low levels during the quiescent or differentiated state, but shows a transient marked increase during the early proliferative response.[14] In a recent report, a transfected ODC gene conferred a transformed phenotype on the 3T3 cell line and also mediated anchorage independence in a soft agar cloning assay.[18] Thus, the ODC gene product, the first regulatory step in the biosynthetic arm of polyamines, may be considered a proto-oncogene. Further evidence supporting this notion came from studies using a mouse myeloma cell line refractory to DFMO treatment. A chromosomal rearrangement between the 5' regulatory promoter of the ODC gene and the switch region of the gamma-1 immunoglobulin gene partially accounted for the enhanced ODC enzyme activity and resistance to DFMO observed in this myeloma cell line.[19,20] The immunoglobulin breakpoint regions have historically only been discovered in the aberrant deregulation of the *c-myc* gene in murine plasmacytomas and human Burkitt's lymphomas.[21] These results suggested that the attenuation of control of the ODC gene was similar in mechanism to *c-myc* activation.[20] Additionally, the increases in gene dosage that can occur with *c-myc* during the progression of some cancers has also been observed with the ODC gene. As a compensatory response to DFMO treatment, an amplified ODC gene has been found in DFMO resistant cell lines.[19]

Another similarity that ODC shares with oncogenes is a protein structure involved in half-life regulation. ODC and the oncogenes c-fos, c-myc, c-fos, v-myb, and the tumor suppressor gene p53, all contain sequences of proline, glutamic acid, serine and threonine amino acids called PEST regions. [22] PEST amino acid motifs are correlated with rapid protein degradation, with ODC possessing the shortest half-life of all these proteins at approximately 10 minutes. The short half-life of ODC also suggests that it will be one of the most rapidly translated proteins in order to maintain high levels during a transitory enzyme response to a proliferation signal.[23] Adding to the complexity of the ODC protein half-life regulation is the presence of an additional protein involved in the degradation of ODC, the ODC antizyme.[24] The ODC antizyme modulates ODC activity by recognizing non-PEST amino acid sequences in the carboxyl end of the protein molecule and is believed to be primarily responsible for the short half-life of ODC.[25]

Further structural similarities of ODC with oncogenes also occur at the mRNA level. ODC is one of several mammalian mRNA transcripts that have 5' leader sequences greater than 200 base pairs.[14] This long 5' nontranslated leader sequence is also found in the *c-myc*,[26] *c-sis*[27] and *lck* oncogenes.[28] A modulatory role in translational regulation has been proposed for the leader sequence of these transforming proteins. Furthermore, the eukaryotic translation initiation factor eIF-4E, which governs the initiation of protein synthesis in mRNA molecules that contain long 5' leader sequences has been shown to fulfill transformation criteria with the standard in vitro and in vivo assays.[29] 3T3 cells overexpressing the eIF-4E protein preferentially maintain abnormally high levels of ODC enzyme activity compared to another polyamine regulatory enzyme with a long 5' mRNA leader sequence, AdoMetDC. Different transcription factor binding motifs in the 5' regulatory DNA sequence of the ODC gene suggest a proto-oncogene role in its expression. The mouse and rat promoter region of the ODC gene both contain DNA cognate sequences for the fos-jun heterodimer AP-1 complex.[30-32] Another transcription binding site located in intron 1 of the ODC gene was hypothesized for *c-myc* and was further identified to have transactivating activity using the chloramphenicol acetyl-transferase (CAT) assay.[33] Assessment of deletion mutants implied that the normal transcriptional activating myc dimerization partner, max, was not participating in the binding complex to the ODC promoter.

Deregulation of ODC activities by oncogenes may occur at multiple regulatory steps. c-Ha-*ras* indirectly causes aberrant ODC activity, partially by increasing the rate of transcription and also by increasing the half-life of the ODC mRNA transcript in c-Ha-*ras* transformed 3T3 cells. A 3-fold increase in the ODC enzyme stability was also detected.[34] Thus, ODC appears, in some cases, to fit the working definition of an oncogene whose regulation can be affected by other oncogenes. However, recent evidence by Clifford et al[109] challenges the hypothesis of ODC as an oncogene and re-emphasizes the particiption of ODC during tumor progression.

C. Inhibitors and Polyamine Analogs

The importance of these cations in growth regulation encouraged the development of inhibitors aimed at the regulatory enzymes essential for polyamine biosynthesis.[35] Metcalf and coworkers first synthesized DFMO, a suicide inhibitor of ODC.[35] Structurally, DFMO was designed as an ornithine analog with two fluorine atoms substituted for the alpha-methyl moiety of the parent molecule. The catalytic activity of ODC is necessary for the activation of the inhibitor, which subsequently covalently binds to ODC and inactivates it (Fig. 8.2).[2,35]

DFMO has demonstrated significant antiproliferative activity against several representative tumor cell types. However, in most cases DFMO treatment is only cytostatic. DFMO depletes putrescine and spermidine levels, but generally has limited effect on spermine levels. The inability of DFMO to completely deplete intracellular polyamines is a result of several compensatory mechanisms in response to depletion. An example of ODC-specific inhibition is a concomitant increase of AdoMetDC activity. The induced AdoMetDC enzyme has been hypothesized to maintain spermine levels associated with DFMO treatment by providing the aminopropyl donating substrate for the formation of spermine from any residual putrescine and spermidine. Although attempts have been made to produce more effective inhibitors of ODC, no compounds which demonstrate clear therapeutic advantages over DFMO have been identified.[5,36]

The development of the polyamine analogs (Fig. 8.2), such as N^1, N^{12} *bis*(ethyl)spermine (BESpm), was based on the premise that they would regulate polyamine intracellular homeostasis, but not substitute for normal polyamine functions. Polyamine analogs are thought to exert their regulation of the polyamine pathway as a response to excess intracellular polyamine concentrations. Included in the action of the polyamine analogs is their intracellular accumulation by the polyamine transporter system with a higher affinity of uptake than the natural polyamines.[10,12] Additionally, the intrinsic antiproliferative qualities of the polyamine analogs can also be manifested without a decrease in polyamine levels, potentiating their effectiveness.[5] These characteristics of the polyamine analogs make them potentially more efficacious chemopreventive agents than DFMO. High concentrations of DFMO are required for antiproliferative activity because it is poorly accumu-

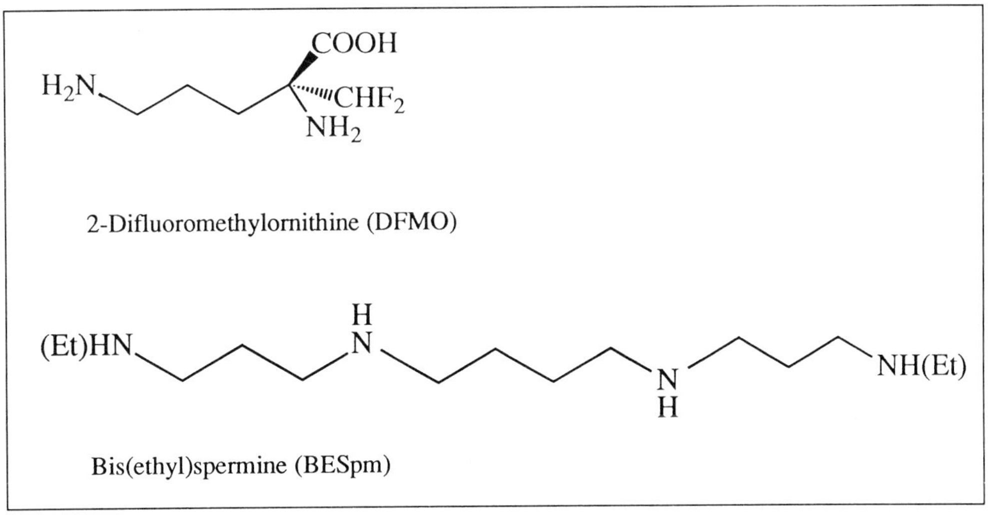

COOH
H_2N CHF_2
NH_2

2-Difluoromethylornithine (DFMO)

(Et)HN — N(H) — N(H) — NH(Et)

Bis(ethyl)spermine (BESpm)

Fig. 8.2. Molecular structures of DFMO and BESpm.

lated by the cell and the efficacy of DFMO also depends on overriding the compensatory nature of the polyamine pathway.

III. POLYAMINES IN CHEMICAL CARCINOGENESIS

A. Mouse Skin Paradigm

The mouse skin model of chemical carcinogenesis has been utilized over a period of decades.[38] Detailed histological and biochemical studies have been performed with this model due to of the ease of chemical application and the ability to monitor the sequence of events accumulating during the intraepithelial neoplastic stages. Histologically, the initial tissue hyperproliferation in response to chemical exposure is a continuum of growth to the benign papillomas formed during tumor promotion.[38]

Chemical carcinogenesis is typically initiated by a single, sub-threshold dose of a carcinogen, followed by the repetitive application of a nongenotoxic, tumor promoter. Clonal expansion of the preneoplastic cell following initiation is propagated by the reversible events characterizing the tumor promotion stage such as alterations in gene expression, increased ODC activity, and enhanced tissue proliferation. The irreversible genetic alterations characterizing the stage of initiation are also prerequisite to the formation of a carcinoma. A chemically generated DNA lesion is manifested into a heritable mutation which can deregulate the function of growth regulatory genes. Conceptually, the multistage mouse skin model provides a fundamental tool for developing an understanding of the myriad of interrelated events participating in the process of chemical carcinogenesis.[38,39]

Association of the induction of ODC activity in mouse skin carcinogenesis, was first delineated by O' Brien et al.[40] Application of 12-O-tetradecanoyl-phorbol-13-acetate (TPA), the tumor promoter used in these experiments, was observed to transiently induce ODC activity 200-fold in normal mouse skin. More importantly, the inducibility of ODC was associated with the dose of TPA and the yield of papillomas. Description of the epithelial target cell of TPA mediated induction of ODC activity was outlined using cultured primary keratinocytes.[41] Epidermal basal cells growing in low calcium concentrations maintained their inducibility of ODC, whereas epidermal cells differentiated with high levels of calcium were refractory to ODC induction by TPA treatment.[41]

The biochemical characterization of papillomas in chemical carcinogenesis protocols has provided insight into the area of the polyamine regulation during tumor progression in the mouse skin. Papillomas, the precursor to carcinoma, display aberrant polyamine metabolic pathway control.[38] Unusually high levels of putrescine and spermidine associated with elevated ODC activity have been detected in papillomas.[42] Exaggerated constitutive levels of polyamines and ODC in these systems suggest deregulation of the proliferation program. AdoMetDC activity was not consistently increased in the papillomas assessed, illustrating the specificity involved in the deregulation of the polyamine pathway with respect to ODC. Additionally, different ODC isoforms were discerned between normal and papilloma tissue, suggesting structural alterations may participate in the elevated enzyme activity.[43] Studies with cultured papillomas further illustrated the uncoupling of control in the normally tightly regulated polyamine cycle.[44] Application of TPA to papilloma cells in vitro demonstrated an induction of ODC along with a concomitant stimulus of DNA synthesis. In further studies describing regulation of ODC by TPA, mouse papilloma cells cultured in high concentrations of calcium were shown to be susceptible to tumor promoter stimulation of ODC activity. These cells were not fully differentiated like normal keratinocytes cells in the presence of high concentrations of calcium, though normal cells were shown to be nonresponsive to TPA induction of ODC in such high levels of calcium. The conclusion drawn from this series of studies was that the transiently elevated levels of ODC induced by TPA in preneoplastic tissue is observed constituitively in papillomas along with abnormally high polyamine levels. These results strongly suggest the deregulation of the polyamine metabolic pathway is required to maintain intraepithelial neoplastic growth.

To help differentiate the molecular mechanisms in the aberrant control of the polyamine metabolic pathway in papillomas, enzyme half-life studies using the protein synthesis inhibitor cyclohexamide were performed.[45] The constitutive half-life of AdoMetDC in papillomas of the mouse skin was similar to the TPA induced AdoMetDC in normal epidermal tissue. ODC activity displayed a bimodal decline, indicative of a radical change in the control of regulation of constitutive enzyme levels. After cyclohexamide treatment in papillomas, ODC initially displayed a decline in half-life activity similar to the decline observed after TPA application to normal epidermal tissue. Surprisingly, after the expected drop in ODC protein activity within 15-20 minutes, ODC exhibited an exaggerated and prolonged spike of activity for greater than 120 minutes. The abrupt increase in ODC activity, occurred during a time when total protein synthesis was inhibited by 90%. The abnormal decline in constitutive ODC activity due to the application of cyclohexamide exemplifies targeted deregulation of the polyamine metabolic pathway in neoplastic tissue. The ODC bimodal half-life decay could be attributed to an alteration of inhibitor proteins, such as with the ODC antizyme. Although the de novo translation of ODC was not specifically addressed, the increase in ODC activity during cyclohexamide treatment suggests that ODC was preferentially synthesized at a more rapid rate than other cellular proteins in neoplastic tissue.[23] Thus, these biochemical studies have helped to identify deregulation of the polyamine metabolic pathway within a mouse papilloma cell system. Collectively, these findings suggest that the aberrations in the control mechanisms governing the polyamine pathway have an important role in the deregulation of proliferation during tumor promotion and are not a mere association with neoplastic growth.

Additional molecular evidence implicating the mutated c-Ha-*ras* proto-oncogene in the formation of papillomas[46] could also be a putative mechanism for the deregulation of constitutive ODC activity in papillomas.[34] Altered ODC control in transformed 3T3 cells with the c-Ha-*ras* oncogene as discussed previously. Mutations to the *ras* genes are associated with aberrant growth in papillomas and squamous cell carcinomas. These data present a potential molecular mechanism of constitutive ODC enzyme deregulation by the mutated ras oncogene relevant to the proliferation of neoplasia.

The rationale for testing DFMO as an inhibitor of mouse skin chemical carcinogenesis was due to the evidence implicating polyamines in the growth of papillomas and their progression to carcinomas. DFMO was consequently shown to be an effective chemopreventive agent in two stage chemical carcinogenesis with the initiating agent 7,12-dimethylbenz[a]anthracene (DMBA) and TPA as the tumor promoter.[47] Oral administration of DFMO (1.0% w/v) during tumor promotion significantly attenuated the incidence of papillomas in mice by 56%. Additionally, a 90% reduction in the multiplicity of papillomas was observed with this treatment regimen. The potency of DFMO was also assessed by using the ODC enzyme assay with TPA. Topically applied DFMO concentrations ranging from 0.01 to 3 mM repressed ODC activity in a graded response with TPA application. Thus, these studies gave impetus to test other tissue examples of chemical carcinogenesis,[47] with DFMO as a chemopreventive agent. Furthermore, direct inhibition of the ODC enzyme activity by DFMO was an efficacious nontoxic inhibitor of mouse papillomas and squamous cell carcinoma, further substantiating a role for polyamines during carcinogenesis.

B. Colon Carcinogenesis

The tumor promoter activity of fatty acids in the rodent colon[48] encouraged the examination of bile salts for tumor promoter activity in colon carcinogenesis in the rodent.[49] Intrarectal instillation of different bile salts exhibited a transient increase of the polyamine biosynthetic enzymes ODC and AdoMetDC with similar kinetics to that of TPA mediated ODC induction in the mouse skin.[40] Substantiating ODC and AdoMetDC association with tumor promotion in the rat colon with bile salts, the activities of each increased in a dose dependent manner.[49] A similar investigation in the rat colon utilized unsaturated fatty acids as the tumor promoting agent. An expected increase of ODC activity correlated with fatty acid exposure but comparable concentrations were also associated with an increase in DNA synthesis.[50]

Continuous oral administration of 1% DFMO in drinking water markedly decreased the incidence of tumors induced by the colon specific carcinogen, dimethylhydrazine. A decline in the levels of putrescine and spermine were associated with a reduction of the proliferative markers such as DNA content and colonic crypt depth in this study.[51] Interestingly, other studies examining the antiproliferative effect of DFMO have shown the agent to be neoplastic tissue specific in the rat colon.[52] A single 100 mg/kg dose of DFMO differentially targeted tumor cell proliferation where normal colonic epithelia was not inhibited. Repetitive doses of 400 mg/kg of DFMO were required to attenuate normal cell proliferation, thus illustrating the specificity of the action of this agent on neoplastic tissue.

Additional studies in the rat assessed the practicality of the transient intervention of DFMO with azoxymethane induced colon carcinogenesis in a 26 week carcinogenesis study.[53] An uninterrupted 26 week exposure of DFMO resulted in the most convincing inhibition of incidence and multiplicity of tumors, with 8 weekly subcutaneous injections of the carcinogen. However, significant reductions were also observed with transient 13 week treatment of DFMO administered at the beginning or later half of the experiment. These studies depicted the efficacy of DFMO with a more realistic short term intervention of the agent.

The genetic delineation of the field effects in human colon cancer has been addressed, though presumably incompletely. A mutated *K-ras* oncogene in human colon cancer is prevalent, and associated with aberrant neoplastic growth.[54] In a controlled chemopreventive setting such as in the rat colon, the p21 *ras* proto-oncogene product was a useful marker for the efficacy of DFMO.[55] With the colon-specific carcino-

QUESTIONNAIRE

Receive a FREE BOOK of your choice

Please help us out—Just answer the questions below, then select the book of your choice from the list on the back and return this card.

R.G. Landes Company publishes five book series: *Medical Intelligence Unit, Molecular Biology Intelligence Unit, Neuroscience Intelligence Unit, Tissue Engineering Intelligence Unit* and *Biotechnology Intelligence Unit.* We also publish comprehensive, shorter than book-length reports on well-circumscribed topics in molecular biology and medicine. The authors of our books and reports are acknowledged leaders in their fields and the topics are unique. Almost without exception, there are no other comprehensive publications on these topics.

Our goal is to publish material in important and rapidly changing areas of bioscience for sophisticated scientists. To achieve this goal, we have accelerated our publishing program to conform to the fast pace in which information grows in bioscience. Most of our books and reports are published within 90 to 120 days of receipt of the manuscript.

Please circle your response to the questions below.

1. We would like to sell our *books* to scientists and students at a deep discount. But we can only do this as part of a prepaid subscription program. The retail price range for our books is $59-$99. Would you pay $196 to select four *books* per year from any of our Intelligence Units–$49 per book–as part of a prepaid program?

   Yes **No**

2. We would like to sell our *reports* to scientists and students at a deep discount. But we can only do this as part of a prepaid subscription program. The retail price range for our reports is $39-$59. Would you pay $145 to select five *reports* per year–$29 per report–as part of a prepaid program?

   Yes **No**

3. Would you pay $39–the retail price range of our books is $59-$99–to receive any single book in our Intelligence Units if it is spiral bound, but in every other way identical to the more expensive hardcover version?

   Yes **No**

To receive your free book, please fill out the shipping information below, select your free book choice from the list on the back of this survey and mail this card to:

R.G. Landes Company, 909 S. Pine Street, Georgetown, Texas 78626 U.S.A.

Your Name ____________________

Address ____________________

City ____________________ State/Province: ____________________

Country: ____________________ Postal Code: ____________________

My computer type is Macintosh______ ; IBM-compatible _____ ; Other _____

Do you own ____ or plan to purchase ___ a CD-ROM drive?

Available Free Titles

Please check three titles in order of preference.
Your request will be filled based on availability. Thank you.

- ❐ Water Channels
  Alan Verkman,
  University of California-San Francisco
- ❐ The Na,K-ATPase:
  Structure-Function Relationship
  J.-D. Horisberger, University of Lausanne
- ❐ Intrathymic Development of T Cells
  J. Nikolic-Zugic,
  Memorial Sloan-Kettering Cancer Center
- ❐ Cyclic GMP
  Thomas Lincoln, University of Alabama
- ❐ Primordial VRM System and the Evolution of Vertebrate Immunity
  John Stewart, Institut Pasteur-Paris
- ❐ Thyroid Hormone Regulation of Gene Expression
  Graham R. Williams, University of Birmingham
- ❐ Mechanisms of Immunological Self Tolerance
  Guido Kroemer, CNRS Génétique Moléculaire et Biologie du Développement-Villejuif
- ❐ The Costimulatory Pathway for T Cell Responses
  Yang Liu, New York University
- ❐ Molecular Genetics of Drosophila Oogenesis
  Paul F. Lasko, McGill University
- ❐ Mechanism of Steroid Hormone Regulation of Gene Transcription
  M.-J. Tsai & Bert W. O'Malley, Baylor University
- ❐ Liver Gene Expression
  François Tronche & Moshe Yaniv,
  Institut Pasteur-Paris
- ❐ RNA Polymerase III Transcription
  R.J. White, University of Cambridge
- ❐ src Family of Tyrosine Kinases in Leukocytes
  Tomas Mustelin, La Jolla Institute
- ❐ MHC Antigens and NK Cells
  Rafael Solana & Jose Peña,
  University of Córdoba
- ❐ Kinetic Modeling of Gene Expression
  James L. Hargrove, University of Georgia
- ❐ PCR and the Analysis of the T Cell Receptor Repertoire
  Jorge Oksenberg, Michael Panzara & Lawrence Steinman, Stanford University
- ❐ Myointimal Hyperplasia
  Philip Dobrin, Loyola University
- ❐ Transgenic Mice as an In Vivo Model of Self-Reactivity
  David Ferrick & Lisa DiMolfetto-Landon,
  University of California-Davis and Pamela Ohashi,
  Ontario Cancer Institute
- ❐ Cytogenetics of Bone and Soft Tissue Tumors
  Avery A. Sandberg, Genetrix & Julia A. Bridge ,
  University of Nebraska
- ❐ The Th1-Th2 Paradigm and Transplantation
  Robin Lowry, Emory University
- ❐ Phagocyte Production and Function Following Thermal Injury
  Verlyn Peterson & Daniel R. Ambruso,
  University of Colorado
- ❐ Human T Lymphocyte Activation Deficiencies
  José Regueiro, Carlos Rodríguez-Gallego
  and Antonio Arnaiz-Villena,
  Hospital 12 de Octubre-Madrid
- ❐ Monoclonal Antibody in Detection and Treatment of Colon Cancer
  Edward W. Martin, Jr., Ohio State University
- ❐ Enteric Physiology of the Transplanted Intestine
  Michael Sarr & Nadey S. Hakim, Mayo Clinic
- ❐ Artificial Chordae in Mitral Valve Surgery
  Claudio Zussa, S. Maria dei Battuti Hospital-Treviso
- ❐ Injury and Tumor Implantation
  Satya Murthy & Edward Scanlon,
  Northwestern University
- ❐ Support of the Acutely Failing Liver
  A.A. Demetriou, Cedars-Sinai
- ❐ Reactive Metabolites of Oxygen and Nitrogen in Biology and Medicine
  Matthew Grisham, Louisiana State-Shreveport
- ❐ Biology of Lung Cancer
  Adi Gazdar & Paul Carbone,
  Southwestern Medical Center
- ❐ Quantitative Measurement of Venous Incompetence
  Paul S. van Bemmelen, Southern Illinois University and John J. Bergan, Scripps Memorial Hospital
- ❐ Adhesion Molecules in Organ Transplants
  Gustav Steinhoff, University of Kiel
- ❐ Purging in Bone Marrow Transplantation
  Subhash C. Gulati,
  Memorial Sloan-Kettering Cancer Center
- ❐ Trauma 2000: Strategies for the New Millennium
  David J. Dries & Richard L. Gamelli,
  Loyola University

gen azoxymethane, DFMO inhibition of tumors strongly correlated with p21 *ras* GTPase activity and protein expression.[55] Further investigations also included bromodeoxyuridine labeling as an index of colonic crypt proliferation. These studies revealed that a carcinogen stimulated increase in cell proliferation was ablated with DFMO application and the concentrations used were associated in other studies with the inhibition of adenoma and adenocarcinoma formation. Furthermore, the antiproliferative effect preferentially targeted carcinogen treated tissue and did not alter the control labeling index of normal tissue treated with DFMO.

Persistent inflammatory disease in the lower bowel increases the risk of colon cancer.[56] Piroxicam, a nonsteroidal antinflammatory drug, has been tested as a chemopreventive agent in the inhibition of prostaglandin biosynthesis associated with cancer. The antiproliferative agent DFMO, in combination with proxicam effectively reduced the incidence of adenocarcinomas in the rat colon carcinogenesis protocol treated with AOM, whereas administration of the same doses of the individual agents did not alter adenocarcinoma growth.[57] These data are in accord with bladder studies[58] where combinations of chemopreventive agents with DFMO, can reduce their effective concentrations and diminish the possibility of a toxicological response from associated polyamine depletion.

C. Field Carcinogenesis

The different rodent models of chemical carcinogenesis enabled researchers to experimentally reproduce the phenomenon of field cancerization, making the studies more relevant to the area of chemoprevention and human disease. Field cancerization, proposed by Slaughter et al,[59] classifies the incidence of tumors and or intraepithelial neoplasm of multicentric origin over a wide area of epithelial tissue noticed in patients with certain cancers. Multifocal tumors of different origins can, in some cases, be indicative of exposure to environmental chemicals such as in the upper areodigestive tract and the bladder.[7,59] Additionally, field cancerization is also prevalent in cancers derived from genetic disorders, such as in familial polyposis of the colon.[58] The genetic alterations occurring during process of human colon cancer has been partially delineated by Vogelstein et al[58] and frequently involves mutations in the growth regulatory gene, *K-ras*. Presently, studies involving the staining of Gal-GalNac carbohydrate moieties in adenocarcinomas of breast tissue also detected putative preneoplastic tissue in the surrounding normal epithelium. Gal-GalNAc expression is a well established indicator of adenocarcinoma tissue in the colon and these results suggest that the human breast tissue exhibited a field effect of possible carcinogen exposure.[60]

D. Other Carcinogenesis Paradigms

Investigations with DFMO as a possible chemopreventive agent in tissues other than the skin and colon included the rodent tongue, bladder and tracheal carcinogenesis protocols. Initiation of tongue neoplasms by the direct acting carcinogen 4-nitroquinoline-1-oxide, were effectively abolished by DFMO treatment in rats. DFMO was administered orally during the 25 week postinitiation phase of the protocol and exhibited a dose dependent response in the inhibition of tongue neoplasms. Analysis by silver staining of nucleolar organization regions to evaluate the antiproliferative effects of DFMO illustrated an abatement of the growth rate of the carcinogen treated tissue.[61] DFMO administration beginning before the initiation phase and continuing for the duration of the experiment significantly diminished tracheal cancers produced by methylnitrosourea exposure in Syrian golden hamster studies.[62] Similarly, concurrent treatment protocols with DFMO and the initiation agent N-butyl-N-(4-hydroxybutyl) nitrosamine that persisted throughout the duration of the study was decisive in inhibiting bladder cancer.[62,63] Rat bladder transitional cell carcinoma was effectively inhibited with oral application of DFMO following and concurrent to N-butyl-N-(4-hydroxybutyl) nitrosamine exposure, with sodium saccharide used as the tumor promoter. Interestingly, DFMO did not have an effect on the preneoplastic hyperplasia and intraepithelial neoplasia in this system, suggesting that main mode of action of DFMO targeted the transition from benign papillary growth to malignant carcinoma. In studies in the mouse bladder, DFMO not only significantly attenuated the incidence and multiplicity of transitional cell carcinoma (TCC), but also the carcinoma exhibited a less invasive phenotype.[58]

Experimental chemopreventive studies with DFMO in target organs where field cancerization has not been extensively characterized clinically include breast and liver cancer. In murine mammary carcinogenesis initiated with methylnitrosourea, DFMO was shown to significantly reduce the incidence of carcinomas during the tumor promotion phase of the experiment.[64] Continuous DFMO application throughout the chemical carcinogenesis protocol did not improve the tumor inhibition effect,[65] when compared to transient exposure during the promotion stage. In comparison, studies with rat hepatocarcinogenesis initiated with diethylnitrosamine demonstrated a antiproliferative effect with administration of DFMO solely during the initiation phase of carcinogen exposure.[66] Hepatocellular carcinogenesis was also diminished when DFMO was administered in the chemical carcinogenesis period of tumor promotion.

E. Conclusion

As summarized in Table 8.1, DFMO is an effective chemopreventive agent in several models of carcinogenesis (Table 8.1). In agreement with initial studies in the mouse skin and the delineation of the importance of the constitutive ODC activity during tumor promotion, DFMO is most efficacious when applied during post carcinogen exposure. Furthermore the antiproliferative effects of DFMO in some cases, targets tissue exposed to the carcinogen. These results are encouraging because of the directed cytostatic action of the agent to the pre- or neoplastic tissue, potentially lessening the toxicity to normal tissue.

IV. POLYAMINE PATHWAY IN CLINICAL CHEMOPREVENTION

Clinical validation of the deregulation of the polyamine metabolic pathway in cancer and the effectiveness of DFMO administration in short term rodent studies enabled researchers to apply these findings

Table 8.1. Chemopreventive studies with DFMO and concomittant treatment with other agents in rodents

Target Organ	Species	Dose/Route	Chemopreventive Agent	Carcinogen	Outcome	Ref.
Skin	Mouse	1% in water 0.3 mg topically	DFMO	DMBA⇒TPA	⇓tumor incidence	47
Colon	Mouse	1 % in water	DFMO	DMH	⇓ tumor incidence	51
Colon	Rat	0.1 % in water	DFMO	AOM	⇓tumor incidence	53
Tongue	Rat	100,1000, 2000 ppm in diet	DFMO	4-NQO	⇓tumor incidence	61
Trachea	Hampster	3.2 g/kg in diet	DFMO	MNU	⇓tumor incidence	62
Bladder	Mice	2 ,4 g/kg in diet	DFMO	OH-BBN	⇓tumor incidence	62
Bladder	Rat	0.5, 0.2% in water	DFMO	OH-BBN	⇓tumor incidence	63
Mammary Glands	Rat	1% in water	DFMO	MNU	⇓tumor incidence	64
Mammary Glands	Rat	0.125, 0.5% in water	DFMO	MNU	⇓tumor incidence	65
Liver	Rat	500 ,1000 ppm in diet	DFMO	DEN	⇓tumor incidence	66
Colon	Colon	400,1000, 2000 ppm in diet 25,75,150 ppm in diet	DFMO + Piroxicam	AOM	⇓tumor incidence	57
Bladder	Bladder	640 mg/kg in diet 100 mg/kg in diet	DFMO + Oltipraz	OH-BBN	⇓tumor incidence	58

DMBA, 7,12-dimethylbenz[a]anthracene;TPA, 12-O-tetradecanoyl-phorbol-13-acetate; DMH, 1,2-dimethylhydrazine; AOM, azoxymethane; NQO, 4-nitroquinoline-1-oxide; MNU, methylnitrosourea; OH-BBN, N-butyl-N(4-hydroxybutyl)nitrosamine; DEN, diethylnitrosamine.

to the prevention of a disease that is potentially latent for decades. Studies with colonic polyposis took advantage of a predisposition to colon cancer and the use of the ODC assay to biochemically survey for neoplasia. Examination of the histological gradation of colonic polyps correlated well with the induction of ODC activity.[67] Dysplastic polyps exhibited the highest ODC activity, while the flat mucosa of patients with colonic polyposis demonstrated a marked reduction of activity. Colonic mucosa of normal controls displayed ODC activity significantly below the preneoplastic normal mucosa. These biochemical studies in the colon were more recently substantiated by measuring levels of putrescine, spermidine and spermine in a similar experimental design in patients with adenomatous polyps. Variations in spermine and spermidine were nonexistent in normal patients and patients with adenomatous polyps. However an increase in levels of putrescine was detected in normal appearing tissue in patients with adenomatous polyps when compared to control colonic mucosa in patients without polyps.[68] A concordant increase in ODC activity appeared to accounted for the abnormal intracellular concentrations of putrescine in the normal appearing mucosa in patients with polyps.

Clinical correlation of polyamines with cancer has been examined in tissues other than the colon, though not with a target of chemoprevention, intraepithelial neoplasia.[69] In studies with breast cancer tissue, an increase in polyamines is associated with the incidence of secondary tumors, indicative of a more virulent reoccurring cancer. The examination of the polyamines as a prognostic indicator of breast cancer by Kingsnorth et al[69] implied that potential fluctuations in the constituents of the polyamine pathway may take place in the dysplastic stages leading to anaplasia in this tissue.

Due to the early chemopreventive studies in the mouse skin carcinogenesis model, the inhibition of TPA induced ODC was well documented[38] and progressed to clinical examinations of the feasibility in using the assay to assess the efficacy of certain chemopreventive agents (Table 8.2).[70] Oral administration of several different chemopreventive agents decreased the inducibility of ODC activity when skin punch biopsies were obtained from subjects and compared to controls. This has been used to delineate the lowest effective concentration of DFMO during a Phase I chemoprevention trial.[71] A maximally tolerated concentration of 0.5 g/m^3 was inferred from initial toxicity studies and given orally over a period of 5 to 12 months. Samples of skin were collected on a weekly basis until the end of the first month of the study, with subsequent samples obtained monthly until the termination of the study after the tenth month. Finally, analysis of the pooled skin samples at all time points achieved a greater than 50% reduction of TPA induction of ODC activity.

Additional clinical trials (Table 8.2) used the reduction of intracellular concentrations of polyamines in the target tissue of the colon

Table 8.2. Clinical chemopreventive studies with DFMO

Type of Study	Route, Dose, Frequency, Duration	Endpoints	Toxicity	Ref.
Phase I	Oral; 0.5 g/m^2 /day for 5-12 months	To assess the inhibition of TPA induced ODC activity and to determine the MTD.	No major toxic effects	71
Phase I	Oral; 0.8, 1.6, 3.1, 6.4g/m^2/day for 26 weeks	To assess the urinary and erythrocyte polyamine levels. To determine the MTD.	Ototoxicity	68,73
Phase I	0.1 to 3.0 g/m2/day for 4 weeks	To assess colorectal mucosa for polyamine levels. To determine the MTD.	No toxicity with effective concentrations 0.25,0.5g/m^2/day	72

MTD, is the most tolerated dose that produces the desired endpoint without the occurrence of toxicity.

as a molecular assay to assess the efficacy of nontoxic concentrations of DFMO.[72] In patients with previous adenomatous polyps removed, concentrations as low as 0.25 gm^2 manifested a significant reduction in putrescine as well as a decrease in the spermidine to spermine ratio. Further versatility of the polyamine pathway as an internal standard for the efficacy of short-term DFMO treatment involved the noninvasive collection of urine samples.[73] Administration of the chemopreventive agent caused an eventual reduction in the urine concentrations of polyamines in all subjects of the study, though not during the same time frame. Interestingly, one patient exhibited a regression of polyp growth in the rectal stump concomitant to a diminishment in urinary polyamine levels during the short term phase I chemopreventive trial.

Potential prospective trials with targeted DFMO intervention in tissues other than the colon include the prostate and the bladder.[74-76] The bladder is a unique organ that can concentrate the chemopreventive agent at the site of the target tissue with the highest incidence of cancer, the bladder transitional epithelium. Furthermore, histological monitoring of the dysplastic stages of the bladder epithelium is feasible. Interestingly, the prostate contains the highest constitutive levels of polyamines in the body and is also considered a rational tissue for DFMO intervention in chemoprevention. Rodent studies examining effects in the ventral prostate have substantiated this possibility.[77] Short-term oral administration of DFMO in male Sprague-Dawly rats have elicited a conspicuous 90% inhibition of ODC activity with a concomitant drop in polyamine levels. The inhibition of the polyamine pathway by DFMO was also associated with a significant preferential diminishment of prostate weight compared to other organs.

V. TOXICITY AND PHARMACOKINETIC STUDIES OF DFMO

Important initial considerations in designing a clinical trial for chemopreventive agents include ascertaining the lowest effective concentration of the agent and the toxicity associated with the treatment.[78] Pharmacokinetic studies demonstrated DFMO to follow a first order linear kinetics that are indicative of rapid absorption, distribution and elimination.[71,72] The half-life of DFMO in the body was determined to be 3.5 hours[71] during a daily oral dosing regimen. DFMO is excreted primarily unchanged in the urine and, with the bioavailabity of the agent being only approximately 55%, readily accumulate in the epithelium of the bladder and colon. These organs are logical targets for DFMO intervention in chemopreventive studies because of this reason.

Long-term administration of DFMO is generally well tolerated. The overt toxicity of ODC inhibition by DFMO is relatively well guarded against by the biochemical compensatory mechanisms that prevent total cellular polyamine depletion. It has been proposed that the reason for the highly regulated nature of the polyamine pathway is to prevent cellular toxicity from their excess production.[79] The untoward effects observed in clinical trials include thrombocytopenia and reversible ototoxicity.[68,72] Hearing loss was characterized in a dose-dependent manner with respect to accumulating concentrations of DFMO and an impairment in hearing at multiple frequency levels.[80] Studies in animals were analyzed to demarcate chronic toxicities associated with extremely high doses of DFMO.[81] Deleterious effects experienced within a duration of 52 weeks included weight loss, gastric inflammation, skin abrasions and liver necrosis. In rats a no-effect dose was observed at 400 mg/kg of DFMO, well above the extrapolated concentrations administered clinically that have elicited a desirable outcome.[71-73,81] These results along with the clinical studies suggested further examination of DFMO as a chemopreventive agent is justified.

Inherent toxicological problems with DFMO may also depend on the health status of the individual. Wound healing and fetal development, which are both biological processes that rely on cellular proliferation, pose a problem with polyamine-related intervention.[82,83] In early tissue maturation studies DFMO caused a delay in development in newborn rat colonic mucosa and was also responsible for the attenuation of tissue repair stimulated by arabinosylcytosine (ara C) exposure in the adult rat.[82] Investigations pertinent to the teratogenic effects of DFMO were also examined in the rodent. DFMO administration was shown to inhibit the sequence of events occurring after implantation into the uterine wall. These postimplantation processes are associated with increased ODC activity and enhanced polyamine levels.[83]

Other possible deleterious effects from long-term DFMO administration include the probability of ODC gene amplification and/or rearrangement analogous to *c-myc* deregulation. This situation was discussed previously and could mimic the growth stimulatory action of a tumor promoter, which has been recently substantiated with the prospect of ODC as a putative oncogene. However, ODC gene amplification has not been observed in any of the rodent models where relatively high concentrations of DFMO were used to inhibit chemical carcinogenesis.

VI. POLYAMINE ANALOGS IN CHEMOPREVENTION

The suitability of the polyamine metabolic pathway as a target for chemotherapeutic intervention has led to the development of several polyamine analogs. The complexity of the ways in which normal and neoplastic cells regulate the pathway suggest a basis for differential sensitivity to these agents.[5,8,9] The potential prophylactic activity of the polyamine analogs satisfies the criteria of a desirable chemopreventive agent in many human cancer cells lines; antiproliferative activity without apparent cytotoxicity.[84-87]

Preliminary studies have determined that the polyamine analogs have a possible application in the prevention of cancer.[37] BESpm exhibited potent antiproliferative effects in two in vitro chemoprevention assays. The spermine analog abolished the formation of TPA stimulated colonies in the JB-6 soft agar cloning assay and markedly slowed the proliferation rate of the PE mouse papilloma cell line. In both instances, BESpm was about 300-fold more potent than DFMO. The two cell lines utilized were derived from chemical carcinogenesis protocols and are cell culture models of tumor promotion and progression.[44,88] Furthermore, of importance to the field of chemoprevention was the absence of toxicity with cytostatic concentrations of BESpm.

An important antiproliferative aspect that is common to agents directly interfering with the polyamine pathway is a significant reduction of ODC activity. While the polyamine analogs reduce ODC activity, the primary cytostatic effect of these agents does not appear to be directly associated with ODC inhibition.[89,90] Eloquent studies by Ghoda et al[89] employed the mutant CHO cell line which are obligate polyamine auxotrophs because of an ODC defect. The difference in antiproliferative activities between the polyamine analogs, N^1,N^{14}, bis(ethyl)homospermine (BE444) and N^1, N^8-*bis*(ethyl)spermidine (BES), was discerned by using a trypanosome ODC construct, whose protein product is refractory to polyamine regulation unlike the mammalian protein. The analog with the more potent cytostatic effect, BE444, diminished the growth rate with either ODC construct in the cells, whereas BES was not effective in reducing proliferation in CHO cells mediated by the trypanosomal ODC construct. The molecular mechanism of the spermine analog efficacy was further delineated in studies by Albanese et al.[90] Attenuation of the cellular proliferative rate in the murine leukemia cell line correlated well with the cellular uptake of the polyamine analogs. Additionally, polyamine depletion produced by the analog augmented the resultant antiproliferative effect. Thus the working hypothesis on the antiproliferative effect of the developing spermine analogs, is that these agents compete with the natural polyamines at intracellular binding sites.[90]

Comprehensive studies with a spectrum of different human lung carcinoma cell lines illustrated yet another important feature related to chemoprevention by the spermine analogs. The differential sensitivity between the toxic or antiproliferative effect of these agents correlated well with the magnitude of induction of the SSAT enzyme.[85] In large cell and adenocarcinoma lung culture lines a cytotoxic response was produced by BESpm, which was associated with >1000 pMol/mg/min SSAT activity level. In comparison, small cell and squamous cell lung carcinoma lines did not induce SSAT levels above 450 pMol/mg/min. Further characterization of the SSAT activity as a marker for spermine analog cytostasis has a potential application in the field of chemoprevention and has already been implemented clinically as a prognostic indicator of tumor sensitivity to BENSpm.

The significance of the enhanced SSAT activity may have use as a marker for clinical studies and yield insight into the molecular mechanism of spermine analog action. Pegg et al[91] has shown that BESpm mediation of polyamine intracellular reduction, included an enhanced SSAT activity that correlated with an efflux of acetyl-spermidine from the cell and the greatest depletion of spermine levels. More importantly, as the natural polyamine intracellular levels decreased, levels of the analogs increased as part of the compensatory mechanism to maintain the homeostatic nitrogen cationic charge.[91,92] Thus, the biochemical activity of the polyamine specific acetylase(SSAT) can mediate the excretion of the acetylated polyamines. The role of the apparent SSAT coupled efflux mechanism is in addition to the ramifications of the reduction of the positive charge on the polyamine molecule when acetylated. Consequently, SSAT-activity can facilitate the removal of the natural polyamines from their molecular targets by stimulating their efflux and reducing their positive charge intracellularly. These actions may subsequently enhance analog uptake into the cell which appears to be the main indicator of their antiproliferative action.[5,10]

The antiproliferative effect of BESpm, which is aimed at molecular processes other than modulation of the polyamine pathway, may involve changes in the expressions of important growth regulatory genes initiated by more global changes in chromatin structure. Numerous biophysical and biochemical binding studies with DNA have determined that polyamines can bind to specific topologies of nucleic acids through hydrogen-bonding and electrostatic interactions. These polycation, nucleic acid interactions have the capacity to alter the tertiary structure of the nucleic acid.[15] Evidence of DNA perturbations at the organelle level have shown that depletion of intracellular polyamine concentrations and uptake of the polyamine analogs have the propensity to alter chromatin structure in the nucleus. Related studies have demonstrated that conformational changes in chromatin associated with histone depletion and alterations in the nuclear matrix attachment sites could be involved in the antiproliferative effects of polyamine depletion and/or analog treatment.[93-95]

Polyamine depletion and treatment with polyamine analogs, has been associated with specific changes in gene expression. More specifically, differential down-regulation of the expression of the immediate early genes have been elicited by antiproliferative agents targeting the polyamine metabolic pathway.[93-96] Although further study is needed, it is possible that these changes in gene expression are mediated by conformation changes at the level of DNA.

VII. INFLAMMATORY PROCESSES AS A TARGET OF INTERVENTION

The effect of polyamine depletion and analog activity in the regulation of oncogenes may suggest a prophylactic role in the prevention of cancers associated with chronic inflammatory disease processes. An increase in oxidative stress accompanies inflammation.[100] Some evidence implicates reactive oxygen species as being mediators of gene expression in some in vitro models providing a molecular connection between the phenomenon of gene regulation by a chronic disease process and a potential regulatory role for polyamine pathway intervention. Furthermore, increases of the constituents of the polyamine metabolic pathway themselves have also been shown to participate in the inflammatory process.[101]

Some of the chronic inflammatory disorders that are associated with an increased risk to cancer include tuberculosis and asbestosis with lung cancer and chronic ulcerative colitis as an epidemiologic factor for adenocarcinoma of the colon.[56] Potential mechanisms provoking the process of carcinogenesis via oxidative stress encompass DNA damage, cellular repair due to cytoxicity,[100] regulation of oncogene expression[102,103] and PKC activation,[104] all of which have been concluded to be important mechanisms in tumor promotion and progression.[105] In studies with the JB-mouse epidermal cells, specifically clone 41, reactive oxygen species generated by a xanthine/xanthine oxidase system were shown to mediate an increase in mRNA expression with the c-*myc* and c-*fos* proto-oncogenes. The increased expression of the c-*fos* was regulated by a preferential enhancement of the rate of transcription.[102] *C-myc* did not exhibit an increased transcription rate and the mechanism of transcript regulation was not further assessed. These studies imply that the induction of the c-*myc* and c-*fos* proto-oncogenes, two genes whose expression are diminished by polyamine depletion,[98] may participate in the ability of the cell to progress to an anaplastic phenotype. In investigations related to the polyamine regulation of gene expression, treatment with inhibitors of the polyamine metabolic pathway demonstrated a preferential decrease in the transcription of the c-*myc* and c-*fos* oncogenes.[97,98] Recently DFMO induced cytostatic treatment of cultured rat mesangial cells resulted in a similar down-regulation of immediate early genes.[99] Furthermore, the antiproliferative effect was attributed to a signal transduction constraint downstream from the initiating signal at the plasma membrane and proximal to the reductions of the mRNA in the nucleus.

Additionally, enhanced ODC activity has been showed to be associated with chronic inflammation.[101] Investigations with parasitic infections in the rodent intestine demonstrated an enhancement of ODC activity along with a concordant increase of DNA synthesis, followed by tissue hyperplasia. Application of DFMO preferentially inhibited both growth related aspects of the chronic inflammatory response without killing the etiologic organism or reducing the levels of neutrophils. This study presented added evidence to the possible preventative benefits of polyamine pathway targeted intervention, aimed at a process associated with certain types of cancers.

VIII. CONCLUSION

The polyamine pathway is an essential part of the proliferative process of carcinogenesis and for this reason can be exploited in the field of chemoprevention. The implementation of the various integral markers of the pathway in the surveillance of intraepithelial neoplasia is being developed and validated. The interference of intraepithelial neoplastic growth when specifically aimed at the polyamine pathway, is relatively safe with the agents discussed because of the compensatory mechanisms that prevent total polyamine depletion. Thus the complexity of the multiple regulatory nature of the pathway can in some cases prevent overt toxicity.

The well-founded basis for the development of the ODC inhibitor DFMO, has included its progression from rodent studies to phase II clinical trials. Even though DFMO has advanced farther than most other chemopreventive agents, there are still problems related to its efficacy involving cellular uptake and targeting a metabolic pathway with multiple compensatory mechanisms. The polyamine analogs are another class of novel agents that target the polyamine pathway and their transport system.

In theory, the antiproliferative effect of chemopreventive agents targeting intraepithelial neoplasia can completely inhibit or prolong the duration for the transformation to a malignant carcinoma. Slowing the growth rate of intraepithelial neoplasia also has profound ramifications on the reduction of the mutation rate in proto-oncogenes and tumor suppressor genes which interfere with the normal growth and differentiation of epithelial cells and tissue.[106-108] Therefore, this gives added impetus in the development of antiproliferative agents because of their duality of effect on the evolution of the neoplasm. Furthermore, evidence strongly suggests that agents directly down-regulating the polyamine metabolic pathway[5] have the ability to nullify proto-oncogenes by inhibiting their expression.

References

1. Heby O. Role of polyamines in the control of cell proliferation and differentiation. Differentiation 1981;9:1-20.
2. Pegg AE. Polyamine metabolism and its importance in neoplastic growth and as a target for chemotherapy. Cancer Res 1988;48:759-774.
3. Seiler N, Sarhan S, Grauffel C, et al. Endogenous and exogenous polyamines in support of tumor growth. Cancer Res 1990;50:5077-5085.
4. Steglich C, Scheffler IE. An ornithine decarboxylase-deficient mutant of Chinese hamster ovary cells. J Biol Chem 1982;8:4603-4609.
5. Marton LJ, Pegg AE. Polyamines as targets for therapeutic intervention. Ann Rev Pharm Toxical 1995;35:55-91.
6. Kelloff GJ, Boone WC, Crowell JA, et al. Chemopreventive drug development: perspectives and progress. Cancer Epidemiol Biom Prev 1994;3:84-98.
7. Kelloff GJ, Boone CW, Malone WF et al. Development of chemopreventive agents for bladder cancer. J Cell Bioch 1992; supp.161:1-12.
8. Wattenberg LW. Inhibition of carcinogenesis by minor dietary constituents. Cancer Res 1992; 52:2085s-2091s.
9. Boone CW, Kelloff GJ, Steele VE. Natural history of intraepithelial neoplasia in humans with implications for cancer chemoprevention strategy. Cancer Res 1992;52:1651-1659.
10. Porter CW, Regenass U, Bergeron RJ. Polyamine inhibitors and analogs as potential anticancer agents. In: Dowling RH, Folsch UR, Loser Chr, eds. Falk Symposium on Polyamines. Dordrecht: Kluwer Press. 1992;301-322.
11. Kelloff GJ, Boone CW, Steele VE, et al. Progress in cancer chemoprevention: Perspectives on agent selection and short-term clinical intervention trials. Cancer Res 1994;54:2015s-2024s.
12. Casero RA, Baylin SB. Concepts for deriving specific inhibitors of polyamine biosynthesis: human lung cancer cells as a model system. In: Palfreman MP, McCann PP, Lovenberg W, Temple JG, Sujoerdsma A, eds. Enzymes as Targets for Drug Design. Academic Press, San Diego, 1989; 185-200.
13. Bergeron RJ, Neims AH, McManis JS, et al. Synthetic polyamine analogs as antineoplastics. J Med Chem 1988;31:1183-1190.
14. Heby O, Persson L. Molecular genetics of polyamine synthesis in eukaryotic cells. TIBS 1990;15:153-158.
15. Feuerstein BG, Williams LD, Basu H S et al. Implications and concepts of polyamine-nucleic acid interactions. J Cell Biochem 1991;46:37-47.
16. Casero RA, Pegg AE Spermidine/spermine N^1-acetyltransferase—the

turning point in polyamine metabolism. FASEB J 1993;7:653-661.
17. Seiler N. Functions of polyamine acetylation. Can J Physiol Pharmaco 1987;65:2024-2035.
18. Auvinen M, Paasinen A, Andersson LC, Holtta E. Ornithine decarboxylase activity is critical for cell transformation. Nature 1992;360:355-358.
19. Kahana CK, Nathans D. Isolation of cloned cDNA encoding mammalian ornithine decarboxylase. Proc Natl Acad Sci USA 1984;81:3645-3649.
20. Katz A, Kahana C. Rearrangement between ornithine decarboxylase and the switch region of the gamma-1 immunoglobulin gene in a-difluoromethylornithine resistant mouse myeloma cells. EMBO J 1989; 8:1163-1167.
21. Ailtalo D, Koshinen P, Makela TP, et al. *Myc* oncogenes: activation and amplification. Biochemica et Biophysica Acta 1987;907:1-32.
22. Rogers S, Wells R, Rechsteiner M. Amino acid sequences common to rapidly degraded proteins: the PEST hypothesis. Science 1986;234:364-369.
23. Schimke RT, Doyle D. Control of enzyme levels in animal tissues. Ann Rev Biochem 1970;39:924.
24. Heller JS, Fong WF, Canellakis ES. Induction of a protein inhibitor to ornithine decarboxylase by the end products of its reaction. Proc Natl Acad Sci USA 1976;73:1858-1862.
25. Li X, Coffino P. Regulated degradation of ornithine decarboxylase requires interaction with the polyamine-inducible protein antizyme. Mol Cell Biol 1992;12:3556-3562.
26. Godeau F, Persson H, Grey HE, Paredee AB. *c-myc* expression is dissociated from DNA synthesis and cell division in *Xenopus* oocyte and early embryonic development. EMBO J 1986;5:3571-3577.
27. Rao CD, Pech M, Robbins KC, et al. The 5' untranslated sequence of the c-sis/platelet-derived growth factor-2 transcript is a potent translational inhibitor. Mol Cell Biol 1988;8:284-292.
28. Marth JD, Overell RW, Meier KE, et al. Translational activation of the *lck* proto-oncogene. Nature 1988;332:171-173.
29. Shantz LM, Pegg AE Overproduction of ornithine decarboxylase caused by relief of translational repression is associated with neoplastic transformation. Cancer Res 1994; 54:2313-2316.
30. Win L, Huang J, Blackshear PJ. Rat ornithine decarboxylase gene. J Biol Chem 1989; 264:9016-9021.
31. Eisenberg LM, Fanne OA. Nucleotide sequence of the 5'-flanking region of the murine ornithine decarboxylase gene. Nucleic Acids Res 1989; 17:2359.
32. Mitchell PJ,Tijian R. Transcriptional regulation in mammalian cells by sequence-specific DNA binding proteins. Science 1989;245:371-378.
33. Bello-Fernandez C, Packham G, Cleveland JL. The ornithine decarboxylase gene is a transcriptional target of c-Myc. Proc Natl Acad Sci 1993;90:7804-7808.
34. Holtta E, Sistonen L, Alitalo K. The mechanisms of ornithine decarboxylase deregulation in c-Ha-*ras* oncogene-transformed NIH 3T3 cells. J Biol

Chem 1988;263:4500-4507.

35. Metcalf BW, Bey P, Danzin C. Catalytic irreversible inhibition of mammalian ornithine decarboxylase by substrate and product analogs. J American Chem Soc 1978;100:8:2551-2553.
36. Mc Cann PP, Pegg AE. Ornithine decarboxylase as an enzyme target for therapy. Pharmaco Ther 1992;54:195-215.
37. Berchtold CM, Casero RA, Kensler TW. In vitro chemopreventive studies with the polyamine analog N^1, N^{12} *bis*(ethyl)spermine(BESpm). Proc Am Assoc Cancer Res 1994;35:3721.
38. Slaga TJ Overview of tumor promotion in animals. Environ Health Perspect 1983;50:3-14.
39. Kensler TW, Trush MA. Oxygen-free radicals in chemical carciogenesis. In: Oberley LW, ed. Superoxide Dismutase Volume III Pathological States. CRC Press, Boca Raton, Florida. 1985;192-236.
40. O'Brien TG, Simsimam RC, Boutwell RK. Induction of the polyamine biosynthetic enzymes in mouse epidermis by tumor promoting agents. Cancer Res 1975;35:1662-1670.
41. Lichti U, Patterson E, Hennings H, et al. The tumor promoter 12-O-tetradecanoylphorbol-13-acetate induces ornithine decarboxylase in proliferating basal cells but not in differentiating cells from mouse epidermis. J Cellular Phys 1981;107:261-270.
42. Koza, RA, Megosh LC, Palmieri M, O'Brien TG. Constitutively elevated levels of ornithine and polyamines on mouse epidermal papillomas. Carcinogenesis 1991;12:1619-1625.
43. O'Brien TG, Madara T, Pyle JA et al. Ornithine decarboxylase from mouse epidermis and epidermal papillomas: differences in enzymatic properties and structure. Proc Natl Acad Sci USA 1986;83:9448-9452.
44. Yuspa SH, Morgan D, Lichti U, et al. Isolation and characterization of cells derived from mouse skin papillomas induced by an initiation-promotion protocol. Carcinogenesis 1986;7:949-958.
45. Astrup EG, Boutwell RK. Ornithine decarboxylase activity in chemical induced mouse skin papillomas. Carcinogenesis 1982;3:303-308.
46. Quintanilla M, Brown K, Ramsden M, et al. Carcinogen-specific mutation and amplification of Ha-*ras* during mouse skin carcinogenesis. Nature 1986;322:78-80.
47. Takigawa M, Verma A K, Simsiman R C, et al. Polyamine biosynthesis and skin tumor promotion: inhibition of 12-O-tetradecanoylphorbol-13-acetate- promoted mouse skin tumor formation by the irreversible inhibitor of ornithine decarboxylase 2-difluoromethylornithine. Biochem Biophys Res Commun 1982;105:969-976.
48. Bull AW, Loulliert BD, Wilson PS. Promotion of azoxymethane-induced intestinal cancer by high fat-diet in rats. Cancer Res 1979;39:4956-4959.
49. Takano S, Matsushima M, Erdogan E, Bryan GT. Early induction of rat colonic epithelial ornithine and S-adenosyl-l-methionine decarboxylase activities N-methyl-N' nitro-N-nitrosoguanidine or bile salts. Cancer Res

1981;41:624-628.

50. Bull AW, Nigro ND, Golembieski WA, et al. In vivo stimulation of DNA synthesis and induction of ornithine decarboxylase in rat colon by fatty acid hydroperoxides, autoxidation products of unsaturated fatty acids. Cancer Res 1984;44:4924-4928.
51. Kingsnorth AN, King WWK, Diekema KA, et al. Inhibition of ornithine decarboxylase with a-difluromethylornithine: reduced incidence of dimethylhydrazine-induced colon tumors in mice. Cancer Res 1983;43:2545-2549.
52. Tutton PJM, Barkla DH Comparison of the effects of an ornithine decarboxylase inhibitor on the intestinal epithelium and on intestinal tumors. Cancer Res 1986;46:6091-6094.
53. Nigro ND, Bull AW, Boyd ME. Importance of the duration of inhibition on intestinal carcinogenesis by α-difluoromethylornithine in rats. Cancer Letters 1987;35:153-158.
54. Fearon ER, Vogelstein B. A genetic model for colorectal tumorigenesis. Cell 1990;61:759-767.
55. Singh J, Kelloff G, Reddy BS Intermediate biomarkers of colon cancer: modulation of expression of ras oncogene by chemopreventive agents during azoxymethane induced colon carcinogenesis. Carcinogenesis 1993;14:699-704.
56. Demopoulos HB, Pietronigro DD, Seligman ML. The development of secondary pathology with free-radical reactions as a threshold mechanism. J Amer Coll Toxicol 1983;2:173-184.
57. Reddy BS, Nayini J, Tokumo K, et al. Chemoprevention of colon carcinogenesis by concurrent administration of piroxicam, a nonsteroidal antiinflammatory drug with α-difluoromethylornithine, an ornithine decarboxylase inhibitor, in diet. Cancer Res 1990;50:2563-2568.
58. Moon RC, Kelloff GJ, Detrisac CJ. Chemoprevention of OH-induced bladder cancer in mice by oltipraz, alone and in combination with 4-HPR and DFMO. Anticancer Res 1994; 14:5-12.
59. Slaughter DP, Southwick HW, Smejkal W. Field cancerization in oral stratified squamous epithelium. Cancer 1953;September:963-968.
60. Shamsuddin AM, Tyner GT, Yang GY. Common expression of the tumor marker Δ-Galactose-β-[1⇒3]-N-Acetyl-D-Galactosamine by different adenocarcinomas: evidence of field effect phenomenon. Cancer Res 1995;55:149-152.
61. Tanaka T, KojimaT, Hara A, et al. Chemoprevention of oral carcinogenesis by DL-α-difluoromethylornithine, an ornithine decarboxylase inhibitor: dose-dependent reduction in 4-Nitroquinoline 1-oxide-induced tongue neoplasms in rats. Cancer Res 1993;53:772-776.
62. Ratko TA, Detrisac CJ, Rao KVN et al. Interspecies analysis of the chemopreventive efficacy of dietary α-difluoromethylornithine. Anticancer Res 1990;10:67-72.
63. Homma Y, Kakizoe T, Samma S, et al. Inhibition of N-Butyl-N-(4-hydroxybutyl)nitrosamine-induced rat urinary bladder carcinogenesis by

α-difluoromethylornithine. Cancer Res 1987; 47:6176-6179.
64. Thompson HJ, Herbst EJ, Meeker LD, et al. Effect of D,L-α-difluoromethylornithine of murine mammary carcinogenesis. Carcinogenesis 1984;5:1649-1651.
65. Thompson HJ, Meeker LD, Herbst EJ, et al. Effect of concentration of D,L-a- difluoromethylornithine of murine mammary carcinogenesis. Cancer Res 1985;45:1170-1173.
66. Kojima T, Tanaka T, Kawamori T, et al. Chemopreventive effects of dietary a-difluoromethylornithine, an ornithine decarboxylase inhibitor, on initiation and postinitiation stages of diethylnitrosamine-induced rat hepatocarcinogenesis. Cancer Res 1993;53:3903-3907.
67. Luk GD, Baylin SB. Onithine decarboxylase as a biologic marker in familial colonic polyposis. New England J Med 1984;July 12:80-83.
68. Creaven PJ, Pendyala L, Petrelli NJ. Evaluation of 2-difluoromethylornithine as a potential chemopreventive agent: tolerance to daily oral administration in humans. Cancer Epidemiol Biom and Prev 1993;2:243-47.
69. Kingsnorth AN, Wallace HM, Bundred NJ, et al. Polyamines in breast cancer. Br J Surg 1984;71:352-356.
70. Loprinzi CL, Verma AD, Boutwell RD et al. Inhibition of phorbol-ester-induced activity by oral compounds : A possible role in human chemoprevention studies. J Clin Onc 1985; 3:751-57.
71. Love RR, Carbone PP, Verma AK et al. Randomized Phase I chemoprevention dose-seeking study of α-difluoromethylornithine. J Nat Cancer Inst 1993;85:732-737.
72. Meyskens FL, Emerson SS, Pelat H, et al. Dose de-escalation chemoprevention trial of α-difluoromethylornithine in patients with colon polyps. J Nat Cancer Inst 1994;86:1994 1122-1130.
73. Pendyala L, Creaven PJ, Porter CW. Urinary and erythrocyte polyamines during the evaluation of oral α-difluoromethylornithine in a phase I chemoprevention clinical trial. 1993:235-241.
74. Loprinzi C, Melssing EM A prospective clinical trial of difluoromethylornithine (DFMO) in patients with resected superficial bladder cancer. J Cell Biochem 1992;Suppl. 16 l:153-55.
75. Bostwick DG. Prostatic intraepithelial neoplasia (PIN): current concepts. J Cell Bioch 1992;Supp.16 H:10-19.
76. Kadmon D. Chemoprevention in prostate cancer: the role of difluoromethylornithine (DFMO). J Cell Bioch 1992;Supp. 16H:122-127.
77. Danin, C., Claverie, N., Wagner, J,. Grove. J, and Kogh-weser, J., Effect of prostatic growth of 2-difluoromethylornithine, an effective inhibitor of ornithine decarboxylase. Biochem J 1982;202:175-181.
78. Plezia, OM, Alberts DS, Peng Y et al. The role of serum and tissue pharmacology studies in the design and interpretation of chemoprevention trials. Preventive Medicine 1989;18:680-687.
79. Morris DR. A new perspective on ornithine decarboxylase regulation: prevention of polyamine toxicity is the overriding theme. J Cell Biochem

1991;46:102-105.
80. Croghan MK, Aickin MG, Meyskens FL. Dose-related 2-difluoromethylornithine ototoxicity. Am L Clin Oncol 1991;14:331-35.
81. Crowell JA, Goldenthal EI, Kelloff GJ, et al. Chronic toxicity studies of the potential cancer preventive 2-(difluoromethyl)-dl-ornithine. Fund Appl Tox 1994;22:341-354.
82. Luk GD, Marton LJ, Baylin SB. Ornithine decarboxylase is important in intestinal mucosal maturation and recovery from injury in rats. Science 1980;210:195-198.
83. Fozard JR, Part ML, Prakash NJ et al. L-ornithine decarboxylase:an essential role in early mammalian embryogenesis. Science 1980;208:505-508.
84. Casero RA, Celano P, Ervin SJ et al. Differential induction of spermidine/spermine N^{-1} acetyltransferase in human lung cancer cells by the bis(ethyl)spermine analogs. Cancer Res 1989;49:3829-3833.
85. Casero RA, Mank AR, Xiao L, et al. Steady-state messenger RNA and activity correlates with sensitivity to N^1,N^{12}-*bis(*ethyl)spermine in human cell lines representing the major forms of lung cancer. Cancer Res 1992;52:1-5.
86. Davidson NE, Mank AR, Prestigiacomo LJ et al. Growth inhibition of hormone-responsive and -resistant human breast cancer cells in culture by N^1, N^{12}-bis(ethyl)spermine. Cancer Res 1993;53:2071-2075.
87. Porter CW, Ganis B, Libby, PR, Bergeron RJ Correlations between polyamine analog-induced increases in spermidine/spermine N^1-acetyltransferase activity, polyamine pool depletion, and growth inhibition in human melanoma cell lines. Cancer Res 1991;51:3715-3720.
88. Colburn NH, Rormer BF, Nelson K, Yuspa SH. Tumor promoter induces anchorage independence irreversibly. Nature 1979; 281:589-591.
89. Ghoda L, Basu HS, Porter CW, et al. Role of ornithine decarboxylase suppression and polyamine depletion in the antiproliferative activity of polyamine analogs. Mol Pharm 1992;42:302-306.
90. Albanese L, Bergeron RJ, Pegg, AE. Investigations of the mechanism by which mammalian cell growth is inhibited by N^1,N^{12}-*bis*(ethyl)spermine. Biochem J 1993;291:131-137.
91. Pegg EA, Wechter R, Pakala R, et al. Effect of $N^{1,}$ N^{12}-*bis*(ethyl)spermine and related compounds on growth and polyamine acetylation, content, and excretion in human colon tumor cells. J Biol Chem 1989; 264:11744-11749.
92. Bergeron RJ, Hawthorne TR, Vinson JRT, et al. Role of the methylene backbone in the antiproliferative activity of polyamine analogs on L1210 cells. Cancer Res 1989;49:2959-2964.
93. Basu HS, Feuerstein BG, Deen DF et al. Correlation between the effects of polyamine analogs on DNA conformation and cell growth. Cancer Res 1989;49:5591-5597.
94. Basu HS, Sturkenboom MCJM, Delcros J, et al. Effect of polyamine depletion on chromatin structure in J-87 MG human brain tumor cells. Biochem J 1992;282:723-727.
95. Basu HS, Wright WD, Deen DF et al. Treatment with polyamine analog

alters DNA- matrix association in HeLa cell nuclei: a nucleoid halo assay. Biochem 1993;32:4073-4076.

96. Celano P, Baylin SB, Giardiello FM, et al.. Effect of polyamine depletion on c-myc expression in human colon carcinoma cells. J Biol Chem 1988;263:5491-5494.
97. Celano P, Berchtold CM, Giardiello FM, et al. Modulation of growth gene expression by selective alteration of polyamines in human colon carcinoma cells. Biochem Biophys Res Commun 1989;165:384-390.
98. Celano P, Baylin SB, Casero RA. Polyamines differentially modulate the transcription of growth associated genes in human colon carcinoma cells. J Biol Chem 1989;264:8922-8927.
99. Schulze-Lohoff E, Fees H, Zanner S. Inhibition of immediate-early-gene induction in renal mesangial cells by depletion of intracellular polyamines. Biochem J 1994;298:647-653.
100. Trush MA, Kensler TW. An overview of the relationship between oxidative stress and chemical carcinogenesis. Free Radical Biol Med 1991;10:201-209.
101. Wang J, Johnson LR, Tsai Y, et al. Mucosal ornithine decarboxylase, polyamines and hyperplasia in infected intestine. Am Physiol Soc 1991;G45-G51.
102. Crawford D, Zbinden I, Amstad P, et al. Oxidant stress induces the proto-oncogenes *c-fos* and *c-myc* in mouse epidermal cells. Oncogene 1988;3:27-32.
103. Shibanum M, Kurodi T, Nose K. Induction of DNA replication and expression of proto-oncogene *c-myc* and *c-fos* in quiescent Balb/3T3 cells by xanthine/xanthine oxidase. Oncogene 1988;3:17-21.
104. Larsson R, Cerutti P. Translocation and enhancement of phosphotransferase activity of protein kinase C following exposure in mouse epidermal cells to oxidants. Cancer Res 1989;4:5627-5632.
105. Guyton KZ, Kensler TW. Oxidative mechanisms in carcinogenesis. British Med Bull 1993;49:523-544.
106. Cohen SM, Ellwein LB. Cell proliferation in carcinogenesis. Science 1990;249:1007-1011.
107. Nowell PC. Mechanisms of tumor progression. Cancer Res 1986; 46:2203-2207.
108. Sporn MB. Approaches to prevention of epithelial cancer during the preneoplastic period. Cancer Res 1976;36:2699-2702.
109. Clifford A, Morgan D, Yuspa SH, et al. Role of ornithine decarboxylase in epidermal tumorigenesis. Cancer Res 1995; 55:1680-1686.

INDEX

Page numbers in italics denote figures (f) or tables (t).

A

AbeAdo,
  inhibition of SAMDC, 34, 178, *179f*
Abrahamsen, MS, 14
AdoDATAD, 172
AdoDATO, 172
AdoMac, 178, 182
  mechanism for inactivation of AdoMet-DC, *181f*, 183-184, *184f*
Agatoxins, 139
Albanese, L, 103, 115, 222
Antizyme,
  role in SAMDC degradation, 63
  ODC enzyme, 209
Anton, DL, 174
Argiotoxins, 139

B

Basic region/leucine zipper (bZip) motif, 13
Bergeron, RJ, 192, 206
Bitoniti, AJ, 190
Boone, CW, 206
Burgoyne, LA, 106

C

Casero, RA, 208
CHENSpm, 196
  influence of polyamine and SSAT levels, *196t*
c-Myc, 17
  Mad, 18
  regulation of ODC transcription, 18
  MXi1, 18
Circular dichroism (CD), 106
Clifford, A, 210
Carginogenesis,
  colon, 214-215
  field, 215
  other paradigms, 216
CPENSpm, 194, *195f*
CpG islands, 84
  content in human SSAT gene, 84
Cyclic AMP response elements (CREs), 12

D

Danzin, C, 178
Davis, RH, 115
Derven, PB, 116
DFMO (2-difluoromethylornithine) *see also inhibitors of ODC*, 205
  antiproliferative activity against tumor cells, 210
  chemopreventive studies, *217t*, *219t*
    polyamine pathway, 217-220
  molecular structure, *211f*
  toxicity and pharmacokinetic studies, 220-221
Diaz, E, 174
DNA alkylating anti-cancer agents,
  Cis-diamminedichloroplatinum (CDDP), 110
  Chloroethylnitrosoureas (CENU), 111
DNA,
  damage, 112
    ionizing radiation induced, 112
  repair, 112
Durand, GM, 145

E

Early growth response (EGR) family, 9
Edwards, ML, 192

F

Foka, M, 111
fos gene, 11, 15, 18, 77

G

GABA (gamma-aminobutyric acid), 129
Garrard, WT, 116
GC factor, 9
Gene regulatory proteins, *see also* individual proteins
  gene repressors, 9-10
Ghoda, LY, 103, 115, 222
Glutamate receptors, 129
  function, 130
  ligand-gated ion channels, 130
    NMDA receptors, 130, 131-134
    AMPA receptors, 130
    kainate receptors, 130
  in synaptic transmission, *131f*
  metabotropic receptors, 130
  subtypes and cloned subunits, *130f*
Guo, JQ, 185

H

Hayashi, S-I, 1
Hewish, DR, 106

I

Internal ribosome entry site (IRES), 45
Inward rectifier K^+ (IRK) channels, 159

J

Jackson, V, 108
jun gene, 10, 15

K

Kameji, T, 35
Kingsnorth, AN, 218
Koenig, H, 116
Kozak, M, 43
Kutny, R, 174

L

Larsson, L-I, 107
Long-term depression (LTD), 132
Long-term potentiation (LTP), 132
Luciferase, 14

M

MAA (methylaminoadenosine), 178
Mach, M, 34
MAR (matrix attachment regions), 104
Maric, SC, 32, 61
Metcalf, BW, 210
MGBG (methylglyoxal *bis*(guanylhydrazone), 174
Morris, DR, 39, 44, 45, 208
mos, 18
Muller, SR, 8
Murakami, Y, 1
myc, 11, 77

N

Nakanishi, N, 140
NF-ODC_1, 8
NMDA (N-methyl-D-aspartate) receptors, 129
  effects of polyamines, 136, 145
    electrophysiological studies, 138-140
    ligand-binding studies, 136, *137f*
  glycine dependent and independent stimulation, 147-153, *151f*
  functional properties, 135,*136f*
  modulation, 134, *152f*
  molecular biology,
    NR1 subunits, 140, *141f*
      heteromeric NR1/NR2 receptors, 147-153
      homomeric NR1 receptors, 145-145
    NR2 subunits, 143-145
  physiology and pathophysiology, 131
  role, 131
    in neurotoxicity and neurodegeneration, 133
  schematic model, *132f*
  topography and subunit arrangement, 142
Nucleosomes,
  defined, 105
  structure,
    "bead and string" model, 105
    DNA incorporation, 105-106
    effects of polyamines, 106

O

O'Brien, TG, 212
Ornithine decarboxylase (ODC), 1, 5, 77, 102, 172
  antizyme, 209
  inhibition by DMFO, 89, 205
  ODC gene, 5
    factors regulating expression, 6, *6f*, 10
      CRE2, 10
      oncogene products (*see also* individual products), 17-19
    mechanism of transcriptional regulation, *11t*
    regulation by cAMP, 12-14
  transformation, 209

P

PAO (polyamine oxidase), 187
Paoletti, J, 111
Palvino, JJ, 14, 16
Pegg, AE, 35, 42, 173, 222
PENSpm, 194
PEST regions, 63, 80, 209
Polyamines,
  analogs in chemoprevention, 221-223
  as anti-neoplastic agents, 110, 113
  BES (*bis*(ethyl)spermidine), 36, 82, 114, 192-197, *211f*, 221
    antitumor activity of BENSpm, *see also individual polyamine analogs,* 82, 194, 210, 222
  development of synthetic analogs, 190, *191f*
  distribution and concentration, 155
  DNA-matrix associations, 104-105
  historical perspectives, 102
  in chemical carcinogenesis, 211-214
  in inflammatory processes, 223-224

interactions with native NMDA receptors, 136-140
summary, 158
metabolism in chemoprevention, 205, *207f*
pathway components, 208
polyamine-DNA interactions, 102, 106, *109f*
alterations in chromatin structure, 109-110
and cell division, 103
structure, *140f*, *189f*
Primary response genes, 10
Protein kinase C (PKC), 12
Proto-oncogenes, 15 (*see also* individual genes)

R

Ransom, RW, 136
ras, 18
Reporter genes,
human growth hormone (hGH), 40
Retinoic acid receptors (RAR), 15
mode of action, 16
Rogers, S, 63
Rosenberg, B, 110
Roti-Roti, JL, 104
Ruan, H, 45, 46

S

S-adenosylmethionine (SAM), 27
S-adenosylmethionine decarboxylase, (AdOMetDC) or (SAMDC), 1, 27, 172, 208
activation by putrescine, 57
degradation, 63
role of polyamines, 63, 102
effect of spermidine, 32
kinetic properties, *58t*, *182t*
inhibitors as antiparasitic agents, 185-186
molecular features and actions, 28
mRNA distribution, 40
non-nucleoside inhibitors, 174-176
nucleoside-based inhibitors, 176, *176f*, *177f*
structure, *180f*
posttranslational modifications, 59-61
primary amino acid sequence of human, yeast and potato SAMDCs, *30f*
proenzyme processing, 50-57
amino acid sequences of pyruvate-containing enzymes, *51t*
effect of dilution, *55f*
putrescine effects, 52-53, *54f*, 56
regulation, 61
RNase treatment, 62, *62f*
structure, function and mechanism, 173, *175f*
transcriptional control, 31-33
translational control of expression, 33-50
model, 47
S-adenosylmethionine decarboxylase 5' UTR
effect of alterations on translational efficiency, *38f*
effects of mutations of the uORF, *41t*
Schantz, LM, 36, 38, 40, 42
Secrist, JA, 177
Serum response element, 17
Slaughter, DP, 215
Snell, E, 50
Snyder, RD, 107
Spermidine, 27
effects on SAMDC activity and mRNA levels, 32
DNA interactions, 106, 113
metabolic pathway, *79f*
modulation of NMDA receptors, 129
Spermine, 27, 129
BESpm (N^1,N^{12}-bis(ethyl)spermine), 82, *83f*
superinduction of SSAT, 85
DNA interactions, 106
metabolic pathway, *79f*
modulation of NMDA receptors, 129, *148f*, *149f*, 153, *154f*, 156
structure, *83f*
Spermidine/spermine N^1-acetyltransferase (SSAT), 1, 77
cloning and characterization of SSAT gene, 80, 82-85, 186-187
human SSAT gene structure, *88f*
SSAT mRNA, 81, *81f*
kinetics and substrate specificity, 187, *188f*
mechanism of action, 78, 206
protein, 79
regulation and deregulation of gene expression, 85-87
in neoplastic growth, 91-92
role of methylation, 87
stress-induced expression, 89-91
transition-state analogue inhibitors, *196f*
Stanley, BA, 208
Stec, NL, 136
Sugihara, H, 142
Suzuki, T, 37

T

Tabor, H, 173
Tew, KD, 111
Topoisomerases, 101, 112
topoisomerase I, 112
topoisomerase II, 112
TPA (12-)-tetradecanoyl-phorbol-13-acetate), 212

Transcription factors,
AP-1 family, 15
CREB/ATF, 6, 10, 13
CREB-binding protein, 13
CAD domain, 14
KID domain, 14
CREM (cAMP-responsive element modulator), 13
E2, 10
EGR (early growth response) family, 9
NF-ODC, 6
Sp1, 6
Sp1-related proteins (Sp2, Sp3, Sp4, BTEB 1, BTEB 2), 7
Translation,
eukaryotic initiation factor (eIF-4E), 43, 209
Trans-retinoic acid (RA), 15
Tumor promoter TPA, 14
TPA-response element (TRE), 15

V

Varmus, HE, 107
Varshavsky, A, 61
Vogelstein, B, 215

W

Williams, JR, 114
Woster, PM, 178, 197
WT1 (Wilms tumor suppressor), 9
Wu, YQ, 178, 184

X

Xiao, L, 208

Molecular Biology Intelligence Unit

Available and Upcoming Titles

- ❐ Organellar Proton-ATPases
  Nathan Nelson, Roche Institute of Molecular Biology
- ❐ Interleukin-10
  Jan DeVries and René de Waal Malefyt, DNAX
- ❐ Collagen Gene Regulation in the Myocardium
  M. Eghbali-Webb, Yale University
- ❐ DNA and Nucleoprotein Structure In Vivo
  Hanspeter Saluz and Karin Wiebauer, HK Institut-Jena and GenZentrum-Martinsried/Munich
- ❐ G Protein-Coupled Receptors
  Tiina Iismaa, Trevor Biden, John Shine, Garvan Institute-Sydney
- ❐ Viroceptors, Virokines and Related Immune Modulators Encoded by DNA Viruses
  Grant McFadden, University of Alberta
- ❐ Bispecific Antibodies
  Michael W. Fanger, Dartmouth Medical School
- ❐ Drosophila Retrotransposons
  Irina Arkhipova, Harvard University and Nataliya V. Lyubomirskaya, Engelhardt Institute of Molecular Biology-Moscow
- ❐ The Molecular Clock in Mammals
  Simon Easteal, Chris Collet, David Betty, Australian National University and CSIRO Division of Wildlife and Ecology
- ❐ Wound Repair, Regeneration and Artificial Tissues
  David L. Stocum, Indiana University-Purdue University
- ❐ Pre-mRNA Processing
  Angus I. Lamond, European Molecular Biology Laboratory
- ❐ Intermediate Filament Structure
  David A.D. Parry and Peter M. Steinert, Massey University-New Zealand and National Institutes of Health
- ❐ Fetuin
  K.M. Dziegielewska and W.M. Brown, University of Tasmania
- ❐ Drosophila Genome Map: A Practical Guide
  Daniel Hartl and Elena R. Lozovskaya, Harvard University
- ❐ Mammalian Sex Chromosomes and Sex-Determining Genes
  Jennifer A. Marshall-Graves and Andrew Sinclair, La Trobe University-Melbourne and Royal Children's Hospital-Melbourne
- ❐ Regulation of Gene Expression in *E. coli*
  E.C.C. Lin, Harvard University
- ❐ Muscarinic Acetylcholine Receptors
  Jürgen Wess, National Institutes of Health
- ❐ Regulation of Glucokinase in Liver Metabolism
  Maria Luz Cardenas, CNRS-Laboratoire de Chimie Bactérienne-Marseille
- ❐ Transcriptional Regulation of Interferon-γ
  Ganes C. Sen and Richard Ransohoff, Cleveland Clinic
- ❐ Fourier Transform Infrared Spectroscopy and Protein Structure
  P.I. Haris and D. Chapman, Royal Free Hospital-London
- ❐ Bone Formation and Repair: Cellular and Molecular Basis
  Vicki Rosen and R. Scott Thies, Genetics Institute, Inc.-Cambridge
- ❐ Mechanisms of DNA Repair
  Jean-Michel Vos, University of North Carolina
- ❐ Short Interspersed Elements: Complex Potential and Impact on the Host Genome
  Richard J. Maraia, National Institutes of Health
- ❐ Artificial Intelligence for Predicting Secondary Structure of Proteins
  Xiru Zhang, Thinking Machines Corp-Cambridge
- ❐ Growth Hormone, Prolactin and IGF-I as Lymphohemopoietic Cytokines
  Elisabeth Hooghe-Peters and Robert Hooghe, Free University-Brussels
- ❐ Human Hematopoiesis in SCID Mice
  Maria-Grazia Roncarolo, Reiko Namikawa and Bruno Péault DNA Research Institute
- ❐ Membrane Proteases in Tissue Remodeling
  Wen-Tien Chen, Georgetown University
- ❐ Annexins
  Barbara Seaton, Boston University
- ❐ Retrotransposon Gene Therapy
  Clague P. Hodgson, Creighton University
- ❐ Polyamine Metabolism
  Robert Casero Jr, Johns Hopkins University
- ❐ Phosphatases in Cell Metabolism and Signal Transduction
  Michael W. Crowder and John Vincent, Pennsylvania State University
- ❐ Antifreeze Proteins: Properties and Functions
  Boris Rubinsky, University of California-Berkeley
- ❐ Intramolecular Chaperones and Protein Folding
  Ujwal Shinde, UMDNJ
- ❐ Thrombospondin
  Jack Lawler and Jo Adams, Harvard University
- ❐ Structure of Actin and Actin-Binding Proteins
  Andreas Bremer, Duke University
- ❐ Glucocorticoid Receptors in Leukemia Cells
  Bahiru Gametchu, Medical College of Wisconsin
- ❐ Signal Transduction Mechanisms in Cancer
  Hans Grunicke, University of Innsbruck
- ❐ Intracellular Protein Trafficking Defects in Human Disease
  Nelson Yew, Genzyme Corporation
- ❐ apoJ/Clusterin
  Judith A.K. Harmony, University of Cincinnati
- ❐ Phospholipid Transfer Proteins
  Vytas Bankaitis, University of Alabama
- ❐ Localized RNAs
  Howard Lipschitz, California Institute of Technology
- ❐ Modular Exchange Principles in Proteins
  Laszlo Patthy, Institute of Enzymology-Budapest
- ❐ Molecular Biology of Cardiac Development
  Paul Barton, National Heart and Lung Institute-London
- ❐ RANTES, *Alan M. Krensky, Stanford University*
- ❐ New Aspects of V(D)J Recombination
  Stacy Ferguson and Craig Thompson, University of Chicago

Neuroscience Intelligence Unit

Available and Upcoming Titles

- ❐ Neurodegenerative Diseases and Mitochondrial Metabolism
  M. Flint Beal, Harvard University
- ❐ Molecular and Cellular Mechanisms of Neostriatum
  Marjorie A. Ariano and D. James Surmeier, Chicago Medical School
- ❐ Ca^{2+} Regulation By Ca^{2+}-Binding Proteins in Neurodegenerative Disorders
  Claus W. Heizmann and Katharina Braun, University of Zurich, Federal Institute for Neurobiology, Magdeburg
- ❐ Measuring Movement and Locomotion: From Invertebrates to Humans
  Klaus-Peter Ossenkopp, Martin Kavaliers and Paul Sanberg, University of Western Ontario and University of South Florida
- ❐ Triple Repeats in Inherited Neurologic Disease
  Henry Epstein, University of Texas-Houston
- ❐ Cholecystokinin and Anxiety
  Jacques Bradwejn, McGill University
- ❐ Neurofilament Structure and Function
  Gerry Shaw, University of Florida
- ❐ Molecular and Functional Biology of Neurotropic Factors
  Karoly Nikolics, Genentech
- ❐ Prion-related Encephalopathies: Molecular Mechanisms
  Gianluigi Forloni, Istituto di Ricerche Farmacologiche "Mario Negri"-Milan
- ❐ Neurotoxins and Ion Channels
  Alan Harvey, A.J. Anderson and E.G. Rowan, University of Strathclyde
- ❐ Analysis and Modeling of the Mammalian Cortex
  Malcolm P. Young, University of Oxford
- ❐ Free Radical Metabolism and Brain Dysfunction
  Irène Ceballos-Picot, Hôpital Necker-Paris
- ❐ Molecular Mechanisms of the Action of Benzodiazepines
  Adam Doble and Ian L. Martin, Rhône-Poulenc Rorer and University of Alberta
- ❐ Neurodevelopmental Hypothesis of Schizophrenia
  John L. Waddington and Peter Buckley, Royal College of Surgeons-Ireland
- ❐ Synaptic Plasticity in the Retina
  H.J. Wagner, Mustafa Djamgoz and Reto Weiler, University of Tübingen
- ❐ Non-classical Properties of Acetylcholine
  Margaret Appleyard, Royal Free Hospital-London
- ❐ Molecular Mechanisms of Segmental Patterning in the Vertebrate Nervous System
  David G. Wilkinson, National Institute of Medical Research-UK
- ❐ Molecular Character of Memory in the Prefrontal Cortex
  Fraser Wilson, Yale University

MEDICAL INTELLIGENCE UNIT

AVAILABLE AND UPCOMING TITLES

- ❒ Hyperacute Xenograft Rejection
  Jeffrey Platt, Duke University
- ❒ Chimerism and Tolerance
  Suzanne Ildstad, University of Pittsburgh
- ❒ Birth Control Vaccines
  G.P. Talwar and Raj Raghupathy, National Institute of Immunology-New Delhi and University of Kuwait
- ❒ Monoclonal Antibodies in Transplantation
  Lucienne Chatenoud, Hôpital Necker-Paris
- ❒ Therapeutic Applications of Oligonucleotides
  Stanley Crooke, ISIS Pharmaceuticals
- ❒ Cryopreserved Venous Allografts
  Kelvin G.M. Brockbank, CryoLife, Inc.
- ❒ Clinical Benefits of Leukpodepleted Blood Products
  Joseph Sweeney and Andrew Heaton, Miriam and Roger Williams Hospitals-Providence and Irwin Memorial Blood Center-San Francisco
- ❒ Delta Hepatitis Virus
  M. Dinter-Gottlieb, Drexel University
- ❒ Intima Formation in Blood Vessels: Spontaneous and Induced
  Mark M. Kockx, Algemeen Ziekenhuis Middelheim-Antwerpen
- ❒ Adult T Cell Leukemia and Related Diseases
  Takashi Uchiyama and Jungi Yodoi, University of Kyoto
- ❒ Development of Epstein-Barr Virus Vaccines
  Andrew Morgan, University of Bristol
- ❒ p53 Suppressor Gene
  Tapas Mukhopadhyay, Steven Maxwell and Jack A. Roth, University of Texas-MD Anderson Cancer Center
- ❒ Retinal Pigment Epithelium Transplantation
  Devjani Lahiri-Munir, University of Texas-Houston
- ❒ Minor Histocompatibility Antigens and Transplantation
  Craig V. Smith, University of Pittsburgh
- ❒ Familial Adenomatous Polyposis Coli and the APC Gene
  Joanna Groden, University of Cincinnati
- ❒ Cancer Cell Adhesion and Tumor Invasion
  Pnina Brodt, McGill University
- ❒ Constitutional Immunity to Infection
  Cees M. Verduin, David A. Watson, Jan Verhoef, Hans Van Dijk, University of Utrecht and North Dakota State University
- ❒ Nutritional and Metabolic Support in Critically Ill Patients
  Rifat Latifi, Yale University
- ❒ Nutritional Support in Gastrointestinal Failure
  Rifat Latifi and Stanley Dudrick, Yale University and University of Texas-Houston
- ❒ Septic Myocardiopathy: Molecular Mechanisms
  Karl Werdan and Günther Schlag, Ludwig-Maximilians-Universität-München and Ludwig-Boltzmann-Instituts für Experimentelle und Klinische Traumatologie
- ❒ The Molecular Genetics of Wilms Tumor
  Bryan R.G. Williams, Max Coppes and Christine Campbell, Cleveland Clinic and University of Calgary
- ❒ Endothelins
  David J. Webb and Gillian Gray, University of Edinburgh
- ❒ Nutritional and Metabolic Support in Cancer, Transplant and Immunocompromised Patients
  Rifat Latifi, Yale University
- ❒ Antibody-Mediated Graft Rejection
  J. Andrew Bradley and Eleanor Bolton, University of Glasgow
- ❒ Liposomes in Cancer Chemotherapy
  Steven Sugarman, University of Texas-MD Anderson
- ❒ Molecular Basis of Human Hypertension
  Florent Soubrier, Collége de France-Paris
- ❒ Endocardial Endothelium: Control of Cardiac Performance
  Stanislas U. Sys and Dirk L. Brutsaert, Universiteit Antwerpen
- ❒ Endoluminal Stent Grafts for the Treatment of Vascular Diseases
  Michael L. Marin, Albert Einstein College of Medicine
- ❒ B Cells and Autoimmunity
  Christian Boitard, Hôpital Necker-Paris
- ❒ Immunity to Mycobacteria
  Ian Orme, Colorado State University
- ❒ Hepatic Stem Cells and the Origin of Hepatic Carcinoma
  Stewart Sell, University of Texas-Houston
- ❒ HLA and Maternal-Fetal Recognition
  Joan S. Hunt, University of Kansas
- ❒ Molecular Mechanisms of Alloreactivity
  Robert L. Kirkman, Harvard University
- ❒ Ovarian Autoimmunity
  Roy Moncayo and Helga E. Moncayo, University of Innsbruck
- ❒ Immunology of Pregnancy Maintenance
  Joe Hill and Peter Johnson, Harvard University and University of Liverpool
- ❒ Protein and Amino Acid Metabolism in Cancer
  Peter W.T. Pisters and Murray Brennan, Sloan-Kettering Memorial Cancer Center
- ❒ Cytokines and Hemorrhagic Shock
  Eric J. DeMaria, Medical College of Virginia
- ❒ Cytokines in Inflammatory Bowel Disease
  Claudio Fiocchi, Case Western Reserve University
- ❒ T Cell Vaccination and Autoimmune Disease
  Jingwu Zhang, Willems Institut-Belgium
- ❒ Immune Privilege
  J. Wayne Streilein, Luke Jiang and Bruce Ksander, Schepens Eye Research Institute-Boston
- ❒ The Pathophysiology of Sepsis and Multi-Organ Failure
  Mitchell Fink, Harvard University
- ❒ Bone Metastasis
  F. William Orr, McMaster University
- ❒ Novel Regional Therapies for Liver Tumors
  Seiji Kawasaki and Masatoshi Makuuchi, Shinshu University
- ❒ Molecular Basis for the Action of Somatostatin
  Miguel J.M. Lewin, INSERM-Paris
- ❒ Growth Hormone in Critical Illness
  Michael Torosian, University of Pennsylvania
- ❒ Molecular Biology of Aneurysms
  Richard R. Keen, Northwestern University
- ❒ Strategies in Malaria Vaccine Design
  F.E.G. Cox, King's College London
- ❒ Chimeric Proteins and Fibrinolysis
  Christoph Bode, Marschall Runge and Edgar Haber, University of Heidelberg, University of Texas-Galveston and Harvard University

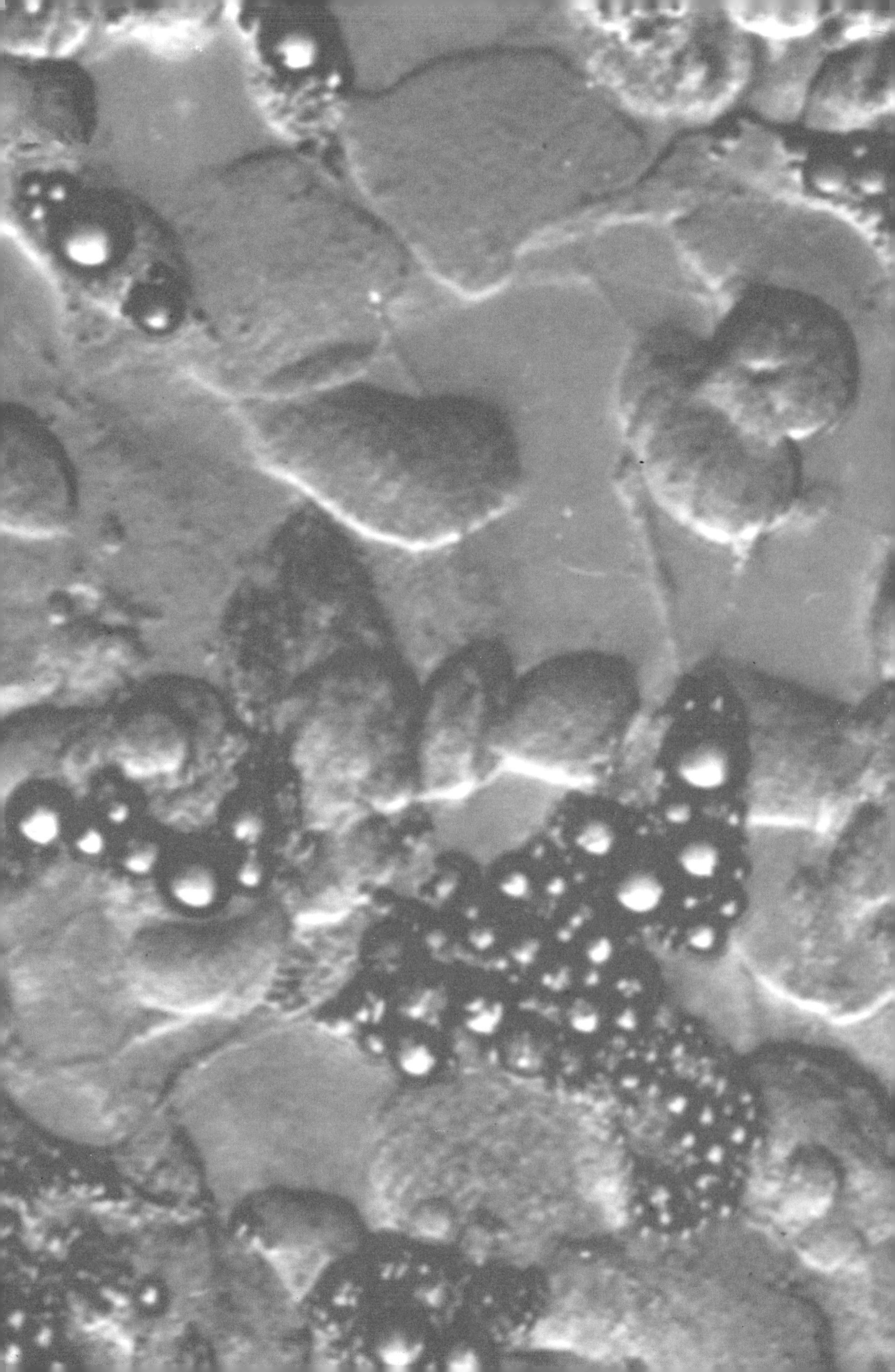